Human Hand Function

Human Hand Function

Lynette A. Jones
Susan J. Lederman

OXFORD
UNIVERSITY PRESS

2006

OXFORD
UNIVERSITY PRESS

Oxford University Press, Inc., publishes works that further
Oxford University's objective of excellence
in research, scholarship, and education.

Oxford   New York
Auckland   Cape Town   Dar es Salaam   Hong Kong   Karachi
Kuala Lumpur   Madrid   Melbourne   Mexico City   Nairobi
New Delhi   Shanghai   Taipei   Toronto

With offices in
Argentina   Austria   Brazil   Chile   Czech Republic   France   Greece
Guatemala   Hungary   Italy   Japan   Poland   Portugal   Singapore
South Korea   Switzerland   Thailand   Turkey   Ukraine   Vietnam

Published by Oxford University Press, Inc.
198 Madison Avenue, New York, New York 10016

www.oup.com

Library of Congress Cataloging-in-Publication Data
Jones, Lynette A.
Human hand function / Lynette A. Jones and Susan J. Lederman.
p. cm.
Includes bibliographical references and index.
ISBN-13 978-0-19-517315-4
ISBN 0-19-517315-5
1. Hand—Physiology. 2. Hand—Anatomy. 3. Hand—Movements.
I. Lederman, Susan J. II. Title.
[DNLM: 1. Hand—physiology. 2. Hand—anatomy & history.   WE 830 J775h 2006]
QP334.J66 2006
611'.97—dc22   2005018742

9 8 7 6 5 4 3 2 1
Printed in the United States of America
on acid-free paper

To all those who have worked to unravel
the mysteries of the human hand

ACKNOWLEDGMENTS

The ideas and initial planning for this book began in September 2001 during Susan Lederman's sabbatical in the Department of Mechanical Engineering at the Massachusetts Institute of Technology. We then collaboratively wrote the book over the next three years, an experience that proved both rewarding and enjoyable. The work for this book was shared equally, and the order of authorship was decided by the toss of a coin.

We wish to thank the following colleagues for providing helpful references and graphic materials, for answering the perplexing questions we posed, or for offering other valuable information: Massimo Bergamasco, Jim Craig, Roz Driscoll, Antonio Frisoli, Yvette Hatwell, Mark Hollins, Peter Hunter, Gunnar Jansson, Charlotte Reed, Hayley Reynolds, and Allan Smith.

A number of individuals have provided support and assistance with the book's preparation, including Aneta Abramowicz, Catalina Tobon Gomez, and Lindsay Eng. We would also like to acknowledge the outstanding assistance provided by Cheryl Hamilton and Monica Hurt.

We thank Catharine Carlin, our editor at Oxford University Press, who has enthusiastically supported this project throughout and provided us with an unusual flexibility, which made writing this book an enjoyable endeavor. We also wish to thank our anonymous reviewers, who offered thorough, thoughtful, and excellent evaluations of the manuscript.

Lynette Jones acknowledges the support of the Massachusetts Institute of Technology and the National Institutes of Health's National Institute of Neurological Disorders and Stroke during the preparation of this book. Susan Lederman acknowledges the support of Queen's University in Kingston, Ontario, Canada; the Canada Council for the Arts, which generously provided a Killam Research Fellowship; the Natural Sciences and Engineering Research Council of Canada; and the Institute for Robotics and Intelligent Systems (Federal Centre of Excellence of Canada).

Finally, we would like to acknowledge the support of our families. Lynette Jones thanks her husband, Ian Hunter, and children, Nicholas and Bridget Hunter-Jones, who enthusiastically supported this project, offered encouragement when needed, and were always willing to assist. Your affection and love are truly sustaining. Susan Lederman wishes to thank her partner, Les Monkman, and his children, Lindsay and Fraser. Each of you has made a profound difference in my life. Your continuing love, support, humor, and irony are gifts offered most generously, and I acknowledge them gratefully.

CONTENTS

Human Hand Function

1

Historical Overview and General Introduction

[W]e must confess that it is in the human hand that we have the consummation of all perfection as an instrument.

BELL, 1833

The human hand is a miraculous instrument that serves us extremely well in a multitude of ways. We successfully use our hands to identify objects and to extract a wealth of information about them, such as their surface texture, compliance, weight, shape, size, orientation, and thermal properties. We demonstrate impressive manual dexterity when reaching for, grasping, and subsequently manipulating objects within arm's reach. Manual gestures, such as those used in sign language and finger spelling, collectively offer valuable forms of communication to those who are deaf or hearing impaired. In addition, raised tangible graphic displays that are explored by hand provide sensory aids to the blind when planning how to navigate their environment without a sighted companion. Finally, the human hand serves as a formidable creative tool in a variety of aesthetic milieus that include dance, sculpture, playing music, and drawing. Depictions of the human hand in art date back to the Upper Paleolithic era (circa 22,000–28,000 years ago) and can still be seen today in caves at Gargas and El Castillo in the Franco-Cantabrian region (Napier, 1993; Piveteau, 1991). There is even a museum dedicated specifically to the hand (La Musée de la Main), which was established by Claude Verdan, a Swiss hand surgeon, in Lausanne, Switzerland.

Historical Roots

Not surprisingly, the study of human hand function derives its historical roots from multiple sources. From archaeology come the very detailed analyses of the changes in the anatomy and structure of the hand as it gradually evolved and became capable of using and, more important, manufacturing tools (Marzke & Marzke, 2000). Zoological and anthropological studies have also focused on the structural aspects of the hand, although this work has primarily concentrated on the functional use of the hand by various primate species, including *Homo sapiens* (Preuschoft & Chivers, 1993). Consideration of these phylogenetic and ontogenetic changes in primate and human hands has led researchers, particularly during the latter half of the 20th century and at the beginning of the 21st century, to examine how such changes relate to the development of human hand function over the lifespan.

Philosophers and, more recently, experimental psychologists have been interested in the sense of touch for centuries, although their attention did not focus on manual sensing per se. As early as Aristotle, those interested in touch recognized that this sensory modality was indeed different from the others (i.e., vision, audition, taste, and smell), a perspective that was maintained throughout the 19th century by individuals such as Weber (1834/1978), Blix (1884), Goldscheider (1886), and Frey (1896).

Two particularly noteworthy developments of the 19th century that have proved to be of significant value to those who study hand function today derive from the work of Weber and, subsequently, Gustav Fechner. It was Weber (1834/1978) who first introduced quantitative techniques to the study of sensory physiology by measuring both absolute and differential thresholds. He also documented a mathematical relation between the size of the change in stimulus magnitude and the just noticeable difference in perceived intensity, which came to be known as Weber's law. Fechner (1860/1966), both a physicist and a philosopher, was interested in offering an alternate interpretation, monism, to Descartes' mind-body dualism. To do so, he argued that the relation between mental and physical events could be described by a general quantitative "psychophysical law" that assumed that all just noticeable differences are subjectively equal and that Weber's law is true. Toward this end, he developed several classical psychophysical methods for measuring sensory thresholds, all of which remain in common use today.

Early tactile concerns centered on issues related to determining the appropriate number of human sensory modalities and submodalities and to the varying sensitivity and spatial acuity of the skin across the entire human body when stimulated by mechanical and thermal forms of energy. During the 19th century, anatomists whose names are well known today (e.g., Meissner, Pacini, Krause) showed much interest in using the new histological techniques of their day to identify the various receptors within the skin. The results of such studies served as the basis for addressing a third important issue, namely, the links between the receptor populations known at that time and the specific human cutaneous sensibility (or sensibilities) that each served. Although the hand was frequently included as a site of stimulation, it was only one among many.

Weber (1834/1978) first introduced the well-known two-point touch threshold as a measure of the skin's spatial acuity, a measure that has subsequently become an important, albeit highly controversial, measure of human hand function. He defined the two-point threshold as the shortest distance between two contact points that could be perceived as two separate contacts, as opposed to just one. Weber made extensive measurements of tactile acuity using his own body and that of his brother. In his work, Weber initially employed an instrument known as the beam compass (or beam trammel), which consisted of a beam with two adjustable connectors that held two steel or pencil points in place at right angles to the beam. To avoid damage to the skin, these points were further enclosed in cork stoppers borrowed from medicine bottles. Eventually, Weber switched to using a pair of metal compasses to stimulate the skin. The two points were applied to many places on the body, including the dorsal and volar regions of the palm and fingers. Weber reported that the normal tactile acuity of the hand improved from the back of the hand to the palm to the fingertips. Indeed, the fingertips, along with the tip of the tongue, were reportedly the most spatially acute parts of the body.

Another important focus during the late 19th century was the skin's sensitivity to applied pressure. The German physiologist Max von Frey (1896) invented two instruments that were used to study pressure sensitivity and the location of "touch" spots on the skin. One device, the limen gauge, was capable of applying force to the skin at any predetermined rate of loading. The other comprised a graduated series of horse and human hairs that were mounted in the ends of sticks and used to measure pressure thresholds (Boring, 1942). These instruments have been employed extensively in studies of pressure sensitivity on the hand over the past 100 years. Indeed, a variant of the von Frey hairs (the Semmes-Weinstein monofilaments) is still in use today.

Much of this early sensory research focused on the tactile sensitivity of a passive subject; however, there was also considerable interest at the same time in understanding the perception of weight (Weber, 1834/1978) and other aspects of what was called the "muscular sense." This term fell into disuse in the early 20th century and was replaced by proprioception, a term coined by Sherrington (1906), and kinesthesis, a concept introduced by Bastian (1887).

Proprioception was defined as the sense subserved by receptors in muscles, tendons, and joints and in the labyrinth, part of the vestibular apparatus. The concept of kinesthesis included sensory information arising from receptors in the skin, muscles, and other deep structures of the limbs, such as tendons and joints. Bell (1826) was the first to postulate the existence of the muscular sense, which he considered to be at parity with the other senses. That is, he regarded it as a sixth sense, although he did not think that muscles were the exclusive site of peripheral sensations (Sherrington, 1900a). The muscular sense was considered to include three classes of sensations: pain and fatigue, weight and resistance, and movement and position. In his 1833 treatise on the hand, Bell noted that "without a sense of muscular action or a consciousness of the degree of effort made, the proper sense of touch could hardly be an inlet to knowledge at all."(p.193). He argued that both muscle and cutaneous sensations must be involved in perceiving the distance, size, form, weight, hardness, and roughness of objects. This indicates a very early recognition of the functional role played by sensors in the skin and muscles in what is now called active haptic processing. These ideas were elaborated further by Sherrington (1900b) in his chapter on cutaneous sensations in Shäfer's *Textbook of Physiology*. In this chapter, Sherrington made the important distinction between active touch, in which the touch sense is related to a series of "muscular sensations," and passive touch, in which an object is allowed to move over a series of tactual points (Phillips, 1986).

The origin of sensations governing the awareness of the forces generated during muscular contractions was fervently debated during the latter part of the 19th century (e.g., Bastian, 1887; Lewes, 1879). In these discussions, a distinction was made between sensations arising in the periphery and sensations of innervation, which were of central origin. Many physiologists and psychologists of the period, including Helmholtz and Wundt, believed that innervation of the motor tracts during voluntary movements produced sensations within the brain. These sensations of innervation were thought to form the basis of the awareness of muscle force and were called a sense of effort. The idea of sensations of innervation declined in popularity, however, with Bastian, Ferrier, and Sherrington all supporting Bell's concept of the muscular sense with its emphasis on peripheral sensations. The notion that internally generated

motor signals might have perceptual consequences did, however, reemerge in the 20th century with the work of Sperry (1950) and Holst (1954), who used the terms corollary discharges and efference copy, respectively, to describe internal signals that arose from centrifugal motor commands and that influenced perception.

From the perspective of human hand function, studies of the muscular sense elucidated two important findings. First, in his research on weight perception, Weber (1834/1978) observed, as was subsequently confirmed by other investigators (e.g., Ferrier, 1886), that although weight discrimination is precise with touch alone (i.e., tactile sensing), it is always more exact if the weight is lifted by the hand (i.e., active haptic sensing). The ability to discriminate between weights of different mass by voluntary muscular exertion was termed a sense of force (Frey, 1914) and was considered one of the essential elements of the muscular sense. The second observation relates to the degree of sensitivity of different joints in the hand and arm in detecting movements. In 1889, Goldscheider published the results from a series of carefully executed experiments in which he measured the movement detection thresholds for various joints while attempting to exclude cutaneous sensations of pressure as much as possible. He found that the thresholds for detecting both active and passive movements were much smaller for more proximal joints, such as the shoulder, as compared to the distal finger joints. This proved especially true for slower movements. In interpreting these findings, he noted that the speed of natural movements of a joint is related to the lever arm over which that joint acts; hence, slower movements are more characteristic of proximal rather than distal joints. Goldscheider's results also showed that there is a well-developed capacity to detect movements passively imposed on the limbs, and so it seemed unnecessary to resort to a concept of sensations of innervation to explain movement perception.

Following the more common examples set by 19th-century investigators interested in internal tactile sensations as opposed to those interested in muscular sensations, the study of touch in the first half of the 20th century focused primarily on tactile sensing by passive (stationary) observers (Boring, 1942; Geldard, 1972). The observer's unmoving hand (and other body sites) was stimulated with a variety of punctuate instruments, some blunt, some sharp, others warm or cool, all in the service of assessing the sensitivity and spatiotemporal acuity of the observer's tactile, thermal, and pain experiences. In keeping with Fechner's significant move away from the dualistic mind-body philosophy of Descartes, these researchers also began to explore the functional relation between numerically assessed sensory magnitudes of perceptual events and their corresponding physical magnitudes. In addition to this behavioral approach, which used psychophysical methodologies, neurophysiologists began to uncover corresponding relationships between the magnitude of the neural activity of cutaneous fibers in nonhuman observers (usually primates) and the corresponding magnitude of the physical stimulus (e.g., Werner & Mountcastle, 1965). Although not the focus of the current book, the study of nonhuman primates has also contributed significantly to our knowledge of how the brain controls the hand.

During the early to mid-20th century, only three individuals—David Katz (1925/1989), Geza Révész (1950), and James J. Gibson (1962, 1966)—stand out for their uncommon emphasis on the critical importance of active, voluntary manual exploration. The focus of their research was on the perception and recognition of external objects and their physical properties rather than on subjective internal tactile sensations. Katz offered a wealth of fascinating and critical observations regarding the sense of touch and how it functions, both in comparison to and in concert with vision and audition. He regarded the hand, as opposed to the skin with its atomistic tactile receptors, as the primary organ for touch. His comprehensive work on the perception of texture, or microstructure, is truly impressive and offers a striking contrast to the historically more common focus on shape, or macrostructure. For example, Révész was particularly interested in form perception by hand as opposed to by eye and to this end proposed a previously unrecognized tactual mode of experience known as haptics. Using his background in philosophy and aesthetics, he offered a theory of haptic form perception, which in turn led him to carefully observe blind individuals as they manually appreciated or created three-dimensional sculptures. He questioned the possibility that the blind can ever possess a meaningful aesthetic appreciation of plastic works of art, an issue we will revisit later. J. J. Gibson continued the pioneering work of Katz and Révész, highlighting the fact that hands are used to explore and to alter the

environment, thus serving both perceptive and executive functions simultaneously. He too critically emphasized the importance of the hand and the haptic system for actively gathering information about the external world. As the works of these three individuals have become better known, particularly those of Gibson, they have led to a surge of scientific and applied interest in haptic sensing as formally defined below, since the early 1990s.

There is a parallel and complementary evolution of research that focuses on the natural actions of the hand. As described above, much of the early experimental research in the motor field focused on the muscular sense. This reflected the belief that it was involved in the regulation of voluntary motor activity and was also probably motivated by the success of psychophysical procedures in the study of vision, audition, and touch (Granit, 1981). Early neurophysiological research on the motor system in the 19th century was also dominated by studies of reflexes, a concept first formulated by Descartes in the mid-17th century. The development of anatomical techniques and methods for physiological experimentation during the 19th century contributed to an understanding of the sensory and motor structures involved in spinal reflex pathways. At this time, the role of sensory inputs in the control of voluntary movement was conclusively demonstrated by Mott and Sherrington (1895). They showed that when the dorsal roots in the brachial region of the spinal cord (C4–T4) of monkeys were unilaterally transected (i.e., dorsal rhizotomy), the animals made no voluntary movements and their affected hands appeared paralyzed, even though electrical stimulation of the motor cortex elicited movements. More recent studies have shown that many aspects of purposive movement can indeed be recovered following dorsal rhizotomy if appropriate care and training procedures are followed (Taub & Berman, 1968). Similar observations had in fact been made by Munk (1909), who described the learning that occurred in monkeys following unilateral deafferentation, provided that their normal limbs were bound. These animals were trained to open and close their hands, to grip, and to produce a variety of movements with their deafferented arms.

During this period in which the focus of motor research was on reflex pathways, there are isolated examples of studies on human movements, such as Bryan and Harter's (1899) analysis of the performance of Morse code operators and the improvement in their skills with training and Woodworth's (1899) and Hollingworth's (1909) monographs on the accuracy of voluntary arm movements. In his research, Woodworth examined the relation between speed and accuracy in goal-directed arm movements, a topic that continued to be of interest throughout the 20th century (e.g., Fitts, 1954; Keele, 1968; Stetson & McDill, 1923). He was also interested in the contribution of feedback (i.e., visual, proprioceptive, and auditory) to movement control and in the movement control processes that result in asymmetries in manual performance. For much of the 20th century, the study of human hand function was limited to the analysis of simple movements similar to those studied by Woodworth, although there were some notable exceptions, such as Bernstein and his colleagues, who started their analysis of movements using tasks such as hammering a nail and striking piano keys (Bernstein, 1967). By the later part of the 20th century, an emphasis on more natural movements emerged, in part stimulated by new methods that enabled the measurement of the kinematic and dynamic features of tasks, such as pointing (Georgopoulos, Kalaska, & Massey, 1981), reaching (Jeannerod, 1984; Wing & Fraser, 1983), and grasping (Westling & Johansson, 1984).

The scientific study of human hand function has always been significantly impeded by the technological demands inherent in systematically producing and displaying objects with multiple attributes and in precisely recording behavioral and neural responses. However, some of these difficulties have been surmounted recently by a third research thrust, which focuses on hand function from the perspective of hardware (e.g., robotics, haptic displays, recording somatosensory neural responses) and software design (e.g., virtual environment technology). The goal of this research is to design and build haptic and multisensory interfaces for remotely exploring and manipulating virtual and real worlds. We note that it is of critical ergonomic

importance that these hardware and software systems be built with the capabilities and limitations of the tactile/haptic and motor systems of the human user carefully considered. For example, software designers who are interested in haptically rendering objects and their properties (e.g., shape, size, surface texture, and compliance) in a virtual environment must deliver the types of tactile and haptic cues that the human user can use effectively. This holds true as well for teleoperator systems that deliver information to the hand(s) of a user who must interact with and act upon a real remote environment, as in the case of performing telerobotic repairs in space.

A New Conceptual Framework for Human Hand Function

The framework that we have adopted in this book conceptualizes hand function along a continuum that ranges from activities that are essentially sensory in nature to those that have a strong motor component. Within this sensorimotor continuum, we delineate four categories in detail, as listed in figure 1.1. Each of these categories was selected because it represents a comprehensive set of primary manual functions.

Tactile sensing serves to effect contact between the person's stationary hand and a surface or object which may or may not be moving. In contrast to the active haptic mode, in tactile sensing, the hand is always passive. This type of mode produces a variety of internal, subjective sensations. Although not typically used to learn about the properties of external objects and surfaces, tactile sensing does provide some information about certain properties (e.g., surface texture, thermal conductivity), especially when the object or surface is moved across the skin.

Active haptic sensing serves to effect contact between the person's hand as it moves voluntarily over a surface or object. The term haptic will be considered in detail later in the book, but in brief, it

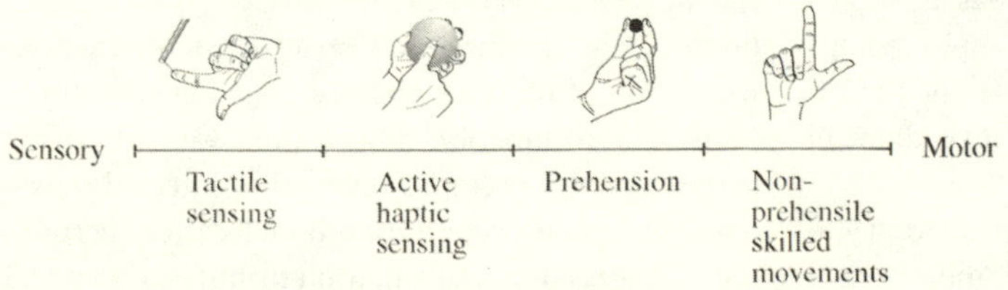

Sensory ├────────────┼────────────┼────────────┼────────────┤ Motor

        Tactile   Active   Prehension   Non-
        sensing   haptic   prehensile
                 sensing   skilled
                                       movements

Figure 1.1. A sensorimotor continuum of human hand function.

involves the use of sensory inputs provided by the stimulation of receptors embedded in skin, muscles, tendons, and joints. In contrast to the tactile mode, the hand is always active. This mode is preferred, if not essential, for identifying objects and extracting more precise information about their properties.

Prehension refers to those activities in which the hand reaches to grasp an object. Most of these activities involve holding an object in the hand. The configuration of the grasp is determined by the task objective and so will often change as the task progresses (e.g., reaching to pick up a pen and then writing with it). Precise and timely completion of the task usually relies on sensory feedback from mechanoreceptors in the hand and in the muscles controlling movements of the fingers.

Non-prehensile skilled movements refer to a diverse class of activities ranging from the gestures made as part of normal speech or as a substitute for speech, to the movements involved in depressing the keys on a keyboard. With the exception of pointing and aiming movements, these activities typically involve all fingers and both hands.

Scope and Contributions

This book surveys normal hand function in healthy humans. It does not attempt to compare tactile and proprioceptive sensory processing with other sensory systems, nor does it examine the vast literature on how damage to the peripheral or central nervous systems influences human hand function. Any comprehensive investigation of human hand function requires the exploration of a wide diversity of disciplines and research areas, including anatomy, neurophysiology, cognitive science, experimental psychology (including psychophysics), developmental psychology and gerontology, kinesiology, hand surgery and rehabilitation medicine, haptic software development and robotics, and human factors/ergonomics. Until very recently, these fields have tended to function relatively autonomously. To the best of our knowledge, there exists no single multidisciplinary source that has attempted to integrate these different but complementary research literatures. We therefore believe the time is now appropriate to provide such a compendium of human hand function.

Our book should be viewed as complementary to several other excellent books on the hand. Most of these may be categorized in terms of one or two of the disciplines listed above. Typically, each has focused in greater depth on a specific aspect of the hand, such as prehension (MacKenzie & Iberall, 1994), handedness and laterality (McManus, 2002), neural control of hand movements (Phillips, 1986; Wing, Haggard, & Flanagan, 1996), and tactile function (Gordon, 1978). In contrast, *Human Hand Function* selectively analyzes and synthesizes the results of fundamental research drawn from a broad range of disciplines that contribute in important ways to our understanding of human hand function. The book further explores how the results of research drawn from more than one of these disciplines may be applied to real-world scenarios that require effective use of the hand. In addition, the book considers ways in which other relevant disciplines might broaden a given field by expanding its current perspectives and methodologies.

Organization

Chapter 2 begins by describing the evolution of the hand. It then examines the hand's anatomy, mechanics, and skin structure. Finally, it considers biomechanical models of the hand.

Chapter 3 considers human hand function from a neurophysiological perspective. It is there that we describe in detail the peripheral receptors that are embedded in the hand's skin, muscles, and joints and the role of corollary discharges that accompany voluntary movements. We further address their associated neurophysiological functions and briefly describe the primary somatosensory pathways that arise in the peripheral receptors and end in the somatosensory cortex.

Chapters 4 and 5 review the performance of a number of manual behavioral tasks that require tactile and active haptic sensing, respectively. Collectively, these tasks span the sensory portion of the sensorimotor continuum of human hand function presented in figure 1.1. Topics in chapter 4 include spatial and temporal tactile sensitivity and resolution, tactile motion, and space-time interactions within somatosensory processing as they pertain to human hand function. In addition, we address the perception of objects (as opposed to internal sensations) and their properties, especially those pertaining to material and geometric attributes. Chapter 5 covers topics that include the critical role that hand movements play in recognizing common objects and their raised 2-D outline representations and in perceiving object properties, and sensory gating with active movement. Finally, we consider the active haptic sensing of objects and their properties when we use a variety of intermediate links (e.g., rigid probes) between the hand and the external environment.

Chapters 6 and 7 address performance related to prehensile and non-prehensile activities, respectively, which relate to the motor portion of the sensorimotor continuum (figure 1.1). In chapter 6, we analyze both reaching and grasping movements and discuss theories of reaching, the role of visual feedback in these movements, the coordination between grip and load force, the adaptation of grip forces to object properties, and anticipatory models of grasping. In chapter 7, we discuss pointing and aiming movements, the keyboard skills of typing and piano playing, issues related to sensory feedback and errors in performance, and the control of movement speed and force in keyboard skills. Finger tapping is also covered in this chapter.

Chapter 8 addresses the complete sensorimotor continuum in terms of the particular constraints imposed on human performance by using different parts of the hand. For tactile and active haptic sensing, topics include site of contact, unimanual versus bimanual sensing, and direct versus remote sensing;

for prehension, topics include site of contact, number of digits, bimanual grasping, remote versus direct action; and for non-prehensile skilled movements, topics include the independent action of individual fingers, the coupling of movements, hand preference, and bimanual movements.

Chapter 9 adopts a developmental perspective in which we consider what is (or needs to be) known about human hand function across the entire lifespan from birth through old age. Up to this point, we have considered human hand function in terms of fundamental research. Chapter 10 explores five application areas for which a sufficient amount of relevant fundamental research has been conducted so that directions and guidelines can be provided about how to approach problems that may arise in these fields. In addition, we consider how such knowledge can be used to guide the ways in which we may potentially improve on manual performance in these domains. The topics covered include evaluation of hand function and rehabilitation; talking and listening with the hand; sensory communication systems for the blind, specifically to display graphic information to the hand and to assist in the manual denomination of banknotes; haptic interfaces; and exploring art by touch.

Chapter 11 draws some conclusions about human hand function and offers a guide to future research on this fascinating topic.

2

Evolutionary Development and Anatomy of the Hand

For all its primitiveness, the human hand is functionally a highly sophisticated organ. The qualities that elevate it to the peerage lie in the degree of differentiation of its musculature, the intricacy of its nerve supply, and its generous representation in the higher centers of the brain.
NAPIER, 1976

In this chapter, the evolutionary development of the hand is first considered in the context of changes in the structure and function of primate hands. The differences between modern human and nonhuman primate hands are discussed with particular reference to tool use and manufacture. We then review the basic anatomy of the human hand as reflected in its bones, joints, muscles, and the structure of the skin surface. The sensory and motor innervation of the muscles and skin is described, and finally we conclude with a brief summary of some of the biomechanical models of the hand that have been developed.

Evolutionary Development of the Hand

From an evolutionary perspective, there are three major developments in the structure and function of the hand as it evolved from a support structure for locomotion to an organ devoted to manipulation. First, there is the development of the pentadactyl (five-fingered) hand with enhanced mobility of the digits, as compared to its predecessors. Second, claws thinned and moved dorsally to become flat

nails and so no longer provided assistance in holding the body when climbing; this was now accomplished by flexing the digits around a branch or by grasping a branch between the digits (Schultz, 1969). Finally, there is the development of highly sensitive tactile pads on the volar (palmar) surface of the digits (LeGros-Clark, 1959). In spite of these advances, the hands of primates have remained primitive in their skeletal structure; indeed, the primate hand is considered to be more primitive than the foot, face, shoulder, knee, or hip (Napier, 1976). The pentadactyl hand has been termed "the absolute bedrock of mammalian primitiveness" (Wood-Jones, 1944, p. 19), as it was established more than 350 million years ago (Mya) in the Devonian period (see figure 2.1) and is seen in the fossil lobefin fish *Eusthenopteron* (Kemble, 1987; Marzke, 1992). The phalangeal structure of 2-3-3-3-3 has also been a mammalian characteristic since the beginning of the Tertiary period (65 Mya), and the only change in the structure of the human hand from this ancient scheme is the absence of the os centrale, a small bone of the carpus (wrist) which has fused with the adjacent scaphoid bone as shown in figure 2.1 (Napier, 1976). The implication of this evolutionary

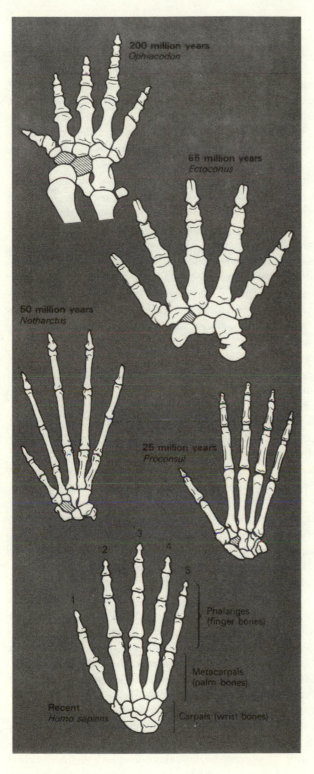

Figure 2.1. Skeletons of the hand from a mammal-like reptile (*Ophiacodon*) to modern *Homo sapiens*. The hatched bones represent the os centrale, which has fused with the adjacent scaphoid bone in *Homo sapiens*. Reprinted from Napier, 1976. Used with permission of the Carolina Biological Supply Company.

record is that the hand evolved early with capabilities that well preceded the cerebral structures required to take advantage of its potential.

A major factor in the evolutionary change in hand use was the development of the opposable thumb, which resulted in a remarkable increase in the versatility of hand function and promoted the adoption of bipedal locomotion. Opposition involves flexion, abduction, and medial rotation of the thumb so that the pulp surface can make contact with the terminal phalanges of one or all of the remaining digits. The articulation of the thumb metacarpal (the proximal bone of the thumb) and the trapezium (a bone of the wrist), in what is known as a saddle joint, enables the thumb to rotate 45° around its longitudinal axis. This movement of opposition is essential for effective handling and exploration of small objects and greatly facilitated primate tool use and construction. Not all primates have an opposable thumb; only Old World monkeys and apes are capable of rotating the thumb, and some primates, such as the colobus and spider monkey, are thumbless (Napier, 1993). The spider monkey compensates for its lack of manual dexterity by using its prehensile tail to pick up small objects. This is possible because the skin on the underside of the distal one third of its tail is hairless and is similar in function to the papillary skin on the hand. Although opposition of the thumb is evident in many Old World monkeys and apes (Napier & Napier, 1985), the extensive area of contact between the compressible pulps of the thumb and index finger is a unique human characteristic, as the thumb is positioned more distally and is longer relative to the index finger in human as compared to nonhuman primates (Napier, 1961). In addition, human hands are distinctive in the relative mass of the thumb musculature, which makes up about 39% of the weight of the intrinsic muscles within the hand, as compared to 33% in gibbons and 24% in chimpanzees and orangutans (Tuttle, 1969). Figure 2.2 illustrates various primate hands, including those of the three larger apes most closely related to humans, namely, the orangutan (*Pongo*), gorilla, and chimpanzee (*Pan*). For these primates, the fingers are so long relative to the thumb that manipulation typically involves the thumb with the lateral surface of the index finger rather than with the tips of the fingers (Marzke, 1992).

In the human hand, unlike those of other primates, the distal phalanges have rough horseshoe-shaped tuberosities (see figure 2.3) to which the soft tissues of the palmar pads are attached. These pads distribute pressure during grasping and allow the fingertips to conform to uneven surfaces (Napier, 1976; Young, 2003). The pad on the volar surface of the distal phalanx of the thumb is broader in humans than in other primates and is distinctive in that it is divided into two wedged-shaped compartments. This permits forces that are applied to the fingerpad to be dissipated more evenly. The proximal compartment is more deformable and mobile than the distal compartment, which allows the soft tissue to mold around an object so that it is held against the distal pulp (Shrewsbury & Johnson, 1983). The tip pinch between the pads of the thumb and index finger that is used to pick up very small objects appears to be a unique human characteristic and is possible because of the human ability to hyperextend the distal phalanx of the index finger passively. In many nonhuman primate species, the fingers must be flexed in order for the tips to meet the tip of the thumb, due to the considerably longer finger lengths relative to that of the thumb, as can be seen in figure 2.2 (Marzke, 1992).

A number of indices derived from the relative projections of various parts of the hand have been used to characterize the structure of different primate hands (Napier & Napier, 1967). These measures, and in particular the opposability index, which expresses the ratio between the lengths of the thumb and index finger (100(total length of thumb/total length of index finger)), have proven useful in characterizing the manipulative capacity of different hands. As the lengths of the index finger and thumb become more similar, the opposability index increases. In *Homo sapiens*, this ratio is 60, in the baboon 57–58 (*Papio* in figure 2.2), but in the orangutan (*Pongo* in figure 2.2) only 40 (Napier, 1993; Schultz, 1956). The baboon has a high ratio because its fingers are short relative to the metacarpals, presumably as part of its adaptation to terrestrial digitigrade locomotion. When it walks, the palmar surfaces of its phalanges make contact with the ground and the metacarpals are hyperextended (Marzke, 1992). Arboreal primates, such as the orangutan and chimpanzee (*Pongo* and *Pan*, respectively, in figure 2.2), have tended to develop a prehensile forelimb for locomotion from tree to tree, at the expense of having pulp-to-pulp contact between the thumb and other fingers (Kemble, 1987). However, in their manual activities, captive chimpanzees

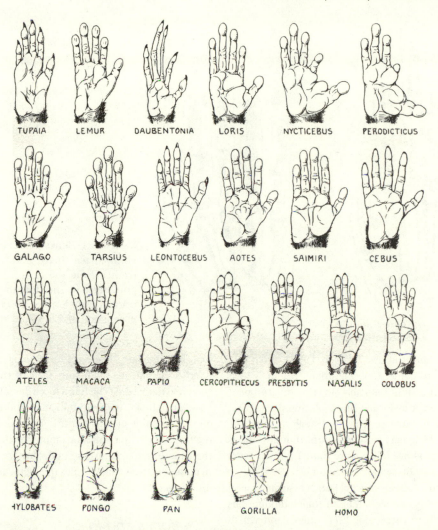

Figure 2.2. Right hands of a range of adult primates, all reduced to the same length. Reprinted from Schultz, 1969, with permission of the Orion Publishing Group.

make extensive use of grips that involve the thumb and lateral surface of the index finger, which does not require a long thumb (Marzke & Wullstein, 1996).

There has been a considerable amount of research on the relation between cranial capacity, the morphology of the hominid hand as it evolved, and the use and construction of stone tools. The discovery in 1960 in the Olduvai Gorge of 1.75 million-year-old fossil hand bones from an early human ancestor (*Homo habilis*) at the same level as primitive stone tools led to a debate that has continued for more than 40 years on the role of tools in the evolution of the human hand (Marzke & Marzke, 2000). The manufacture of specific tools was preceded by a

period of unmodified tool use in which objects found in the environment, such as stones, bones, or pieces of wood, were used for obtaining food or providing protection. It seems likely that tool use, which is evident in many animals, including primates, was followed by a period in which tools were modified or adapted to improve their performance. This type of behavior is evident in present-day chimpanzees, who "fish" for termites by poking long slender twigs down flight holes to a termite nest and then retracting the twig that is now covered in termites. Chimpanzees have been observed collecting twigs prior to arriving at a termite nest and removing any small offshoots or branches that would impede the twig's

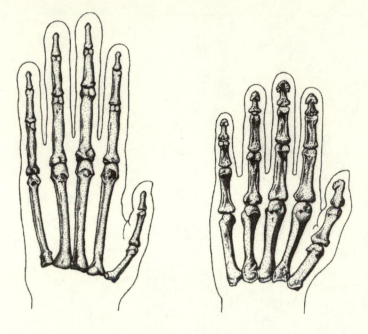

Figure 2.3. Bones of the hands of a chimpanzee (left) and a human (right). Note the tufts on the distal phalanges in the human hand. Reprinted from Young, 2003, with the permission of Blackwell Publishing.

movement into the nest (Napier, 1993). From tool modification came tool manufacture, in which objects occurring naturally in the environment (stones) were transformed into tools (hand axes) that had a specific purpose. This final stage required the intellectual capacity for conceptual thought and is considered to have taken more than 1 million years to emerge in tool users (Wiesendanger, 1999).

It is from analyses of fossil hands associated with various prehistoric tools that an understanding has arisen of how these tools were made and which species were capable of making them. It appears that a primitive tool culture existed at least 2 million years ago in Tanzania and that *Homo habilis* was capable of fashioning stones into useful shapes (Leakey & Lewin, 1977). Studies of contemporary stone tool-making behavior in both human (New Guinea stone tool makers) and nonhuman primates reveal the patterns of hand use required to make different types of tools. They indicate that a number of precision grips are used in tool manufacture, but that a fine precision grip between the tips of the thumb and index finger is not essential (Marzke, 1997).

Anatomy of the Hand

The basic structure of the human hand can be considered in terms of its bones and associated joints,

muscles, and skin. Standard terminology is used to refer to the various digits and parts of the hand as shown in figure 2.4. The hand has a dorsal surface, a volar or palmar surface, and radial and ulnar borders. The palm of the hand is divided into three regions: the thenar mass or eminence that overlies the thumb metacarpal, the midpalm area, and the hypothenar mass that overlies the metacarpal of the little finger.

Bones and Joints

There are 27 bones in the hand, with 8 carpal bones constituting the wrist, 5 metacarpal bones in the palm, and 14 phalangeal bones that make up the digits (2 in the thumb and 3 in each finger). The carpal bones are arranged in two rows, with the more proximal row articulating with the radius and ulna in the forearm. The proximal row comprises the scaphoid, lunate, triquetrum, and pisiform, and the distal row, which articulates with the metacarpals, comprises the trapezium, trapezoid, capitate, and hamate. The metacarpal bones in the palm of the hand articulate closely with the adjacent carpal bones in the distal row, and these carpometacarpal joints are capable of flexion/extension movements and radial and ulnar deviation. With the exception of the first metacarpal of the thumb, independent motion of these joints is very limited, but the range of movement increases

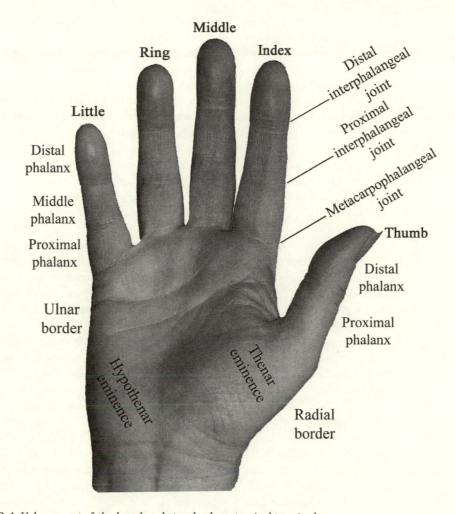

Figure 2.4. Volar aspect of the hand and standard anatomical terminology.

from the second to the fifth metacarpal (Kapandji, 1970; Taylor & Schwarz, 1970). The five metacarpophalangeal joints (biaxial or condyloid type) are universal or saddle joints capable of both flexion/extension and abduction/adduction movements (movements away from and toward the midline of the hand), whereas the nine interphalangeal joints in the digits are hinge joints capable of only flexion and extension. The three bones in the fingers are known as the proximal, middle, and distal phalanges, and each finger has three joints, the metacarpophalangeal (MP), the proximal interphalangeal (PIP), and the distal interphalangeal (DIP) joints.

    The total active range of motion of a typical finger is 260°, which is the sum of active flexion at the MP (85°), PIP (110°), and DIP (65°) joints (Ameri-

can Society for Surgery of the Hand, 1983). The range of active extension at the MP joint varies between people but can reach 30–40° (Kapandji, 1970). Passive and active flexion of the MP joint increases linearly from the index to the little finger, and the total active range of motion of the fingers also increases from the index to the little finger (Mallon, Brown, & Nunley, 1991). Although there is no passive extension beyond 0° at the PIP joint, at 30° it is appreciable at the DIP joint, a uniquely human feature, as noted previously. With the exception of the thumb, the index finger has the greatest range of abduction/adduction movements at 30°. These movements become difficult, if not impossible, when the MP joint is flexed due to tautness in the collateral ligaments of the joint. In contrast to the other digits, the thumb does not have a second phalanx and so has only two phalangeal bones and much greater mobility in the carpometacarpal joint, as described earlier. The carpometacarpal joint of the thumb is described by many authors as a saddle joint with two degrees of freedom. Although there is considerable axial rotation in addition to flexion/extension and abduction/adduction movements, this is constrained and so it is not considered a true third degree of freedom (Cooney, Lucca, Chao, & Linscheid, 1981). In total, the human hand, including the wrist, has 21 degrees of freedom of movement.

Muscles

There are 29 muscles that control movements of the hand, although some of these muscles are divided into distinct parts with separate tendons, such as the flexor digitorum profundus, which sends tendons to the distal phalanx of all four fingers. If these subdivisions of multitendoned muscles are counted separately, the number of muscles controlling the hand increases to 38 (Alexander, 1992). Most of the muscles that control movements of the hand are in the forearm and are known as the extrinsic hand muscles (see figure 2.5). These long flexor and extensor muscles of the wrist and fingers take their origin from the bones in the arm, and as they approach the wrist the muscle bellies are replaced by tendons (Kapandji, 1970). The flexor muscles arise from the medial epicondyle of the humerus in the upper arm or from the volar surface of the radius and ulna, and they then travel down the inside of the forearm. At the wrist, the flexor tendons pass through the carpal tunnel bounded dorsally by the carpal bones and on the volar surface by the transverse carpal ligament. The main flexor muscles are the flexor pollicis longus that flexes the distal joint of the thumb, the flexor digitorum profundus that flexes the DIP joint of each finger, and the flexor digitorum superficialis that flexes the PIP joint of each finger. Flexion of the wrist involves the flexor carpi ulnaris, flexor carpi radialis, and palmaris longus. It is of interest to note that the palmaris longus muscle is absent in about 15% of the population (Taylor & Schwarz, 1970), apparently without any functional consequences. It is sometimes used, however, as a donor of tendon graft material in correcting hand deformities (Taylor, Raj, Dick, & Solomon, 2004).

The extensor muscles take their origin from the ulna and course down the dorsal side of the forearm

(see figure 2.5). The dorsal carpal ligament guides the extensor tendons at the wrist, where they are arranged in six tendon compartments. The tendons of one of the extensors (extensor pollicis brevis) and abductors (abductor pollicis longus) of the thumb, which originate in the mid and distal parts of the radius, are in the first dorsal compartment. The second compartment contains the tendons of the extensor carpi radialis longus and extensor carpi radialis brevis muscles, which insert into the dorsal base of the index and middle metacarpals, respectively. The extensor pollicis longus, which is in the third dorsal compartment, inserts into the dorsal base of the distal phalanx of the thumb, and the extensor indicis proprius and extensor digitorum communis, which are the MP joint extensors of the fingers, are in the fourth dorsal compartment. The extensor digiti minimi, which extends the MP joint of the little finger, is in the fifth dorsal compartment, and the sixth compartment contains the tendon of the extensor carpi ulnaris. This muscle inserts into the dorsal base of the fifth metacarpal (American Society for Surgery of the Hand, 1983).

Muscles that take their origin and insertion within the hand are known as the intrinsic hand muscles. There are four groups of these, as shown in figure 2.6. The three thenar muscles (i.e., the abductor pollicis brevis, opponens pollicis, and flexor pollicis brevis) cover the thumb metacarpal and are involved in pronating the thumb, that is, raising the thumb up straight to form a 90° angle with the palm, and in opposing the thumb to the fingertips. The three muscles of the hypothenar eminence (abductor digiti minimi, flexor digiti minimi, and opponens digiti minimi) abduct (move the finger away from the midline of the hand) and flex the little finger. The four lumbrical muscles flex the MP joints of the fingers and extend the interphalangeal joints together with the four dorsal and three palmar interosseus muscles, which also adduct and abduct the fingers. The lumbrical muscles are unique in the human body in that they originate from tendons and not from bone.

Motor and Sensory Innervation

The muscles and skin of the hand are innervated by the radial, median, and ulnar nerves. The radial nerve innervates the extensors of the hand and the thumb abductor (abductor pollicis longus), and so its primary

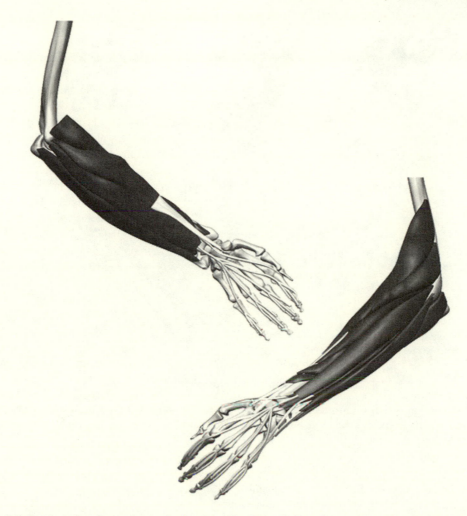

Figure 2.5. Extrinsic hand muscles in the forearm. The flexor (left) and extensor (right) muscles are shown separately. Geometric models of the muscles were created using finite element geometries that were fitted to anatomical data digitized from the Visible Human data set. Adapted from from Reynolds, Smith, & Hunter, 2004.

motor function is to innervate the muscles that extend the wrist and metacarpophalangeal joints and that abduct and extend the thumb. The median nerve innervates the flexors of the wrist and digits, the abductors and opponens muscle of the thumb, and the first and second lumbrical. The ulnar nerve innervates all other intrinsic muscles of the hand. Each of these major nerves branches extensively. The wrist and hand receive their blood supply from the radial and ulnar arteries, which run parallel to the bones and enter the hand through the flexor tunnel.

The sensory innervation of the hand is also provided by the radial, median, and ulnar nerves. The radial nerve innervates the radial three quarters of the dorsum of the hand and the dorsal surface of the thumb. It also supplies sensibility to the dorsal surfaces of the index and middle fingers and the radial half of the ring finger, as far as the proximal interphalangeal joint of each finger. The median nerve innervates the volar surfaces of the thumb, index, and middle fingers and the radial side of the ring finger. Dorsal branches of the nerve arise to innervate the dorsal aspect of the index and middle fingers distal to the proximal interphalangeal joint and the radial half of the ring finger. The ulnar nerve innervates the little finger and the ulnar half of the ring

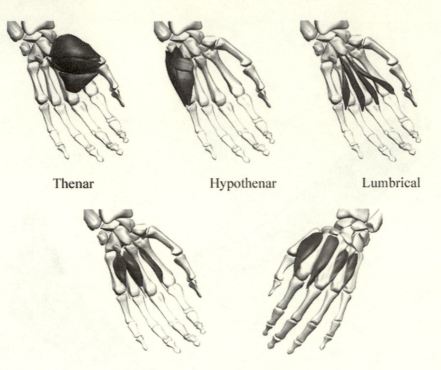

Thenar   Hypothenar   Lumbrical

Palmar and Dorsal Interossei

Figure 2.6. Four main groups of intrinsic muscles in the hand. Geometric models of the muscles were created using finite element geometries that were fitted to anatomical data digitized from the Visible Human data set. Adapted from Reynolds et al., 2004.

finger on the palmar surface as well as the dorsal aspect of the hand over the ring and little finger metacarpals, the dorsum of the little finger, and the dorso-ulnar half of the ring finger (American Society for Surgery of the Hand, 1983).

The exact pattern of sensory and motor innervation does vary considerably across individuals. For example, the entire ring finger and ulnar side of the middle finger may be innervated by the ulnar nerve, or the whole ring finger may be innervated by the median nerve. The index finger appears to be the only digit whose volar surface is always supplied by the median nerve. Similarly, the little finger is invariably innervated by the ulnar nerve (Önne, 1962). These variations need to be taken into account when there is injury to a major nerve innervating the hand. From the perspective of the functional use of the hand, the median nerve is the most important in that it conveys information from a large area of skin on the palmar surface of the hand and innervates the intrinsic muscles controlling the thumb.

Structure of the Skin

The skin on the volar surface of the hand is described as "glabrous," in contrast to hairy skin, which is found on the dorsal surface of the hand. The skin on the volar aspect is relatively thick and is capable of bending along the flexure lines of the hand when objects are grasped. These folds enhance the security of the grasp and demarcate the areas of the hand where the skin is mobile as compared to the adjacent areas that are tightly bound to the underlying tissue and bones. The skin is tethered by fibrillar tissue that connects the deep layers of the dermis with the sheaths of the flexor tendons. Papillary ridges are found on the palm of the hand and occur on those areas involved in grasping, where it is assumed that they act as friction pads (Napier, 1976). In contrast to the volar surface, the skin on the dorsum of the hand is thin, soft, and pliable. When the fingers are flexed, the skin opens out along the tension lines (Langer's lines), and when the hand is relaxed the skin recoils and accumulates as the transverse wrin-kles evident over the phalangeal joints.

All skin consists of two major subdivisions: epi-dermis and dermis. The epidermis is the outermost layer, which is responsible for the appearance and health of the skin. It also provides a protective bar-rier to prevent the loss of moisture and intrusion of bacteria and houses free (unencapsulated) nerve endings. The skin of the hand contains numerous eccrine sweat glands that keep the skin damp and assist in controlling body temperature through evaporative heat loss.

We show the structure of glabrous skin and the location of the mechanoreceptor endings in figure 2.7, although we reserve discussion of the structure and function of these mechanoreceptors until chap-ter 3. On the glabrous skin of the hand, the epider-mis is composed of five sublayers: from outer to inner layer, these are the stratum corneum, the stratum lucidum (particularly evident in the thick palmar skin), the stratum granulosum, the stratum spin-osum, and the stratum germinativum (figure 2.8).

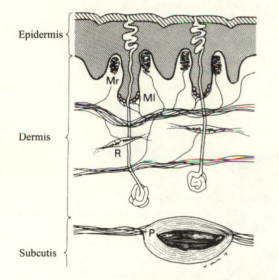

Figure 2.7. Vertical section through the glabrous skin of the human hand schematically depicts the two major divisions of the skin (epidermis and der-mis) and the underlying subcutaneous tissue. The locations of the organized nerve terminals, which are discussed in detail in chapter 3, are also shown. Mr = Meissner corpuscle; M1 = Merkel cell neurite complex; R = Ruffini ending; P = Pacinian corpuscle. Reprinted from Johansson & Vallbo, 1983, with per-mission of Elsevier Biomedical Press.

The cells in the stratum germinativum reproduce by subdividing. They subsequently migrate toward the surface while progressively dehydrating and flatten-ing in shape. With increasing age, this layer thins, making it more difficult to retain water. In young people, the newly generated cells take about 28 days (37 days for persons over the age of 50) to travel from the stratum germinativum to the stratum corneum. During cell migration, the cells fill with a granular substance known as keratin. The outer-most layer, known as the stratum corneum or corneum layer, which varies in thickness from 0.2 to 2 mm, contains flattened, dry cells of soft keratin that are tightly bound together and sloughed off. Indeed, it has been suggested that about 80% of all house dust consists of these dead skin cells.

The second major layer of skin, the dermis, is shown in figure 2.7. It consists of only two sublayers (not shown): the papillary (upper) and reticular (lower) layers (Thomine, 1981). The dermis is nor-mally five to seven times thicker than the epidermis and makes up 15–20% of the total body weight (Montagna, Kligman, & Carlisle, 1992). The dermis serves two primary functions: first, it nourishes the epidermis through its extensive network of blood vessels and capillaries; second, the dermis is com-posed of collagen and elastin and reticular fibers which form a strong supportive structure. The elastin fibers, which are concentrated in the lower layer of the dermis, give the skin its elasticity. Elastic-ity declines with increasing age. Within the dermis's fibrillar network are found encapsulated nerve end-ings, such as the Meissner corpuscles and Ruffini endings that are described in chapter 3, as well as blood vessels and sweat glands.

The epidermis and dermis are interconnected via a series of papillary folds, which give rise to the exter-nally observable fingerprint structure. It is interest-ing to note that during contact, droplets of sweat form evenly along the fingerprint ridges, as shown in figure 2.9. This has led to the intriguing speculation that during contact, the fingertip skin is softened, rendering it more compliant (Quilliam, 1978). The sweat also increases friction between skin and sur-face. In both ways, then, the production of sweat may help to stabilize an object when it is grasped.

A third layer of subcutaneous tissue consisting of fatty tissue lies beneath the dermis (figure 2.7) and houses the largest of the four mechanoreceptors, the Pacinian corpuscles, described in chapter 3. This layer,

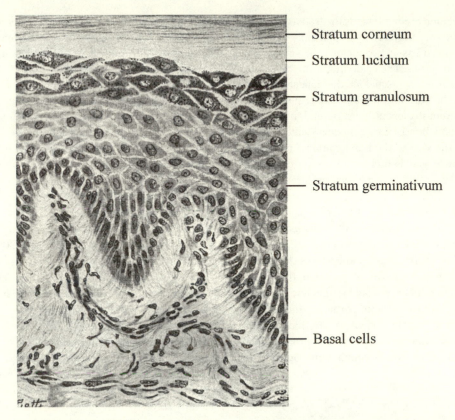

Stratum corneum

Stratum lucidum

Stratum granulosum

Stratum germinativum

Basal cells

Figure 2.8. Detail of the structure of the epidermis. Reprinted from Geldard, 1972, with permission.

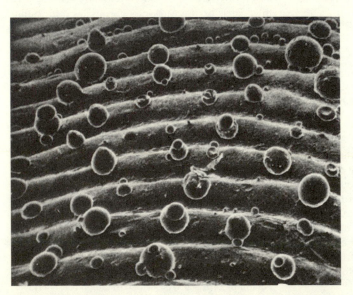

Figure 2.9. Scanning electron micrograph (SEM) of fingertip skin. The papillary ridges and the intervening sulci recur at regular intervals. At short intervals, globules of sweat exude from the summits of the ridges. Quilliam notes that it is possible that the shape and position of the sweat globules, which are seemingly permanently poised on the ridges, may be artifacts due to the method used to obtain the SEM (human index finger; araldite replica, field of view 5 mm). Reprinted from Quilliam, 1978, with permission of Pergamon Press.

known as the subcutis or hypodermis, is not formally considered to be a part of the skin and consists of a loose network of connective tissue bundles and septa, which connects with the dermis and muscles. The primary functions of the hypodermis are thermoregulation, protecting the underlying tissues against mechanical trauma, and providing a source of energy (Montagna et al., 1992).

The hairy skin on the dorsal surface of the hand is histologically similar to palmar skin, in terms of the layered structure of the epidermis (Montagna et al., 1992). The epidermis is thinner on the dorsal as compared to the palmar surface and the stratum corneum is only 0.02 mm thick (Thomine, 1981). The dermis is also thinner and more mobile in dorsal skin, as it lacks the connective and elastic elements that anchor the palmar skin to the underlying fascial planes. This allows the dorsal skin to stretch considerably with finger movements, and so as the middle finger moves from extension to full flexion, the skin can be lengthened by up to 30 mm. As discussed in chapter 3, it is this stretching of the dorsal skin during finger movement that provides an important input to cutaneous mechanoreceptors.

Models of the Hand

Biomechanical models of the forces and moment arms of the tendons and muscles controlling the fingers have been developed from anatomic and radiographic experiments on cadaverous specimens (Biryukova & Yourovskaya, 1994; Chao, An, Cooney, & Linscheid, 1989; Poznanski, 1974) and from measurements of the external dimensions of the hands (Garrett, 1971). Given the complex anatomical structure of the hand with 29 independent muscles, ligaments, and tendons and its anatomical and kinematic redundancy, a number of simplifying assumptions have been made in developing these models. Nevertheless, they have been used to estimate the centers of joint rotation, determine the muscle-tendon moment arms during angular rotation of the joint (Buford & Thompson, 1987), predict the posture of the hand as it grasps different sizes of objects (Buchholz, Armstrong, & Goldstein, 1992), and assist in selecting muscles for tendon transfer surgery (Brand, 1985).

The mechanical behavior of the fingers has also been studied in terms of their mechanical imped-

ance, which specifies the static and dynamic relation between limb movement and an externally applied force. The focus of this research has been to understand what factors influence the mechanical impedance of the hand (e.g., joint angle, muscle activation level, velocity of joint movement) and how this in turn affects the stability and performance of the haptic interfaces that are used for teleoperation (Colgate & Brown, 1994). For the index finger, the relation between force and displacement has been well characterized by a second-order linear model for both rapid (Hajian & Howe, 1997) and longer-term transients applied to the hand (Becker & Mote, 1990). In both of these studies, the stiffness and damping components of the identified models increased with fingertip force; for extension movements over the force range of 2–20 N, Hajian and Howe (1997) found that the damping ratio doubled (on average from 2.2 N s/m to 4.0 N s/m), and the stiffness increased fourfold (from 200 N/m to 800 N/m). This experiment involved very brief displacement perturbations, and the damping ratios identified were considerably higher than those reported in other studies with longer perturbations (e.g., Becker & Mote, 1990), during which reflex responses could have contributed to the system's response. It is therefore predicted that with longer time scales, the finger would be underdamped, consistent with the behavior of other joints. The fingertip is essential for haptic exploration of the external world and is the point of contact in the hand for many motor activities, such as grasping and typing. Any force that is applied at the fingertips is transmitted to the muscles, tendons, and other soft tissues in the hand and arm. There has therefore been interest in determining how the mechanics of the fingertip contribute to tactile perception and the control of grasping. Measurements of the contact force and fingertip displacement as individuals tap on a flat, rigid surface (Serina, Mote, & Rempel, 1997) or grasp an object between the thumb and index finger (Westling & Johansson, 1987) indicate that most of the displacement of the fingertip pulp occurs at forces of less than 1 N. As can be seen in figure 2.10, this relation is nonlinear. Due to the curvature of the fingertip, as the compression increases, there is a larger contact area, and so even at these low forces, the contact area increases rapidly with displacement as illustrated in figure 2.10. On the index finger, the contact area at 1 N is almost 70% of that at 5 N (Serina, Mockensturm, Mote, &

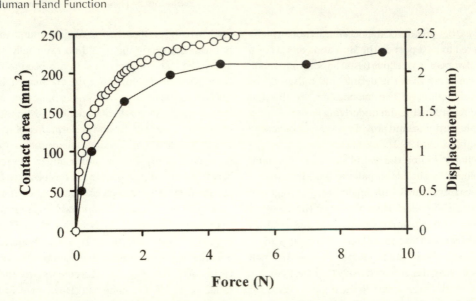

Figure 2.10. Relation between the force applied at the fingertip and the area of contact on the tip of the index finger during grasping (filled circles) and between finger force and the displacement of the index fingertip pulp during contact with a rigid surface (open circles). The force-contact area data are from Westling & Johansson, 1987, and come from a single participant, and the force-pulp displacement data are from Serina et al., 1998, and are also from a single participant. Redrawn from Serina et al., 1998, with the permission of the authors and Elsevier.

Rempel, 1998; Westling & Johansson, 1987). At higher forces, the pulp stiffens rapidly from around 3.5 N mm^{-1} at 1 N to over 20 N mm^{-1} at 4 N, which mechanically protects the distal phalanx from impact (Serina et al., 1997). These data suggest that it is the characteristics of the pulp tissues, and not the underlying hard tissues, that determine the force-displacement relation in the fingertip. They also indicate why the fingertip functions effectively as a tactile sensor at low forces (less than 1 N) in that small changes in force result in a considerable enhancement in tactile mechanoreceptor activity, due to the increase in contact area. At higher forces, there is only a modest increase in contact area, and so little additional tactile information is gained (Serina et al., 1998). The effect of fingertip force on tactile perception is discussed further in chapter 8.

The response of the finger pulp to forces applied tangential to the skin surface has also been investigated. These tangential forces play an important role in the control of grasping, particularly when an object begins to slip between the fingers (see chapter 6). Nakazawa, Ikeura, and Inooka (2000) measured the deformation of the tip of the thumb and middle and little fingers as forces were applied tangentially

to the skin surface, while the contact force was maintained constant at 1.9 N. Using a Kelvin model to represent the dynamics of the fingertips in the shearing direction, they estimated the stiffness and viscosity as a function of shear force amplitude. At small shear forces of around 0.6–1.2 N, they found that the stiffness and viscosity were large in comparison to the values obtained at higher shear force amplitudes and that at shear forces greater than 1.7 N, they remained relatively constant. Nakazawa et al. (2000) also reported that the shear stiffness of the thumb was the largest of the three digits tested, when the data were normalized for contact area. These findings suggest that when an object grasped in the hand is light, the shear deformation of the fingertip is small and that changes in shear stiffness will occur with compression of the fingertip.

A number of models of the fingertip pulp have been developed to predict the force-displacement and force-contact area responses of the human fingertip during contact with rigid objects. Serina et al. (1998) modeled the fingertip as an inflated, ellipsoidal membrane that was assigned the properties of skin and filled with an incompressible fluid. They showed that this structural model of the fingertip, which incorporated both its inhomogeneity (i.e., the layered structure) and geometry, predicted the nonlinear force-displacement data from the fingertip well (Serina et al., 1997). In addition, the model's predictions of contact area closely matched those measured experimentally. However, the model is limited in that the loading on the fingertip must be quasi-static and applied at a fixed angle. At other angles, the contact area is not circular, which violates one of the model's assumptions. Srinivasan (1989) modeled the pulp as a two-dimensional, infinite channel filled with incompressible fluid that was covered with a flat, elastic membrane. This model predicted the surface deformation of the fingerpad that matched experimental data from the primate finger to within 3 mm of the load, but it was limited to line loads. More recently, Dandekar, Raju, and Srinivasan (2003) have developed a multilayered, 3-D, finite-element model of the fingertip based on realistic external geometries, which is able to predict the behavior of mechanoreceptors in the skin. Nonlinear viscoelastic models of soft tissue (Fung, 1993) have also been used to describe the force-displacement response of the fingertip pulp to a range of compression velocities (Pawluk & Howe, 1999). This viscoelastic model, which incorporates a force-relaxation function, was able to predict fingertip force from pulp compression during rapid finger tapping, such as occurs when typing (Jindrich, Zhou, Becker, & Dennerlein, 2003). Fung's model's parameters were similar to those estimated by Pawluk and Howe (1999), despite the much greater dynamic loading conditions in the finger-tapping experiments.

These models of the dynamic response of the human fingerpad to loads have generally been evaluated within fairly restricted experimental conditions. Due to the complexity of the skin and underlying tissues, which are nonlinear, inhomogeneous, and anisotropic, many assumptions have been made in developing these models. Nevertheless, from a biomechanical point of view, they have achieved some success in identifying the critical elements involved in prehension and in maintaining a stable grasp. This research would clearly benefit from better in vivo models of the glabrous skin of the hand.

The evolution of the hand and its present anatomical configuration provide a framework for understanding the versatility and limitations of the human hand. Clearly, the development of an opposable thumb resulted in a dramatic increase in the types of tasks that the hand could now undertake, from exploring small items within the hand to grasping objects between the fingers and thumb. These functions are described in more detail in chapter 8. However, it is not only the structural elements of the hand that have contributed to its versatility, but also the soft tissues within the hand and the properties of the glabrous and hairy skin that covers the palmar and dorsal surfaces, respectively. The sensory mechanoreceptors within the skin are uniquely positioned to provide the central nervous system with information about finger movements and the material and geometric properties of objects held within the hand. These receptors are the subject of the next chapter.

3

Neurophysiology of Hand Function

An important conceptual advance of recent years is the understanding that the study of sensory performance by quantitative methods, called Psychophysics, and the study of neural events that follow sensory stimuli by the method of electrical recording, called Sensory Neurophysiology, represent only different approaches to what are generically the same set of problems. The methods and the concepts of the two can be combined to yield greater insight into those problems than is possible with either alone.

MOUNTCASTLE, 1975

In this chapter, we commence with a discussion of the structure and function of peripheral sensory mechanisms that are relevant to all four subclasses that collectively define the hand-function continuum (figure 1.1). This includes mechanoreceptors and thermoreceptors in the skin, proprioceptive inputs from muscles and joints, and centrally originating signals known as corollary discharges. Peripheral motor neurophysiology is not covered in this chapter, as the general principles of motor unit organization and muscle activation are similar in skeletal muscles throughout the body (see McComas, 1996; Vrbova, 1995). We subsequently describe the primary sensory pathways that project to the cortex and the principal cortical areas that are involved in processing sensory and motor information involving the hand. We close the chapter by discussing the nature of cortical sensory and motor representation and plasticity.

Mechanoreceptors: Glabrous Skin

Areas of Controversy

Before describing the different mechanoreceptor-afferent populations and their presumed functions,

it is important to highlight several critical areas of discussion or controversy. First, a good deal of what is known about hand function originally derived from two separate but related scientific disciplines, namely, psychophysics and single-unit neurophysiology. It was Mountcastle (1967) who provided the critical impetus to combine these two experimental perspectives simultaneously within a single organism because they served the same purpose: to understand sensory performance and its underlying neural mechanisms. At that time, however, it was not possible to perform both techniques within the same subject. Hence, monkeys were used in the neurophysiological studies and humans in the corresponding psychophysical research. It was assumed that "what monkeys and humans feel with their hands is in principle the same, and that neurophysiological observations made in the one may with some validity be correlated with psychophysical measures in the other, given the precise experimental design in the two cases" (Talbot, Darian-Smith, Kornhuber, & Mountcastle, 1968, p. 301).

It is important to highlight the fact that while the structure and function of their nervous systems are highly similar, there are also differences that may alter perceptual function in humans and monkeys in meaningfully different ways. In the next section, we will discuss in detail the four mechanoreceptor populations that are present in the glabrous skin of humans. For now, we merely note that while one of these mechanoreceptor populations, slowly adapting type II (SA II), has been reliably identified in the human hand using microneurography (Vallbo & Johansson, 1978), it has never been observed in neurophysiological studies with monkeys (Johnson, 2001). Another difference between human and monkey hands relates to the location of the Pacinian corpuscle mechanoreceptors. These are concentrated near nerves and blood vessels in the metacarpophalangeal (MP) region of the human cadaver hand (Stark, Carlstedt, Hallin, & Risling, 1998), as opposed to the thenar and hypothenar areas in monkey hands (Kumamoto, Senuma, Ebara, & Matsuura, 1993). The functional consequences, if any, of differences such as these have rarely been investigated.

Second, for well over a century now, investigators have argued over the relationship between various cutaneous afferent fibers and their receptor endings. A number of different methods have been used to address this issue, including histology, single-unit recordings, immunofluorescence labeling, and even to some extent simple logic. Given incomplete evidence pertaining to afferent fiber-receptor associations in human skin, investigators have assumed the existence of several associations based on the available histological and neurophysiological evidence in animals and humans. Conclusive evidence has only been obtained for the association between the considerably larger Pacinian corpuscles and fast adapting type II (FA II) afferents.

Third, over the past several decades, researchers in the fields of neurophysiology and psychophysics have become interested in determining the nature of the codes that the nervous system uses to represent human somatosensory sensations and percepts. Until about the middle of the 20th century, the two disciplines worked separately but in parallel on complementary facets of this problem. Whereas neurophysiological studies focused on the relation between the physical stimulus (physical events) and the responses of single afferent units (neural events), psychophysical studies focused on the nature of the psychophysical law, that is, the mathematical function describing the relation between the physical stimulus (physical events) and human tactual sensory experiences (mental events). More recently, investigators have tried to relate these two branches in their attempt to understand how the somatosensory system represents human perceptual experience. However, problems do arise because in many cases the neurophysiological research has been based on animal models and often on the responses of anesthetized animals (e.g., monkey, cat, raccoon, mouse), whereas the psychophysical work has been based on the responses of alert humans. One interesting attempt to bridge the gap was developed in Sweden in the late 1960s, when investigators combined two techniques, microneurography (recording from a single afferent unit) and microstimulation (stimulation of the same afferent), within an attentive human observer (e.g., Knibestöl & Vallbo, 1970; Macefield, Gandevia, & Burke, 1990; Vallbo, 1995).

The manner in which the somatosensory system represents sensory experiences is a matter of some controversy. One approach advocates a feature-detector model, in which a specific population of tactile units is responsible for encoding a specific feature, such as roughness, softness, curvature, orientation, slip, skin stretch, or vibration (see Connor & Johnson, 1992; Dodson, Goodwin, Browning, & Gehring, 1998; Goodwin & Wheat, 1992; Johansson & Westling, 1987; LaMotte & Srinivasan, 1987a, 1987b; Talbot et al., 1968). An alternate class of model has argued for a multichannel approach to human sensation in which multiple information-processing channels with overlapping sensitivities, each one mediated by its own neural mechanism, simultaneously encode the physical stimulus continuum (e.g., the multichannel model of vibratory sensation proposed by Bolanowski, Gescheider, Verrillo, & Checkosky, 1988). We will discuss some aspects of these approaches in greater detail.

Types of Mechanoreceptors

Tactile sensing activates a variety of tactile units, each consisting of an afferent fiber and its (presumed) ending. There are four different types of endings in the glabrous skin of the human hand: Merkel cells, Meissner's corpuscles, Ruffini endings, and Pacinian corpuscles. Merkel cells are located in clusters within the basal layer at the tip of the deep epidermal folds that project into the dermis. Meissner's corpuscles are ovoid structures that are also somewhat superficially located, specifically at the epidermal–dermal junction in the dermal papillary ridges. Ruffini endings are spindle-shaped structures that are located more deeply within the connective tissue of the dermis. Pacinian corpuscles are multilayered, onion-shaped structures that are large enough to be visible to the naked eye. They are located deep within both the dermis and the subcutaneous fat layer that lies below the dermis. The four types of endings can be seen in figure 3.1.

Four separate populations of tactile afferents have been documented in the glabrous skin of the human hand (e.g., Knibestöl & Vallbo, 1970; for valuable summaries, see Johansson & Vallbo, 1983; Johnson, 2001; Vallbo & Johansson, 1984). These are known as SA I (slowly adapting type I; described as SA in the monkey), FA I (fast adapting type I; in the monkey described as RA, or rapidly adapting, and in the older literature, as QA, or quick adapting), FA II (fast adapting type II; also known as PC or Pacinian in the monkey), and SA II (slowly adapting type II; never reported in the monkey) afferents. In this book, we adopt the afferent nomenclature used by the author(s) of each study under discussion. Although all populations respond to mechanical stimulation, they may be differentiated using several criteria. First, each type of afferent fiber is (presumed to be) associated with a specific type of ending. In human glabrous skin, SA I afferents end in Merkel cell neurite complexes, SA II afferents in Ruffini end-

ings, FA I afferents in Meissner's corpuscles, and FA II units in Pacinian corpuscles.

Microneurography has been used to differentiate and identify the physiological responses of the four types of human cutaneous afferents (e.g., Knibestöl & Vallbo, 1970). In this procedure, the nerve is first located by palpation or low-intensity electrical stimulation. A recording microelectrode (an insulated tungsten needle) is then manually inserted into the skin, usually with a pair of isolated forceps. The microelectrode penetrates the underlying tissue until a nerve fascicle (skin or muscle) is found and impaled. The electrode is then moved in very small increments until the appropriate activity from a single unit in the nerve is recorded, which may take from a few minutes to an hour (Prochazka & Hulliger, 1983). The receptive field of the specific tactile unit that has been engaged by the electrode is initially mapped out by exploring the skin of the hand with a probe. The relative physiological response of the single afferent unit to sustained stimulation is then determined. These two parameters, relative receptive-field size and relative response to sustained stimulation, are the criteria used to identify the type of unit, as shown in figure 3.2.

Figures 3.3 and 3.4 should also be consulted in conjunction with the following description of the four mechanoreceptor populations. Figure 3.3 shows the innervation density for each of the four mechano-

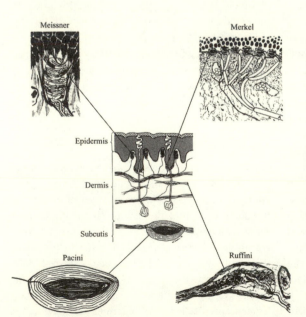

Figure 3.1. Vertical section through the glabrous skin of the human hand (center) schematically demonstrating the structure and locations of organized nerve terminals found in this area. The Meissner corpuscle and the Merkel cell neurite complex are shown in the expanded drawings (top left and top right, respectively). Reprinted from Andres & Düring, 1973, figure 16; and from Iggo & Muir, 1969, figure 1, with the permission of the authors, Springer, and the Physiological Society and Blackwell Publishing, respectively. The Pacinian corpuscle and Ruffini ending are shown in the expanded drawings (bottom left and bottom right, respectively). Reprinted from Chambers, Andres, Düring, & Iggo, 1972, figure 8, with the permission of the Physiological Society and Blackwell Publishing. The combined figure was adapted from Vallbo & Johansson, 1984, with the permission of Springer-Verlag.

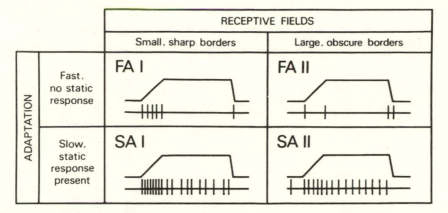

Figure 3.2. Types of mechanoreceptive afferent units in the glabrous skin of the human hand, classified on the basis of adaptation and receptive field properties (Knibestöl, 1973, 1975; Knibestöl & Vallbo, 1970). Graphs show the impulse discharge (lower trace) to the perpendicular ramp indentation of the skin (upper trace) for each unit type. The FA I units are preferentially sensitive to the rate of skin indentation, whereas the FA II units are highly sensitive to acceleration and higher derivatives. The fast adapting units, particularly the FA II units, respond not only when the indentation is increased, but also when the stimulus is retracted. The slow adapting units exhibit a sustained discharge during constant skin indentation in addition to their discharge during increasing skin indentation. The SA I units have a high dynamic sensitivity and often a rather irregular sustained discharge. The SA II units, on the other hand, have a less pronounced dynamic sensitivity and a very regular sustained discharge. Often they show a spontaneous discharge in the absence of tactile stimulation. Reprinted from Johansson & Vallbo, 1983, with permission of Elsevier Biomedical Press.

receptor-afferent populations in glabrous skin on the palm and at two different sites on the finger (see Paré, Smith, & Rice, 2002, for mechanoreceptor distributions in the distal fingerpads of the monkey). Figure 3.4 shows the receptive-field sizes of the mechanoreceptors across the palmar surface of the hand.

The SA I units are assumed to end in Merkel cell neurite complexes (Figure 3.1) and are densely distributed (i.e., about 1 per mm²; figure 3.3). They are particularly sensitive to features of the local stress-strain field (Phillips & Johnson, 1981), which render them strongly sensitive to fine spatial details (e.g., points, edges, bars, corners, and curvature) as opposed to overall indentation. Their receptive fields are relatively small, about 2–3 mm in diameter (see figure 3.4), with local areas within the receptive field that are highly sensitive. Relative to the receptive fields of the FA I units, their size is minimally affected by increasing the depth of penetration. This implies that human tactile pattern recognition is independent of contact forces ranging from 0.2 to 1.0 N (Johnson, Yoshioka, & Vega-Bermudez, 2000). SA I spatial resolution is minimally influenced by the scanning velocity up to at least 80 mm/second (Vega-Bermudez, Johnson, & Hsiao, 1991). The response of

SA I units to repeated skin indentation is invariant; that is to say, the variability is around 1 impulse/trial, independent of the number of action potentials elicited (Vega-Bermudez & Johnson, 1999). Given these response characteristics, humans can discriminate form and texture very well. As SA I units constitute the only population that responds linearly to skin indentations up to about 1500 μm (Blake, Johnson, & Hsiao, 1997), they also code curvature very well (Bisley, Goodwin, & Wheat, 2000). Humans are able to differentiate curvature without regard to contact force (normal) (Goodwin, John, & Marceglia, 1991) or contact area (Goodwin & Wheat, 1992), which suggests that like other forms of spatial pattern, curvature is neurally represented by a spatially distributed, as opposed to an intensive, code. Based on psychophysical data, it has been suggested that the SA I units are maximally sensitive to very low temporal frequencies of vibrotactile stimulation ranging from about 0.4 to 3 Hz, which usually result in the sensation of "pressure" (Bolanowski et al., 1988). Single-unit recording data have also been obtained with respect to the SA I unit responses to sinusoidal stimulation across a frequency range of 0.5–400 Hz (Johansson, Landström, & Lundström,

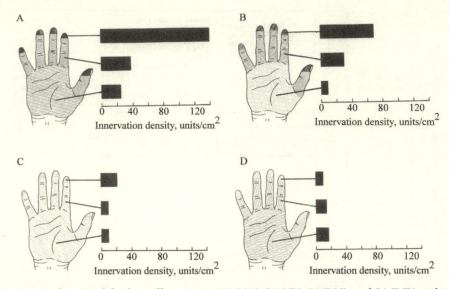

Figure 3.3. Average density of the four afferent units: FA I (A), SA I (B), FA II (C), and SA II (D) within different glabrous skin areas. Each dot in the drawings of the hand represents a single sensory unit innervating the skin area. Histograms give the density of innervation in the following three skin regions: the tip of the finger, the rest of the finger, and the palm. Reprinted from Vallbo & Johansson, 1984, with permission of Springer-Verlag.

1982). When these responses, which were originally reported in terms of number of impulses/cycle, are replotted in terms of the intensity characteristics of the mechanoreceptor, the low-frequency range of sensitivity of the physiological data resembles the human psychophysical data (Bolanowski et al., 1988). SA I units in the human fingertip respond to the application of normal force (e.g., Westling & Johansson, 1984). In addition, when a force is applied to a site on the human fingertip at a 20° angle from the normal, all afferents that are reliably activated (i.e., SA I, SA II, and FA I, but not FA II units) respond; however, in most units the responses are broadly tuned to a specific direction. With respect to the SA I population, response sensitivity is greatest to tangential force components applied in the distal direction (Birznieks, Jenmalm, Goodwin, & Johansson, 2001; Goodwin & Wheat, 2004). SA units in the primate fingerpad also show sustained responses to lateral stretch (Srinivasan, Whitehouse, & LaMotte, 1990).

The FA I units are presumed to terminate in Meissner's corpuscles (figure 3.1). They are even more densely distributed than the SA I units, about 1.5 units per mm^2 in the fingertip (figure 3.3). Like the SA I units, their receptive fields are relatively small, about 3–5 mm in diameter (figure 3.4). However, because they respond evenly across their entire recep-

tive field, they are poorer at discriminating very fine spatial details. Although they are insensitive to static contact, they respond well to transient deformation and particularly to the low-frequency vibration (Johnson et al., 2000) that occurs during initial contact and periods of relative motion between skin and object. FA I units respond to the application of normal force (e.g., Westling & Johansson, 1984) and are most sensitive to tangential force components in the proximal and radial directions (Birznieks et al., 2001). In addition, FA I units provide critical feedback for precise grip control (Johansson & Westling, 1984), in that they detect both actual slip between skin and object and local micro-slips. FA I units also signal forces that act suddenly on an object grasped in the hand (Macefield, Häger-Ross, & Johansson, 1996). Based on psychophysical data, it has been suggested that FA I units are maximally sensitive to vibrotactile temporal frequencies ranging from about 3 to 40 Hz, which produce the sensation of "flutter" (Talbot et al., 1968). As noted above, the shape of the psychophysical curve for threshold amplitude as a function of sinusoidal frequency is similar to that obtained for FA I unit responses to sinusoidal stimulation (0.5–400 Hz) by Johansson et al. (1982), when the latter data are replotted in terms of intensity characteristics (figure 3.5). We further note, how-

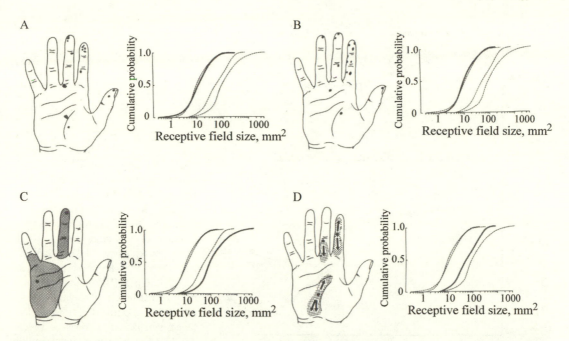

Figure 3.4. Receptive field size of four types of afferent unit. Each graph indicates the cumulative distribution curves of receptive field sizes for all four unit types, with the relevant function of interest highlighted (dark line). (A, B) The black patches on the drawings of the hand indicate receptive fields of 15 FA I units (A) and 15 SA I units (B), as measured with von Frey hairs that provide a force of 4–5 times the threshold force of the individual unit (based on a total sample of 255 units). Dark curves refer to the FA I (A) and SA I (B) units, respectively. Most of the fields are circular or oval, and they usually have a size between 3 mm^2 and 50 mm^2, which correspond to circular areas of 2–8 mm diameter. (C, D) The shaded areas in the drawings of the hand illustrate the receptive fields of two FA II units (C) and four SA II units (D), as determined by manually delivered taps with a small glass rod and skin stretch, respectively. The locations of the zones of maximal sensitivity are indicated by dots. With respect to the SA II units (D), the arrows indicate directions of skin stretch, which give rise to an increased discharge. Dark curves on the graphs indicate cumulative distributions of receptive field sizes of the FA II (C) and SA II (D) units, respectively. Combined figure adapted from Vallbo & Johansson, 1984.

ever, that the psychophysical function is more sensitive than the neurophysiological function. For a detailed discussion of the cause of the discrepancy, the reader should consult Bolanowski et al. (1988).

Each FA II unit ends in a single Pacinian corpuscle (figure 3.1). These units are less densely distributed in the hand than are either SA I or FA I units (about 350 per finger and 800 in the palm; figure 3.3). They also have extremely large receptive fields, which render the FA II units poor at discriminating spatial details (figure 3.4). However they are exquisitely sensitive to transient stimulation, including vibration. For example, Talbot et al. (1968) showed that the thresholds of some FA II units for sinusoidal stimulation were as low as ~1 μm at 200–300 Hz. Sensitivity dropped steeply below about 150 Hz, presumably the result of the low-frequency temporal filtering properties of the Pacinian corpuscle with its multilayer structure. FA II units are also very important for detecting more remote events, for example, those that occur with hand-held objects (Johnson, 2001). Based on psychophysical data, it has been suggested that FA II units are maximally sensitive to vibrotactile temporal frequencies ranging from 40 to >500 Hz, with a maximum sensitivity around 300 Hz, and that they produce the sensation of "vibration" within that frequency range (Bolanowski et al., 1988). One notable difference between the psychophysical function and the FA II neurophysiological function directly obtained by Johansson et al. (1982) is that the decrease in sensitivity over the lower frequency range is steeper for the former than for the latter.

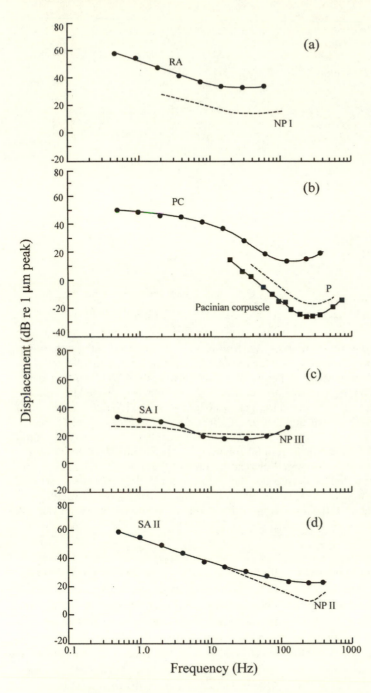

Figure 3.5. Relationship between physiologically measured frequency characteristics for different fiber types—(a) RA; (b) PC; (C) SA I; and (d) SA II—and psychophysically obtained vibrotactile thresholds as a function of frequency for (a) NP I (Bolanowski et al., 1988: 2.5–12 Hz; Verrillo & Bolanowski, 1986: 12–150 Hz); (b) P (Bolanowski & Verrillo, 1982); (C) NP III (Bolanowski et al., 1988); and (d) NP II (Gescheider, Sklar, Van Doren, & Verrillo, 1985: 15–150 Hz; Verrillo & Bolanowski, 1986: 100–500 Hz) channels. Neurophysiological data points are interpolations and extrapolations of the average results presented by Johansson et al., 1982, for selected response criteria: (a) 1 impulse/stimulus; (b) 4 impulses/stimulus; (C) 0.8 impulses/s; and (d) 5 impulses/s. For the P channel (b), an additional physiological curve (Pacinian corpuscle), which was derived from unpublished results of Bolanowski, 1981, has been plotted. This curve is the average response of six excised Pacinian corpuscles maintained at 33°C and for a response criterion of four impulses occurring during the central 200 ms of a 300-ms stimulus burst. Reprinted from Bolanowski, Gescheider, Verrillo, & Checkosky, 1988, with the permission of the Acoustical Society of America and the authors.

SA II units have never been found in monkey skin (Darian-Smith & Kenins, 1980; Lindblom, 1965), although they have been reliably detected in the skin of humans (Knibestöl & Vallbo, 1970). Like FA II units, their density is relatively low compared to either the SA I or FA I units (figure 3.3; Wu, Ekedahl, & Hallin, 1998). They have long been associated with

Ruffini endings that are situated within the dermis (figure 3.1). However, this traditional view has been challenged by a recent study on the distribution of Ruffini endings in the distal phalanx of the human finger in three human donors (Paré, Behets, & Cornu, 2003). Using immunofluorescence labeling, Paré et al. found only one presumptive Ruffini corpuscle. They concluded, "[I]t seems unlikely that Ruffini corpuscles in human glabrous skin account for all but a small proportion of physiologically identified SA II afferents" (p. 265). Paré, Smith, and Rice (2002) have suggested that "clumped" Merkel endings are the terminals of the SA I units, and further that the elongated "chain" Merkel endings may in fact be the source of the SA II afferent responses. The receptive fields of SA II units are about five times larger and their sensitivity to skin deformation about six times poorer than those of the SA I units (figure 3.4). Hence, they are not useful in detecting fine spatial details. Based on human psychophysical data and by a process of elimination, Bolanowski et al. (1988) have suggested that SA II units are most sensitive to vibrotactile frequencies ranging from 100 to >500 Hz and produce the sensation of "buzzing" (personal observation; Bolanowski, Gescheider, & Verrillo, 1994). SA II units are two to four times more sensitive than SA I units to skin stretch (particularly in the proximal direction), as described later in this chapter (Birznieks et al., 2001; Johnson et al., 2000). They contribute to the perception of the direction of object motion and force when an object produces skin stretch (Olausson, Wessberg, & Kakuda, 2000). Together with muscle spindles (and perhaps joint afferent units), SA II units also play an especially important role in perceiving hand configuration and finger position (Edin & Johansson, 1995).

Mechanoreceptors: Hairy Skin

There are five main types of mechanoreceptors with large myelinated afferents that have been identified in hairy skin, two of which are slowly adapting, namely, Merkel cells and Ruffini endings, and three that are rapidly adapting. The latter include non-encapsulated hair-follicle receptors that innervate hairs, field units whose histological structure remains unclear, and Pacinian corpuscles (Vallbo, Olausson, Wessberg, & Kakuda, 1995). These mechanoreceptors and their associated afferent fibers have properties similar to the tactile units described in glabrous skin (Greenspan & Bolanowski, 1996; Järvilehto, Hämäläinen, & Soininen, 1981). In hairy skin, the SA I receptors are Merkel cells that are found within the epidermis and are grouped under visible touch domes with each cell receiving a single afferent that innervates many receptors. The Ruffini endings (SA II receptors) are usually found around hair follicles in the dermis. Hair-

follicle receptors are usually identified by their brisk response to movements of individual hairs and light air puffs, whereas field units are not sensitive to hair movements and are classified on the basis of their rapidly adapting responses and the multiple spots with high sensitivity found within their receptive fields (Vallbo et al., 1995). In contrast to glabrous skin, Pacinian corpuscles are rare in hairy skin and when found are usually in the vicinity of joints, tendon sheaths, and interosseous membranes (Gandevia, 1996; Zimny, DePaolo, & Dabezies, 1989). There are also unmyelinated (conduction velocities of about 1 m/s) tactile units in hairy skin, some of which have low thresholds to mechanical stimuli. Their effects within the central nervous system (CNS) and the role they play in perception have not been fully elucidated as yet (Vallbo, Olausson, & Wessberg, 1999), although there is some evidence that they are involved in the affective aspects of touch, sometimes termed "limbic touch" (Olausson et al., 2002).

Neurophysiological studies of large myelinated afferents indicate that RA units on the dorsum of the hand have discrete receptive fields with distinct borders and are primarily found close to joints, whereas SA units are more uniformly distributed across the dorsal surface (Edin & Abbs, 1991). The relative number of SA units is higher in the hairy skin on the back of the hand than in the glabrous skin on the palm (Järvilehto et al., 1981), and the distinction between SA I and SA II units is not as clear as in glabrous skin (Edin & Abbs, 1991). In addition, the properties of some SA units in hairy skin differ from those in glabrous skin (Järvilehto et al., 1981). One difference between SA I units in hairy and glabrous skin is that in hairy skin they are responsive to levels of stimulation that are below human perceptual thresholds (Harrington & Merzenich, 1970), and they can discharge continuously without eliciting any perceptual response (Järvilehto, Hämäläinen, & Laurinen, 1976). However, the force thresholds for activating tactile units in hairy skin are similar to those reported for glabrous skin, with SA II units having the highest thresholds (5 mN) and the field units the lowest (0.1 mN) (Edin & Abbs, 1991).

Bolanowski et al. (1994) have suggested that tactile sensations resulting from mechanical (i.e., vibratory) stimulation of hairy skin are determined by activity in three psychophysically defined channels, each neurally controlled by a separate mechanoreceptor population. Based on the threshold-frequency characteristics recorded from afferent fibers arising from mechanoreceptors in the hairy skin of the forearm and the changes in their responses with variations in skin temperature and contact area, Bolanowski et al. proposed that three channels mediate mechanoreception in hairy skin: the $NP_{h\,low}$ channel (non-Pacinian, hairy skin, low frequency), the $NP_{h\,mid}$ (non-Pacinian, hairy skin, middle frequency), and the P (Pacinian) channel. The $NP_{h\,low}$ is mediated by SA II fibers that innervate Ruffini endings and is the neural substrate for low-frequency responses (0.4–4 Hz). The $NP_{h\,mid}$ channel, which is probably mediated by rapidly adapting hair-follicle receptors, responds to stimulation frequencies in the midrange (4–45 Hz or 4–150 Hz), depending on contactor area. Finally, the P channel mediates responses to high-frequency vibrotactile stimulation ranging from about 45 to more than 500 Hz. The responses in this channel are probably mediated by Pacinian corpuscles located in the subcutaneous tissue. The above results suggest that the capacity of hairy skin to signal mechanical events differs from glabrous skin, which probably uses four channels, with one channel (P) being considerably more sensitive than any of the channels seen in hairy skin. The threshold functions for vibratory stimuli measured on the hairy skin of the forearm are an order of magnitude higher than those recorded on the palm of the hand (Merzenich & Harrington, 1969). This is not only due to variations in the innervation density of the two types of skin, but also reflects differences in the response properties of the mechanoreceptors and possibly in the mechanical coupling of the skin and stimulator.

Mechanoreceptors in hairy skin respond to mechanical deformation of the skin and also discharge in response to finger movements, which typically stretch the loosely connected skin on the dorsum of the hand. The majority of these receptors respond to movements of more than one joint, as shown in figure 3.6, with the pattern of skin stretch during finger movements determining which units respond. SA II units are particularly responsive to different types of movements around a joint (Edin & Abbs, 1991). Their responses to joint movements are similar to those of muscle spindle afferents with firing rates up to 50 impulses/second, which is considerably lower than their discharge rates for tactile inputs, which can be greater than 200 impulses/second (Grill & Hallett, 1995).

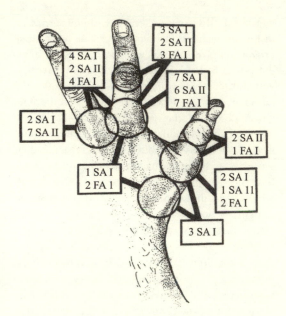

Figure 3.6. Number of slowly adapting (SA I and SA II) and fast adapting (FA) units in the dorsal skin of the human hand that responded to movements of one or more than one joint. The boxes indicate the number and type of units that discharged in response to movement of the joint indicated. For example, 20 units responded only to movement of the MP joint of the index finger, whereas 8 units responded to movements of both the MP and PIP joints; a further 10 units responded to movements of the MP joints of both the index and middle fingers; and 3 units responded to movements of the MP joint of the index finger and of the thumb. Reprinted from Edin & Abbs, 1991, with the permission of the American Physiological Society.

The proprioceptive information that SA II mechanoreceptors provide the CNS is ambiguous with respect to movement direction and amplitude. However, the ensemble response from a population of SA II receptors on the back of the hand could provide a sensory population vector that may reliably differentiate individual finger movements, as Edin (1992) has argued. When these inputs are eliminated following anesthesia of the skin, there is an impairment in proprioception (Ferrell & Milne, 1989). It has also been shown that when mechanical strain patterns are imposed on the skin overlying a joint in the absence of any finger movement, subjects can perceive these as indicating an actual movement of the joint (Collins & Prochazka, 1996;

Edin & Johansson, 1995). Afferent inputs from the skin influence most levels of the motor system from motoneurons in the spinal cord (Garnett & Stephens, 1980) to the motor cortex.

In closing this section on mechanoreceptors, we raise the possibility (A. Smith, personal communication, June 2005) that neurophysiologists may have oversimplified receptor innervation and function in cases where a one-receptor/one-fiber association has been assumed. Electrophysiological studies have tended to ignore the multiple large- and small-fiber innervations of Meissner corpuscles and hair follicles.

Thermoreceptors

During active haptic exploration and object manipulation, information concerning hand movements and the objects contacted is derived from mechanoreceptors in skin and muscle. Thermal information about objects comes from thermoreceptive afferent units in the skin. There are two kinds of thermoreceptors, known as cold and warm receptors, both of which are free nerve endings located in the epidermal and dermal skin layers (Spray, 1986). Cold receptors are more numerous than warm receptors, respond to decreases in temperature over a range of 5–43°C, and discharge most vigorously at skin temperatures of 25°C. In contrast, warm receptors discharge with increasing skin temperature and reach a maximum around 45°C (Stevens, 1991). When the skin is maintained at a temperature of 30–36°C, no thermal sensation is noted, although both types of receptor exhibit spontaneous firing. During the course of daily activities, the skin temperature of the hand can span a range of 20–40°C, but typically remains between 30 and 35°C. The relative distribution of thermoreceptors in glabrous and hairy skin has not been studied, and so it is not known to what extent differences in thermal sensitivity in different regions of the hand reflect variations in innervation.

There is a fairly abrupt change in the character of thermal sensation to one of pain if the temperature rises above 45°C or falls below 13°C (Darian-Smith & Johnson, 1977; Havenith, van de Linde, & Heus, 1992; Johnson, Darian-Smith, & LaMotte, 1973). At these extreme temperatures, thermal information is conveyed by thermal nociceptors, which signal tissue damage. The onset of cold pain, which is associated with skin temperatures below 14°C is slow, as damage from freezing usually takes time. This is very different from the rapid withdrawal of the affected body part when the skin temperature rises to 45°C, and damage is imminent.

It has been estimated that the innervation density of cold and probably warm fibers in the area of glabrous skin innervated by the median nerve in the hand is approximately 50–70 fibers/cm^2 (Darian-Smith, 1984; Johnson et al., 1973). Afferent fibers innervating warm receptors have conduction velocities in the range of 1–2 m/s, which identifies them as unmyelinated fibers. In contrast, the conduction velocities of afferent fibers innervating cold receptors are much faster at 10–20 m/s, indicating that they are small-diameter myelinated fibers (Darian-Smith, 1984).

Nociceptors

The skin also possesses myelinated and unmyelinated afferent units known as nociceptors that are selectively sensitive to high-intensity stimulation of several different energy forms (e.g., electrical, mechanical, chemical, or thermal) that can damage tissue. These units mediate pain sensations (Lynn & Perl, 1996) and will not be covered in this book.

Muscle Receptors

There are three types of mechanoreceptor in muscle that provide the central nervous system with information about muscle length and force. Two of these are stretch receptors known as the primary and secondary spindle receptors. They are located in muscle spindles, which are elongated structures ranging from 4 to 10 mm in length, that are composed of bundles (up to 14) of small intrafusal fibers (Swash & Fox, 1972). The spindles lie in parallel to the extrafusal muscle fibers, the force-generating component of muscle, and attach at both ends either to extrafusal fibers or to muscle tendons. Each spindle normally has one or two primary sensory receptors innervated by Group Ia (large myelinated) afferent fibers and one to five secondary sensory receptors innervated by Group II afferent fibers, as illustrated in figure 3.7. Muscle spindles have their own motor innervation through the fusimotor, or gamma, system, which regulates the sensitivity of the receptors

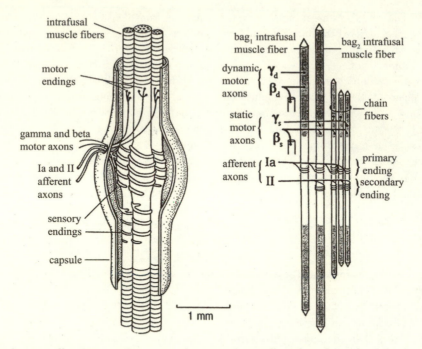

Figure 3.7. Schematic illustrations of a mammalian muscle spindle. When the ends of the intrafusal fibers that compose the spindle contract in response to gamma or beta motor axon stimulation, the central regions are stretched, which causes the primary (Ia) and secondary (II) endings to discharge. Reprinted from Prochazka, 1996, which was adapted from Kandel, Schwartz, & Jessel 1991. Reprinted with the permission of McGraw-Hill and Oxford University Press.

and is activated with the skeletomotor system. Some spindles are also innervated by skeletofusimotor, or beta, fibers that project to both extrafusal and intrafusal muscle fibers (Hulliger, 1984).

The number of spindles in human muscles varies depending on the function of the muscle, and for the hand has been estimated to range from 12 for the abductor digiti minimi, one of the hypothenar muscles, to 356 for the flexor digitorum superficialis (Voss, 1971). When expressed in terms of the number of spindles per gram of mean weight of adult muscle, higher spindle densities are generally found in muscles involved in fine movements, such as the distal finger muscles, and in the maintenance of posture. In contrast to the tactile sensory system, in which higher densities of mechanoreceptors are clearly associated with superior tactile acuity, for the proprioceptive system, higher spindle densities do not appear to be associated with superior sensory acuity. There is no evidence indicating a superior acuity for detecting movements and changes in limb position as one goes from proximal to distal joints.

When expressed in terms of the absolute angular rotation of the joint, the ability to detect movements is in fact better for more proximal joints, such as the elbow and shoulder, than for the distal joints of the hand, as Goldscheider first noted in 1889. This superior performance of more proximal joints is not surprising because they move more slowly than distal joints, and rotation of these joints results in a larger displacement of the end-point of the limb (the fingertip in the case of the arm) than the same angular rotation at a more distal joint. If the extent of movement is expressed in terms of changes in muscle fascicle length during joint rotation, then proximal and distal joints actually give comparable proprioceptive performance, and the distal to proximal gradient in sensitivity disappears (Hall & McCloskey, 1983).

Muscle spindle receptors are specifically responsive to changes in muscle length because of their location in muscles. Both primary and secondary afferents respond to changes in muscle length, but primary afferents are much more sensitive to the velocity of the contraction and respond more vigorously as the velocity of the stretch increases. Primary afferents are, however, highly nonlinear, and their discharge rates depend on several factors, including the recent contractile history of the muscle, the length of the muscle, the velocity with which the muscle changes length, and the activity of the fusimotor system. Secondary afferents demonstrate much less dynamic responsiveness and have a more regular discharge rate than do primary afferents at a constant muscle length (Prochazka, 1996). They also display much greater (around 50%) position sensitivity than do primary muscle spindle afferents (Edin & Vallbo, 1990). The higher dynamic sensitivity of the primary spindle receptors is consistent with the idea that they signal the velocity and direction of muscle stretch or limb movement, whereas the secondary spindle receptors provide the central nervous system with information about static muscle length or limb position. Figure 3.8 provides a schematic summary of the discharge rates of primary and secondary spindle afferent fibers during changes in muscle length, with and without fusimotor stimulation. As can be seen, the discharge rates of these receptors are not simply a function of changes in muscle length but also reflect the activity of the fusimotor system. In order to decode these afferent signals, the central nervous system must have access to information regarding the level of fusimotor activity so that it can distinguish between changes in discharge rates that are proprioceptively significant from those that are the result of fusimotor activity. It has been proposed that this is accomplished with corollaries of the motor command, which are transmitted to the sensory centers in the brain (McCloskey, 1981).

The third class of mechanoreceptor found in muscle is the Golgi tendon organ, an encapsulated receptor about 0.2–1 mm long and 0.1 mm in diameter, normally found at the junction between muscle tendon and a small group of extrafusal muscle fibers. This receptor is therefore "in series" with a small number (usually 10–20 in human muscles) of extrafusal muscle fibers. It is selectively responsive to the forces these fibers develop and has little or no response to the contraction of other muscle fibers (Jami, 1992). Because tendon organs are very sensitive to the in-series forces, most tendon organs in a muscle discharge in all but the smallest contraction. A single Group Ib axon innervates each tendon organ. In primate muscles, Golgi tendon organs are less numerous and more variable in number than spindle receptors, and some muscles, such as the lumbrical muscles of the hand, do not appear to have any tendon organ receptors (Devanandan, Ghosh, & John, 1983).

The distribution of tendon organs in a muscle is generally wide enough that every motor unit is included in the activity of the receptors at least once.

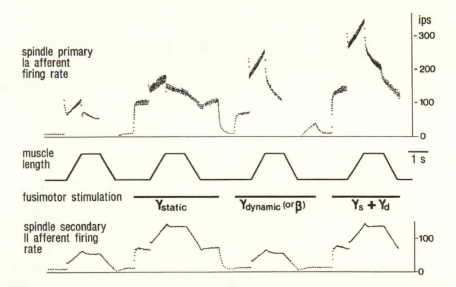

Figure 3.8. Schematic summary of the discharge rates of primary and secondary spindle afferent fibers to muscle length changes that occur either with or without concomitant fusimotor stimulation. Reprinted from Prochazka, 1996, with permission of Oxford University Press.

This suggests that the summed output of the Ib afferents from the tendon organs provides the central nervous system with an overall measure of muscle tension (Jami, 1992). The relation between the activity of a single motor unit and an isolated tendon organ afferent in human muscles is usually nonlinear, and may even be discontinuous, with steps in the frequency of discharge reflecting the contraction of individual motor units (Vallbo, 1974).

Microneurographic recordings have provided valuable information about the activity of muscle receptors during voluntary movements and have had an important impact on hypotheses regarding muscle spindle function. Recordings from human subjects have shown that the position sensitivity of spindle receptors in finger flexor muscles is remarkably low (less than 1 impulse/mm muscle length) and that both primary and secondary afferents have a low resting discharge and are silent at the muscle length that corresponds to the comfortable resting position of the human hand (Vallbo, 1974). It has also been shown that the discharge rates of primary and secondary afferents from the finger extensor muscles do not differ significantly when the joint is actively maintained at different positions, as can be seen in figure 3.9. The absence of any position response in the discharge rates of muscle afferents suggests that changes in fusimotor activity compensate for variations in muscle length (Hulliger, Nordh, & Vallbo, 1982).

There are limitations to the use of microneurography in muscle nerve fascicles, particularly with respect to the range of movement speeds, amplitudes, and forces that can be studied. Generally, the movements and forces generated are restricted to the lower end of the physiological range in order to prevent the microelectrode from dislodging and to limit interference from other active units. The low background firing rates of human muscle spindle afferents (<20 impulses s^{-1}) and the small changes in discharge rates attributed to fusimotor action (<30 impulses s^{-1}) as compared to recordings made in alert cats have been ascribed to the restricted experimental conditions used in human studies and the heightened level of alertness of animals in the laboratory environment (Prochazka & Hulliger, 1998).

In summary, muscle spindle receptors and Golgi tendon organs convey information about the length of muscles, the rates at which muscle lengths are changing, and the forces generated by muscles. This information provides the basis for the perception of the amplitude and velocity of finger movements and the static position of the hand. It also enables us to estimate the weight and compliance of objects supported by the hand. Information about changes in limb position and movement arises from other sensory sources, such as receptors in the skin and joints. Cutaneous afferents from the hand can reflexively affect fusimotor neurons innervating forearm extensor muscles and so make an indirect, as well as direct, contribution to proprioception (Gandevia, Wilson, Cordo, & Burke, 1994). In addition to these peripheral sources of proprioceptive information, there is a considerable body of evidence that suggests that central feedback, such as corollary discharges of the motor outflow, provide information that is used both to decode spindle afferent signals and to perceive force (McCloskey, 1981).

Joint Receptors

Several types of receptor are found in and around joint capsules, including Ruffini endings, Golgi end-

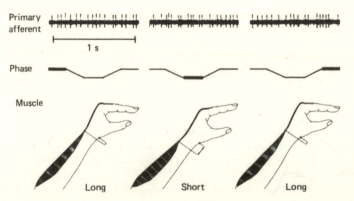

Figure 3.9. Response of the primary muscle spindle afferent in the extensor digitorum communis muscle during 1-s periods (upper trace) when the muscle was held at either a long (muscle length long, finger flexion) or short (muscle length short, finger extension) length (bottom trace). Reprinted from Hulliger et al., 1982, with permission of Blackwell Publishers.

ings, encapsulated Paciniform endings, and free nerve endings. Paciniform endings are probably rapidly adapting, like their counterparts in skin, whereas Ruffini and Golgi endings are probably slowly adapting (Zimny, 1988). The distribution of these receptors is nonuniform in a joint, which may reflect the location of stresses during limb movements (Gandevia, 1996). Most articular receptors discharge near the extremes of joint movement and often are active during both extreme flexion and extension and so do not provide an unambiguous signal related to joint position (Proske, Schaible, & Schmidt, 1988). Other joint afferents discharge during the midrange of joint movement, and estimates of the percentages of this class of receptor range from 5% to 18% in various animal species (Ferrell, 1980).

These midrange joint receptors do respond to finger movements throughout the physiological range of joint rotation and so provide the central nervous system with some kinematic information (Edin, 1990). However, their capacity to signal the direction of joint movement is limited, and so it has been proposed that joint afferent input may only be significant when muscle spindle afferents cannot contribute to proprioception (Burke, Gandevia, & Macefield, 1988). Studies of patients with rheumatoid arthritis who have undergone prosthetic replacement of the metacarpophalangeal joints in the hand suggest that whatever information is conveyed by joint receptors is adequately duplicated by other sensory sources. These patients have normal position sense in the absence of joint receptors (Cross & McCloskey, 1973) and are able to reproduce accurately the amplitude of movements made by fingers with prosthetic joints (Kelso, Holt, & Flatt, 1980).

An anatomical peculiarity of the hand has provided a preparation that has been used in numerous experiments to study the contribution of joint and cutaneous receptors to proprioception. When the fingers surrounding the middle finger are all maximally extended and the middle finger is flexed, the distal interphalangeal joint of the middle finger can no longer be voluntarily controlled as its tendon is slack because the muscles are decoupled from the joint (Gandevia & McCloskey, 1976). When the finger is positioned this way, only joint and cutaneous feedback can contribute to proprioception. The ability to detect the direction of applied flexion and extension movements within the midrange of joint position is impaired with the muscles disengaged,

and the residual ability to detect these movements does depend on joint afferent feedback. Clark, Grigg, and Chapin (1989) found that a local anesthetic block of the finger joint resulted in a further impairment in the ability to detect movements imposed on the joint. This indicates that when movement signals from muscle receptors are absent, articular receptors can provide cues that signal joint movement.

Corollary Discharges

The central nervous system appears to monitor its own activity in producing muscle contractions, and these correlates or copies of the motor command have been referred to historically as sensations of innervation (Helmholtz, 1866/1925) and more recently as corollary discharges or efference copies (McCloskey, 1981). The results from a number of experiments suggest that the perception of force is derived centrally from these internal neural correlates of the motor command and is not based on the activity of peripheral receptors in the muscle (Gandevia, 1996). Evidence to support this proposition comes from experiments in which it has been shown that as the motor command sent to a muscle increases, the perceived magnitude of the force of contraction also increases, even when the force exerted by the muscle remains constant (Jones & Hunter, 1983). This should not be interpreted as dismissing a role for peripheral receptors, such as Golgi tendon organs, in the perception of force as peripheral afferent activity is still required to indicate that the muscle force generated in response to the motor command is adequate to perform the task. A further role for corollary discharges is in the decoding of muscle spindle responses. The discharge rates of muscle spindle receptors are not only a function of changes in the length of the muscle but also reflect the activity of the fusimotor system. The central nervous system therefore needs to distinguish the spindle responses that are a consequence of fusimotor activity from those that are a result of an external perturbation and that have proprioceptive significance.

Sensory Cortical Projections

Sensory information from receptors in the hand and from muscles controlling finger movements is conveyed via afferent nerve fibers to the dorsal root ganglion neurons which lie on the dorsal root of a spinal nerve. The cell bodies of these neurons have two branches, one that projects to the periphery and the other to the central nervous system. The nerve terminals of dorsal root ganglion neurons determine the function of the neuron. The area of skin innervated by a single dorsal root is known as a dermatome, which can be identified in animals by stimulating the skin with different probes and measuring the response from afferent fibers in the dorsal root. Dermatomal maps are often used as a diagnostic aid for locating the site of injury in humans with damage to the spinal cord or dorsal roots.

The large-diameter axons mediating touch and proprioception diverge from the smaller-diameter afferents subserving temperature and pain in the spinal cord, and they follow different pathways to the brain. Figure 3.10 depicts these two ascending pathways on a series of brain slices. The small-diameter fibers terminate on second-order neurons in the dorsal horn of the spinal cord, and the axons of these neurons cross the midline to form the anterolateral system (Gardner, Martin, & Jessell, 2000). The axons serving these two modalities are segregated and are arranged somatotopically as they ascend the spinal cord. The anterolateral system has both direct and indirect connections to the thalamus via the reticular formation in the medulla and the pons. From the thalamus, there are projections to the primary somatosensory cortex, the dorsal anterior insular cortex, and the anterior cingulate gyrus (E. G. Jones, 1985).

Tactile and proprioceptive information is transmitted to the cerebral cortex via the central axons of dorsal root ganglion cells, the first-order neurons, that enter the spinal cord. These axons ascend directly to the medulla through the ipsilateral dorsal columns, in what is known as the dorsal column—medial lemniscal system. The two sensory modalities are segregated anatomically, with axons from proprioceptors being located more ventrally in the dorsal columns than are axons from tactile receptors, which are positioned dorsally. Not all of these primary afferent fibers reach the medulla; some 50% terminate at spinal levels and are known as propriospinal fibers. At higher spinal levels, the dorsal columns are divided into two fascicles: the gracile fascicle and the cuneate fascicle. The gracile fascicle ascends medially and includes fibers from the ipsilateral sacral, lumbar, and lower thoracic segments, whereas the cuneate fascicle ascends laterally and contains fibers from the upper thoracic and cervical segments. Axons in these fascicles terminate in the second-order neurons in the gracile nucleus and cuneate nucleus in the medulla (Florence, Wall, & Kaas, 1989). From there, they cross to the other side of the brain stem and ascend to the ventral posterior lateral nucleus of the thalamus in a fiber bundle known as the medial lemniscus.

The third-order neurons in the thalamus that receive these inputs from the dorsal column—medial lemniscal system send axons to Brodmann's areas 3a, 3b, 1, and 2 in the primary somatosensory cortex in the postcentral gyrus of the parietal lobe (Paul, Merzenich, & Goodman, 1972). Most of the thalamic input terminates in areas 3a and 3b, and the cells in these areas then project to areas 1 and 2. There is also a direct, but much smaller, connection between thalamic neurons and areas 1 and 2. The four areas of the primary somatosensory cortex differ in terms of the inputs they receive from the thalamus. Sensory information from receptors in the skin is received in areas 3b and 1, whereas proprioceptive information is transmitted to areas 3a and 2 (Mima et al., 1997). Area 2 sends a significant amount of efferent axons to the motor cortex, and loss of this cortical input has a striking effect on hand function, with impairments in coordinating the fingers when grasping, even though there is no weakness of the hand nor change in simple finger movements (Hikosaka, Tanaka, Sakamoto, & Iwamura, 1985). There are extensive interconnections between the primary sensory areas, and they in turn send connections to the secondary somatosensory cortex on the superior bank of the lateral sulcus (Burton, 2002).

Each of the four areas within the human somatosensory cortex contains a full representation of the body but with different types of information depicted (Kurth et al., 2000). In addition, parts of the body with higher innervation densities, such as the fingertips and the thumb, have disproportionately larger areas of cortex devoted to processing their sensory signals (Kaas, Nelson, Sur, & Merzenich, 1981). These cortical maps of the body surface are dynamic, not static, and are modified as a function of experience (Buonomano & Merzenich, 1998; Kaas, 1991). It has been shown in monkeys that extensive use (over 3 months) of the fingertips in a

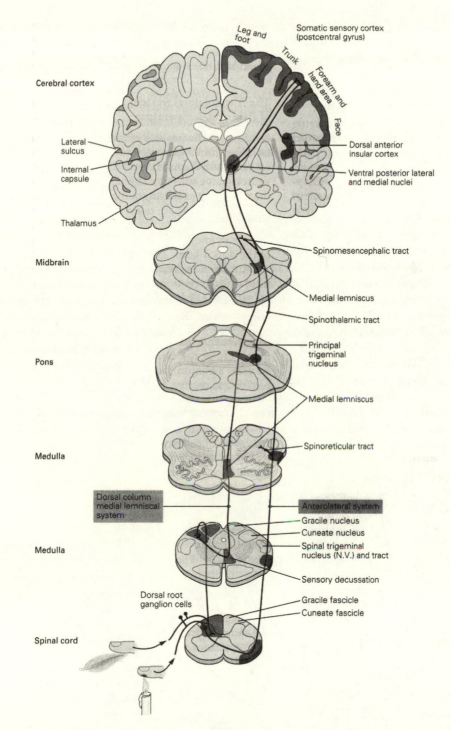

Figure 3.10. Sensory pathways involved in transmitting information from the hand to the cerebral cortex. The dorsal column medial lemniscal system transmits proprioceptive and tactile information from the periphery to the thalamus, whereas pain and thermal sensations are transmitted to the thalamus via the anterolateral system. Reprinted from Gardner et al., 2000, with permission of McGraw-Hill.

task involving touching a rotating disk with the middle fingers to obtain food pellets, results in a significant enlargement of the cortical representation of the stimulated fingers (Jenkins, Merzenich, Ochs, Allard, & Guic-Robles, 1990). Figure 3.11 shows the enlargement of the cortical territory of the stimulated finger as compared to its normal territory prior to differential stimulation. In a related human study of the cortical hand representation of proficient Braille readers, Pascual-Leone and Torres (1993) used transcranial magnetic stimulation (TMS) to show that the cortical representation of the right index (reading) finger was significantly larger than that of the left index finger of Braille readers and of the right index finger of control subjects. Moreover, representations of parts of the hand that were not involved in reading Braille were differentially smaller than those of control subjects. These results were confirmed in another study that used magnetoencephalography (MEG) to image the cortical representations of the fingers of string-instrument players (Elbert, Pantev, Wienbruch, Rockstroh, & Taub, 1995). When playing, these musicians are continuously fingering the strings with the second through fifth digits of the left hand, while the thumb grasps the neck of the instrument and so is not as active as the other digits. Elbert et al. (1995) found that the cortical representation of the digits of the left hand of string players was larger than that of controls and that this difference was smallest for the thumb. In addition, there was a negative correlation between the size of the cortical representation of the fingers of the left hand (as inferred from dipole strength) and the age at which the person began to play the musical instrument.

In contrast to this expansion of the cortical representation of the fingers with training, when the hand area in the human somatosensory cortex becomes deafferented as a consequence of extensive peripheral deafferentation (spinal cord injury) or amputation, there is shrinkage in the area of the cortex devoted to processing information from the hand. In these cases, the cortical representation of adjacent areas expands to occupy the deafferented cortex (Kew et al., 1997; Ramachandran, 1993). The reorganization that results from deafferentation and training appears to takes place at a cortical and not a thalamic or spinal level (Garraghty & Kaas, 1991; Wang, Merzenich, Sameshima, & Jenkins, 1995), although there is evidence of alterations in representational maps in dorsal column nuclei after deafferentation (Jain, Catania, & Kaas, 1997). The

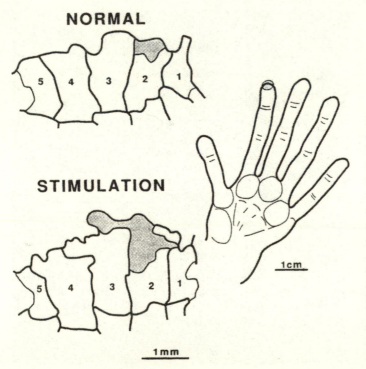

Figure 3.11. Cortical representation of the finger in the adult owl monkey before (normal) and after (stimulation) differential stimulation of the tip of the index finger. Outlines of the territories for each of the digits are shown in area 3b of the somatosensory cortex, together with the area on the fingertip that was stimulated. After stimulation, there was a substantial enlargement of the territory representing the surface of the index finger that was stimulated, as shown by the filled area in the digit 2 representation. Reprinted from Jenkins et al., 1990, with the permission of the American Physiological Society.

mechanisms involved in dynamically allocating cortex to selected peripheral inputs are not fully understood, and many issues, such as the site of plasticity and the cellular mechanisms that determine synaptic competition, require further study (Buonomano & Merzenich, 1998; E. G. Jones, 2000).

An interesting illusion that is related to the plasticity of cortical representation has been reported during acute anesthesia of the thumb or lips. Gandevia and Phegan (1999) observed that following complete anesthesia of the digital nerves that innervate the thumb, the perceived size of the thumb increased in magnitude by about 60–70%. In addition, the perceived size of the lips also increased (by 55%), even though the adjacent unanesthetized index finger did not change in perceived size. A similar result was found when the lips were anesthetized, with the effect on the thumb being much smaller (an increase in size by 5%), although still significant. The latter finding may reflect the fact that the lips were anesthetized using topical anesthesia, which resulted in a marked but not complete loss of sensory function, unlike the digital nerve block that eliminated all sensory feedback from the thumb. This experience of increased size following anesthesia is often reported by people who have had dental anesthesia and feel that their lips and lower face are "swollen."

The sensory representations of the thumb and lips are anatomically close in subcortical and cortical sites, and some cells respond to inputs from both the thumb and lips, so it is not surprising that changes in sensory inputs affected both body parts. The absence of any effect of thumb anesthesia on the perceived size of the index finger is surprising, however, and suggests that the boundaries between the thumb and lip representations are less distinct than those between the thumb and index finger (Gandevia & Phegan, 1999). The increase in perceived size with anesthesia has been interpreted as resulting from the unmasking of sensory inputs to the primary somatosensory cortex and the enlargement in the size of the receptive fields of cortical cells representing skin areas adjacent to the anesthetized region, as has been shown to occur in monkeys (Calford & Tweedale, 1991). These illusory increases in size appear to be related to the extent of the area that is anesthetized, as they do not occur when the whole arm is anesthetized (Melzack & Bromage, 1973). In this situation, the somatosensory cortex devoted to the anesthetized region becomes "silent."

Motor Cortical Areas

The areas of the cerebral cortex that contribute directly to the control of hand movements include the primary motor cortex (M1, or area 4), the supplementary motor area (SMA, or medial part of area 6), the presupplementary motor area (pre-SMA, or part of area 6), and the premotor cortex (PM, or lateral part of area 6) as shown in figure 3.12. There is considerable debate about the number of motor areas in the human brain, and estimates range from a conservative count of four areas to a more liberal estimate of ten areas (Roland & Zilles, 1996). The motor cortical areas receive inputs from subcortical motor areas, such as the basal ganglia and cerebellum, via the thalamus. Intracortical microstimulation studies in primates have shown that a large area of the primary motor cortex is involved in controlling the distal muscles of the contralateral arm (Asanuma & Rosén, 1972; Donoghue, Leibovic, & Sanes, 1992). These studies have also revealed that there are multiple representations of each muscle and type of movement, and so the same muscle can be activated by stimulation at different foci within the hand area of the primary motor cortex. In addition, single-unit recordings from neurons in M1 in monkeys show that there is considerable spatial overlap in the cortical territories for movements of different fingers. This indicates that many cortical motor neurons influence more than one finger muscle and that there are widespread horizontal connections throughout the M1 hand area (Donoghue et al., 1992; Schieber & Hibbard, 1993).

Similar results have been obtained from imaging studies of motor cortex organization in the human brain. Geyer et al. (1996) found two sets of representations for the thumb and index and middle fingers within primary motor cortex, and Sanes, Donoghue, Thangaraj, Edelman, and Warach (1995) reported that movement of a single finger gave rise to several fields of activation within M1. In the latter study, the representations of the thumb and index and ring fingers overlapped each other (as defined by the volume of activation) by 40–70%. These findings are at variance with the classic motor homunculus derived from electrical stimulation of the cortical surface (Penfield & Rasmussen, 1950; Schott, 1993) and indicate that somatotopy in primary motor cortex may hold only for major divisions of the body, such as the arm, face, and leg.

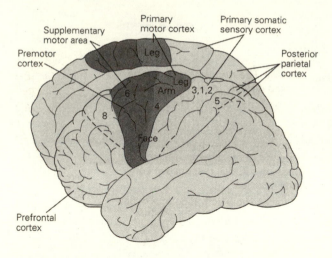

Figure 3.12. Major motor areas of the human cerebral cortex, including the primary motor cortex, the supplementary motor area, and the premotor cortex. The somatotopy of the primary motor cortex is shown with the hand and arm occupying a relatively large area on the lateral surface. Reprinted from Krakauer & Ghez, 2000, with the permission of McGraw-Hill.

There are additional representations of the hand in the supplementary motor and premotor areas. These nonprimary motor areas are believed to control different aspects of hand function but to operate cooperatively (Rouiller, 1996). In both human and nonhuman primates, the pre-SMA is now regarded as an area distinct from the caudally adjacent SMA on both anatomical and physiological grounds (Picard & Strick, 1996; Tanji, 1996). The two areas differ in their anatomical connectivity in that only the pre-SMA receives substantial input from the dorsolateral prefrontal cortex, and only the SMA projects directly to the primary motor cortex (Tanji, 1996). Studies of neuronal activity in the pre-SMA in monkeys suggest that this area plays a role in updating the information that is required to perform an upcoming sequence of movements, and so the neurons in this area are more active during the preparation for upcoming movements (Matsuzaka, Aizawa, & Tanji, 1992; Shima, Mushiake, Saito, & Tanji, 1996). Similar studies in the SMA reveal that it contributes to planning and controlling sequential movements that have multiple spatial components and in particular to arranging multiple movements in the correct temporal order (Shima & Tanji, 1998). The results from imaging studies in human subjects are consistent with this distinction in that they have shown that there is sharply localized activity in the pre-SMA while participants are learning a finger-sequencing task, whereas the SMA becomes more active during the performance of these sequential movements (Boecker et al., 1998; Hikosaka, Sakai, Miyauchi, Takino, Sasaki, & Pütz, 1996).

The PM appears to be involved in the planning and control of movements that are guided by sensory information (Roland & Zilles, 1996). Several lines of evidence from nonhuman primate studies suggest that the ventral premotor cortex can be distinguished from the dorsal premotor cortex both functionally and structurally. In particular, the ventral premotor cortex participates in the visual guidance of hand movements, such as those involved in reaching for and manipulating objects (Ochiai, Mushiake, & Tanji, 2005; Wise, di Pellegrino, & Boussaoud, 1992). Behavioral and imaging studies in both human and nonhuman primates suggest that one role of this cortical area is the transformation of information about the location of objects in space, which is specified in an extrinsic visual frame of reference, into an intrinsic motor reference frame (Kakei, Hoffman, & Strick, 2001). In contrast, the dorsal premotor cortex is involved in movement planning and the mental rehearsal and observation of motor tasks (Ramnani & Miall, 2004). Many task-related neurons in this cortical area exhibit the same activity patterns during observation as during performance of a task (Cisek & Kalaska, 2004, 2005).

All of these cortical motor areas project to the spinal cord, and in human and nonhuman primates, a subset of the neurons in the primary motor cortex, known as corticomotoneuronal cells, send their axons directly to the distal motoneurons that innervate hand and finger muscles. About 40% of the axons in the human corticospinal tract originate in the primary motor cortex, and these axons descend through the subcortical white matter, the internal capsule and cerebral peduncle, until they form the medullary pyramids. This entire projection is therefore often called the pyramidal tract. Only those fibers constituting the corticospinal tract itself terminate in the spinal cord, and the other (approximately 50%) pyramidal tract fibers terminate or send collateral branches to a number of supraspinal structures (Davidoff, 1990). The corticospinal tract is the least mature of the descending motor pathways and, as described in chapter 9, undergoes considerable changes during the early postnatal years.

The contribution of the pyramidal tract, and in particular the corticospinal system, to skilled manual activities has been subject to much debate and experimentation (Lemon & Griffiths, 2005). Early animal experiments emphasized the loss of independent finger movements that occurred after complete transection of the pyramidal tract (Lawrence & Kuypers, 1968; Tower, 1940), although other studies showed that, following intensive retraining, discrete finger movements were possible after unilateral and bilateral lesions of the bulbar pyramids in monkeys (Chapman & Wiesendanger, 1982; Hepp-Reymond, Trouche, & Wiesendanger, 1974). More recently, Sasaki et al. (2004) have shown that after complete transection of the direct (monosynaptic) corticomotoneuronal pathway (at C4/C5), monkeys could still grasp small morsels of food by using independent finger movements. These animals did show persistent deficits in preshaping the hand when grasping the food and had difficulties in controlling the force between the index finger and thumb. These findings indicate that indirect as well as direct corticomotoneuronal pathways are involved in prehension and in the control of independent finger movements.

As was noted for the somatosensory cortex, the somatotopic organization of the motor cortex is dynamic and can change following injury or when learning new motor skills (for a review, see Sanes & Donoghue, 2000). Studies of the motor cortex of people who have had an upper limb amputated, either traumatically or surgically, or who were born without the limb have revealed that muscles adjacent to the stump can be activated (using TMS) from a larger area of cortex than those on the contralateral side. In addition, the threshold for activating muscles is lower on the amputated side (Cohen, Bandinelli, Findley, & Hallett, 1991; Hall, Flament, Fraser, & Lemon, 1990). In this situation, the muscles in the limb adjacent to the stump are controlled by a larger area of cortex that presumably became silent following amputation.

The effects of motor learning on the cortical representation of the hand in human primary motor cortex have been examined using finger-sequencing tasks. These studies have shown that changes in M1 can occur within 20–30 minutes of learning a task (Classen, Liepert, Wise, Hallett, & Cohen, 1998) but can revert back to baseline levels when the task is no longer performed, even for well-practiced skills, such as reading Braille (Pascual-Leone, Wassermann, Sadato, & Hallett, 1995). Using a simple motor task that involved touching the thumb to the tip of each finger in a prescribed sequence, Karni et al. (1995) studied the short- and longer-term (5 weeks) effects of motor learning on the cortical representation of the hand. Functional magnetic imaging (fMRI) scans revealed that after 3 weeks of daily training (20 minutes), at which point performance was asymptotic in terms of speed and accuracy, the area of the motor cortex activated during performance of this finger-movement sequence was larger than that activated when performing an untrained sequence. This enlargement in the cortical activation area was not the result of differences in the speed with which the two sequences were executed and was specific both to the motor task learned and to the hand being trained. It appears that over the time that the skill was acquired, synaptic connections between neurons in primary motor cortex were unmasked and strengthened, so that a larger cortical network of neurons became involved in the task performance.

The continuous modifications of the motor cortex that occur with different manual experiences have important implications for understanding the acquisition and retention of motor skills. As described in chapters 7 and 9, fine manual skills, such as typing and piano playing, are highly dependent on practice for both their acquisition and retention and are very specific in terms of the skill sets acquired. The thousands of times with which finger movements are repeated by skilled typists and pianists no doubt reinforces particular networks in the motor cortex. These connections remain strong with practice, which presumably accounts for the retention of these skills with age, while other motor skills deteriorate (see chapter 9).

4

Tactile Sensing

Despite the fact that the skin is, from the evolutionary standpoint, the oldest of the sensitive tissues of the body, it has yielded up its secrets reluctantly.

GELDARD, 1972

In chapters 2 and 3, we considered the anatomy and sensory neurophysiology of the hand. In the next four chapters, we turn to the research literature on manual performance. Tasks are categorized in terms of how they differentiate hand function along the sensory-motor continuum introduced in chapter 1 (figure 1.1). Chapters 4 and 5 address the sensory performance of the hand based on tasks that involve tactile sensing and active haptic sensing, respectively. Chapters 6 and 7 address motor performance based on tasks involving prehension (reaching and grasping) and non-prehensile skilled activities, respectively. The primary objective of these chapters is to review the extensive, but highly diverse and often uneven, literature on human hand function. We highlight issues and concerns that we believe to be critical to any discussion of hand function, describe selected studies in each of the four categories along the sensory-motor continuum together with the methodologies used to study them, note important control parameters that may affect performance, and provide additional references for the reader to pursue a specific topic in greater depth.

Within the domain of tactile sensing, it is important to distinguish between tasks that involve stimu-lating the stationary hand with a stationary object ("passive static") and tasks that involve an external agent moving an object across the stationary hand ("passive movement"). Passive-static tasks activate fast-adapting mechanoreceptor units at the beginning of the contact period, and sometimes at the end as well. Slow-adapting units respond to initial contact and continue to respond throughout the contact period. Passive-movement tasks activate the slow-adapting units even more vigorously (Johnson & Lamb, 1981); in addition, the fast-adapting mechanoreceptor populations are activated by vibrations produced by the relative motion between skin and object.

Sensitivity and Resolution

Any discussion of human hand function must consider both the hand's sensitivity to various forms of mechanical, thermal, and electrocutaneous stimulation and its spatial and temporal resolving capacities. These questions pertain to threshold-level stimulation and have been addressed by applying different types of probes to the observer's stationary hand, usually on the volar surface. Investigators have considered a number of dimensions that are relevant to mechanical, thermal, and electrical stimuli. The reader should consult Sherrick and Cholewiak (1986) and Greenspan and Bolanowski (1996) for more detailed discussions of this topic.

Sensitivity

Pressure

The pressure threshold is the most common intensive measure of absolute tactile sensitivity (i.e., the smallest intensity that is just detectable). Nylon monofilaments of varying diameters are most often used to apply controlled pressure to a designated site on the stationary hand. Typically, some variant of the psychophysical method of limits is used to determine the absolute threshold. Observers must indicate whether or not they detect the presence of a filament applied to the skin. On alternating series of trials, the pressure is decreased or increased until a threshold is reached (for further methodological details, see Gescheider, 1997). Body site and gender both influence sensitivity to pressure: for men, normal mean threshold values average about 0.158 g on the palm and about 0.055 g on the fingertips; the corresponding values for women are consistently lower, namely, 0.032 g and 0.019 g (Weinstein, 1968). In keeping with the literature on cutaneous pressure sensitivity, we report the values in grams. It should be noted that figures 10.2 and 10.3 in the original Weinstein article both represent force in units of $\log_{10} F_{0.1mg}$. The mean force values for men and women above, which were derived from these two figures, were converted to grams by dividing by 10^4. Although Weinstein's monofilaments were calibrated in terms of force, Levin, Pearsall, and Ruderman (1978) have argued that stress (force per unit area) is the more appropriate variable to use when measuring pressure sensitivity.

Intensity difference thresholds (i.e., the smallest perceptible difference in intensity between two stimuli) for taps generated by 2-ms square waves have also been reported (Craig, 1972). The taps were controlled using a pulse generator and imposed on a vibrator contacting the right index finger. Four standard intensity values (14, 21, 28, and 35 dB SL, where sensation level at threshold is measured in voltage and defined as 0 dB SL) were evaluated. The procedure used to determine the difference thresh-

olds for intensity combined a two-interval, temporal forced-choice paradigm with a blocked up-and-down method (Campbell, 1963). Observers were required to choose which tap felt more intense. The results are shown in figure 4.1. The unit of measure for the Y axis is the difference in attenuator settings for the standard and comparison stimuli, measured in dB and plotted as a function of the sensation level in dB, rather than the more traditional Weber fraction. If these ratio values were plotted in terms of the Weber fraction ($\Delta I / I$) as a function of intensity (amplitude), they would confirm what has been commonly found with respect to Weber's law, namely that, except for low intensity levels, the Weber fraction for intensity remains relatively constant as the sensation level of the standard stimulus is increased (Craig, 1974).

The results of many studies collectively show that the cutaneous system effectively resolves normal forces applied to the skin and that humans are capable of scaling the magnitude of such forces (e.g., Burgess, Mei, Tuckett, Horch, Ballinger, & Poulos, 1983; Cohen & Vierck, 1993; Hämäläinen & Järvilehto, 1981; Knibestöl & Vallbo, 1980). Until recently, less attention was paid to the influence of tangential forces on neural coding and on the corresponding human sensation. Most people are indeed capable of scaling the perceived magnitude of tangential, as well as normal, forces (0.15–0.70 N) applied without slip to the distal pad of the index finger (Paré, Carnahan, & Smith, 2002; Smith & Scott, 1996). The perceived force magnitude is unaffected by the rate at which the force is applied. Sensitivity of the human fingerpad to tangential force is lower than that to normal force, presumably due to the high impedance of the fingerpad to tangential stimulation (Biggs & Srinivasan, 2002a). As we shall see presently, tangential forces play a role in the perception of surface texture and shape.

Temperature

When the hand is thermally stimulated, people report that their skin appears to become warmer or cooler. These reports have been quantified in terms of intensity and duration. J. C. Stevens and Choo (1998) reported that young adults could detect fingertip changes from a baseline temperature of 33°C as small as 0.16°C and 0.12°C for warmth and cold, respectively. Corresponding values for the thenar eminence (volar base of the thumb) were still lower,

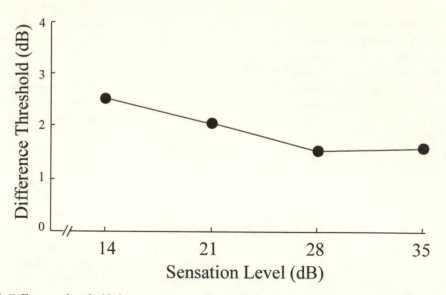

Figure 4.1. Difference thresholds for taps, expressed as the difference in attenuator settings between standard and comparison stimuli in dB, plotted as a function of sensation level (SL), where absolute threshold is defined as 0 dB SL. Two observers participated. Each data point is based on 10 measurements, with approximately 100 trials per measurement. If these data were to be replotted in terms of $\Delta I/I$ as a function of I (in dB above absolute threshold), the function would be relatively flat except for low-intensity values, as discussed in the text. Reprinted from Craig, 1972, with permission of the Psychonomic Society.

namely, 0.11°C and 0.07°C. Harju (2002) used a simplified version of the method-of-limits procedure (discussed earlier) to determine absolute threshold from a baseline skin temperature of 33°C on the thenar eminence. Observers were asked to indicate when they first detected a warm or cold event. Healthy males and females of two different age ranges (20–30 years and 55–65 years) served as observers. At this site, neither gender nor age differences were noted for the warm thresholds (34.4°C); however, cold thresholds for men (28.8°C) were higher than those for women (28.1°C). It is likely that the discrepancies with the results obtained by Stevens and Choo (1998) derive primarily from the relative precision of the methodologies used. Nevertheless, both studies confirm that the hand is more sensitive to cold than to warmth.

Relative sensitivities to warmth and cold have also been investigated by measuring the difference thresholds (at 75% accuracy) for resolving a change in intensity between a pair of warming or cooling pulses sequentially applied to the thenar eminence (Johnson, Darian-Smith, LaMotte, Johnson, & Old-field, 1979). To investigate relative sensitivity to thermal warming, the skin was maintained at base-

line temperatures of 29, 34, or 39°C for at least 10 min before testing. A range of standard amplitude steps for the warming pulse relative to the baseline temperature was evaluated. Calculation of the optimal difference threshold for warming pulses was based on the five lowest values obtained at each standard intensity level. The difference thresholds for baseline temperatures of 34°C and 39°C were very small, that is, rarely more than 0.050°C, while those for a baseline of 29°C were only a little higher, no more than 0.075 °C. An earlier study (Johnson et al., 1973) used a similar approach to evaluate relative sensitivity to cooling pulses on the thenar eminence. Regardless of baseline temperature (29–41°C), the difference thresholds for cooling pulses varied linearly from a mean of 0.072°C for a 0.5°C step to 0.206°C for an 8°C step.

The rates of growth of warm and cold sensations have also been examined with a psychophysical procedure known as "magnitude estimation." The observer is required to assign the number that best matches the perceived magnitude of the percept being judged, in this case, warmth or cold. Harju (2002) obtained power functions with exponent values of 1.6 and 1.1, respectively, and neither of these values was affected by gender or age. However, the exponents of the magnitude estimation functions for warmth do vary considerably with both stimulus size and site of stimulation (Stevens & Marks, 1971). The thermal growth function obtained by magnitude estimation potentially offers clinicians a new diagnostic tool for assessing thermal sensitivity on the hand.

When skin temperature changes very slowly (i.e., < 0.5°C/min), an alteration of less than 5–6°C may not be detected by an observer, provided that the temperature remains within the neutral zone of 30–36°C. However, when the change occurs more rapidly (e.g., 0.1°C/s), small increases and decreases in skin temperature are perceived (Kenshalo, Holmes, & Wood, 1968).

Spatial features of the thermal stimulus (e.g., the area and its shape) are poorly resolved, and changes in intensity within the region of stimulation are poorly detected. Low spatial resolution occurs because the skin summates thermal intensity over space, a process labeled spatial summation. Spatial summation occurs at absolute threshold, but may also be experienced at levels above threshold, as shown by the following simple demonstration. First, press the edge of a coin that has been warmed (or cooled) into the skin; then, repeat the process using one of its flat faces. It is important to avoid altering the thermal experience by touching the coin with the other fingers. By switching from the edge to the flat region of the coin, the magnitude of warmth (cold) experienced markedly increases, as noted by Sherrick and Cholewiak (1986). More recently, Green and Zaharchuk (2001) have reported that about two thirds of the spatial summation of warmth may be attributed to the spatially nonuniform thermal sensitivity of the skin. Increasing the size of the stimulus thus increases the likelihood of stimulating regions with greater sensitivity. Accordingly, neural summation may play only a modest role in producing the overall amount of summation observed in psychophysical studies. Almost complete temporal summation (i.e., the summation of warmth over time) occurs on the forehead over a range of temporal durations from 0.5 to 1.0 s, with no further effect up to 10 s (Stevens, Okulicz, & Marks, 1973). We know of no scientific studies that have parametrically assessed either temporal or spatial summation effects on the hand. Although both summation effects are likely to occur, there may well be differences between the hand and other body regions.

In closing this section, we note that when the skin is warmed or cooled as a result of touching an object, the location and extent of the thermal stimulus can be determined on the basis of the tactile input (e.g., Green, 1977). We will discuss this phenomenon in greater detail when we consider spatial acuity.

Electrocutaneous Stimulation

Electrocutaneous thresholds describe the sensitivity of the skin to electrical stimulation. Note that the skin's receptors do not sense electrical stimulation because the nerve is stimulated directly. The absolute threshold for electrical stimulation is strongly influenced by the type of electrode and its configuration, as well as by the locus of stimulation, wave form, repetition rate, and the individual observer. To date, very few studies have specifically focused on the human hand. Hahn (1958) presented square-wave electrocutaneous pulses (pulse width range: 0.1–7.0 ms; pulse frequency: 60, 100, 200, 500, and 1,000 pulses/s) to the right index finger resting on the active electrode and with the right thenar eminence on the indifferent electrode. The results are shown in figure 4.2. For frequencies ranging from 60 to 1,000 pulses/s, sensitivity (measured as the inverse of the observer's absolute threshold relative to baseline response) increases with increasing unidirectional pulse width, indicating the occurrence of temporal summation effects for electrocutaneous stimulation of the hand.

Subsequent work by Higashiyama and Tashiro (1983) showed that partial summation of single constant-current pulses occurs at absolute threshold for durations between 0.1 and 1.0 ms; however, beyond 1 ms, temporal summation does not occur. This result was subsequently confirmed using double electrical pulses (Higashiyama & Tashiro, 1988). Electrocutaneous spatial summation has also been investigated. For example, greater-than-complete summation, or supersummation, occurs at low sensation levels, whereas complete spatial summation occurs at high sensation levels (Higashiyama & Tashiro, 1990, 1993a).

Investigators have also determined the difference threshold for intensity of electrocutaneous sinusoidal stimulation. For example, Hawkes (1961) reported that the Weber fraction (i.e., difference threshold/intensity of the standard stimulus) for the

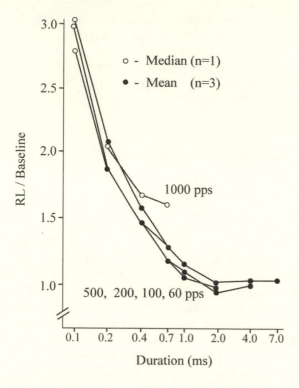

Figure 4.2. Strength-duration curves at several pulse rates with duration plotted on a logarithmic scale. RL/baseline is the ratio of the absolute threshold (RL), measured in milliamperes using an ascending method of limits, to the mean RLs at pulse durations of 2 ms and longer, since these values were essentially the same for a given observer. The parameter is pulses per second (pps). The means of data from three observers are indicated by solid circles; at those points for which the RLs of two observers were above maximum stimulator output, the medians of the data for the remaining observer are indicated by open circles. Reprinted from Hahn, 1958, with permission of the American Association for the Advancement of Science.

index fingerpad remains constant at approximately 3% for frequencies ranging from 100 to 1500 Hz. The dynamic range from threshold to pain for electrical stimulation is only about 15 dB, which is considerably narrower than the corresponding 60–80 dB dynamic range measured for vibrotactile stimulation (Geldard, 1972).

The rate of growth of electrocutaneous sensation has typically been assessed by using the magnitude-estimation procedure to estimate the perceived magnitude of suprathreshold stimuli. A power function is usually obtained, although the value of the exponent has varied. The results of an early experiment by S. S. Stevens, Carton, and Shickman (1958) are shown in figure 4.3. The 60 Hz stimuli, which were either 0.5 or 1 s in duration and varied from 0.25 to 1.10 mA, were delivered through electrodes to the index and ring fingers. The rate of growth of perceived current amplitude for the data combined across duration levels was exceptionally high (compared to other sensory tasks) as revealed by the exponent of the power function, which was 3.5. Since then, other researchers have obtained exponents that are considerably lower and that vary somewhat across studies (e.g., Babkoff, 1978; Cross, Tursky, & Lodge, 1975; Rollman & Harris, 1987).

The results of a study by Higashiyama and Tashiro (1989) suggest that the diversity in exponent values may be attributed to differences in current level inasmuch as the value of the exponent is higher for low current values than for high current values. To the extent that this suggestion is true, the relation between perceived intensity and current cannot be described by a simple power function.

Vibration

There is a considerable body of psychophysical research that has evaluated the sensitivity of the hand to vibrotactile stimulation. Although much of this research has been performed on the thenar eminence, data do exist for the fingertip (e.g., Gescheider, Bolanowski, Pope, & Verrillo, 2002; Verrillo, 1971) as shown in figure 4.4. Typically, vibration is delivered via a contactor mounted on a mini-shaker. A rigid surround with a small gap separating the contactor from the surround (e.g., 1 mm) is often used to prevent any traveling-wave effects on the skin. Verrillo (1963) first described a two-limbed psychophysical function for vibrotactile thresholds as a function of frequency (25–700 Hz). At lower frequencies (25–40 Hz), the psychophysical function was relatively flat;

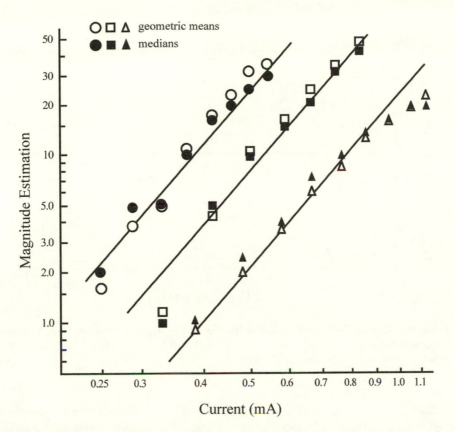

Figure 4.3. Three experiments (indicated by circles, squares, and triangles) on the estimation of the apparent intensity of electric current. The filled points are medians; the unfilled points are geometric means. Reprinted from S. S. Stevens, Carton, & Shickman, 1958, with the permission of the American Psychological Association.

that is, the skin was not sensitive to stimulus frequency. However, beyond that frequency range the psychophysical function was U-shaped, with the threshold progressively declining up to approximately 250 Hz, at which point it began to increase up to 700 Hz. These initial psychophysical results led Verrillo to propose a two-channel theory of vibrotactile sensitivity, with each channel (non-Pacinian, or NP, and Pacinian, or PC) presumably mediated by a different population of tactile units, as suggested later by Talbot et al. (1968). The low-frequency, non-Pacinian channel showed no spatial or temporal summation, in contrast to the high-frequency, Pacinian channel, which demonstrated both forms of summation.

Converging psychophysical paradigms (selective adaptation, masking, and enhancement) were subsequently used to uncover the full frequency response of each vibrotactile channel, since Verrillo and his colleagues suspected that only the more sensitive channel mediates the psychophysical threshold responses. Selective adaptation was used to show selective-fatigue effects within, but not across, the NP and PC channels. With this paradigm, the skin is adapted to high-intensity stimulation at the channel's most sensitive frequency in order to reveal the response of the normally less sensitive channel (Verrillo & Gescheider, 1977). Temporal masking has also been used to confirm the hypothesis that vibrotactile detection is mediated by at least two receptor systems that do not mask one another (e.g., Gescheider, Verrillo, & Van Doren, 1982). Briefly, this psychophysical paradigm involves determining thresholds for a range of vibrotactile frequencies in the presence of a masking stimulus that varies in intensity. The masking frequency is also manipulated so that the masking and test stimuli activate either the same or different channels. Predictions regarding the masking functions were based on the two-channel theory and depended on which

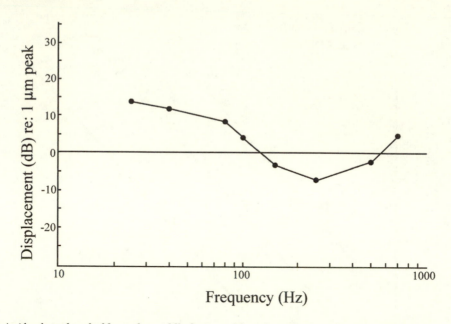

Figure 4.4. Absolute thresholds on the middle fingerpad for vibration plotted as a function of vibrotactile frequency. The contractor size was 0.005 cm². The data are based on five subjects. Reprinted from Verrillo, 1971, with permission of the Psychonomic Society.

vibrotactile channel the masking stimulus activated and on whether the masking intensity exceeded the threshold of the channel under consideration. In keeping with the predictions, within-channel masking effects were observed; however, no cross-channel effects occurred, as evident in figure 4.5A. Additional support has been obtained using the enhancement paradigm, which consists of presenting three brief (20 ms) sinusoidal bursts in turn: a conditioning stimulus followed after a variable interval by a test stimulus (10 dB SL), followed 750 ms later by a matching stimulus, whose intensity is controlled by the observer. The observer must match the perceived intensity of the test stimulus in the presence of the conditioning stimulus (Verrillo & Gescheider, 1975). The effect of the conditioning stimulus is determined by calculating the difference between this match and a control match that does not include the conditioning stimulus. The frequencies of the conditioning and test stimuli were chosen from the same vibrotactile channel (e.g., 300–300–300 Hz) or from different channels (e.g., 25–300–300 Hz). In the same-channel condition that was designed to test for sensory interactions within the PC system, the magnitude of sensation was increasingly enhanced as the time interval between the conditioning and test stimuli

was reduced. However, in the different-channel condition, which involved using conditioning and test stimuli from the separate NP and PC systems, there was no enhancement effect (figure 4.5B).

This early psychophysical work assessed the influence of a number of stimulus parameters on absolute vibrotactile thresholds, including, for example, contactor size (e.g., Verrillo, 1963), stimulus duration (e.g., Verrillo, 1965), skin-surface temperature (e.g., Bolanowski & Verrillo, 1982;), age (e.g., Verrillo, 1977; and chapter 9 in this volume) and body region (e.g., Löfvenberg & Johansson, 1984; Verrillo, 1971; also see chapter 8). On the basis of additional psychophysical work, a three-channel model was then proposed (Capraro, Verrillo, & Zwislocki, 1979), in which one population of tactile afferents, the SA II units, was not a contributor.

Subsequently, Bolanowski, Gescheider, Verrillo, and Checkosky (1988) used extensive anatomical, neurophysiological, and psychophysical data to propose a multichannel model of tactile sensation in glabrous skin that involves the contribution of four psychophysical channels, each with its own physiological substrate, as shown in figure 3.5. Psychophysical thresholds for vibrotactile stimulation of the glabrous skin were obtained using a probe that deliv-

A

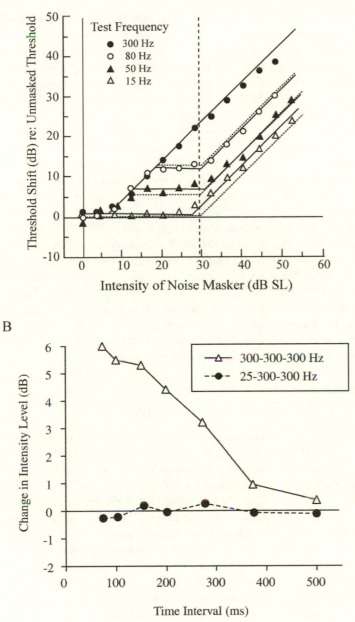

Figure 4.5. (A) Predictions (dotted lines) and experimental masking results using a narrow-band noise masker centered at 275 Hz and four test frequencies (15, 50, 80, and 300 Hz). The solid lines were determined by using a least-squares method of best fit to the data points. Masking occurred only when masker and test stimuli were presented within the same vibrotactile system. Reprinted from Gescheider, Verrillo, & Van Doren, 1982, with permission of the Acoustical Society of America. (B) Enhancement of the subjective magnitude of the test stimuli by the presence of a preceding conditioning stimulus at time intervals between 75 and 500 ms. All three bursts were set at a frequency of 300 Hz, which lies within the frequency range of maximal response for the Pacinian system (open triangles). The second experiment was the same as just described, but the conditioning stimulus was set at a frequency (25 Hz) where the NP system determines the response, and the test and matching bursts were set at a frequency (300 Hz) where the PC system dominates the response. There appears to be no enhancement effect across the two systems (filled circles). Reprinted from Verrillo, 1985, with permission of the Acoustical Society of America.

ered vibration perpendicular to the skin surface. A variety of different psychophysical techniques were used to uncover the relation between vibrotactile threshold and vibratory frequency for each of the four psychophysical channels. Across experiments, the stimulus parameters included stimulus frequency and amplitude, probe/contactor area, stimulus duration, and skin-surface temperature. Vibrotactile masking

procedures were used as well to uncover the vibrotactile response characteristics of the psychophysical channels. The psychophysical results were then correlated with existing human anatomical and neurophysiological data (Johansson et al., 1982; Johansson & Vallbo, 1979).

The model proposes that each of the four psychophysical channels for vibration is mediated by the population of tactile units that is most sensitive to a particular range of vibrotactile frequencies. Stimulation of any one mechanoreceptor population results in a unique "unitary" sensation, similar to the scotopic (rods) and photopic (three populations of cones) systems in vision. Two assumptions are made. First, the channel with the greatest sensitivity processes threshold sensations. Second, more than one activated psychophysical channel is responsible for producing suprathreshold sensations, although it can be as few as two. The outputs of all activated channels are presumably combined to produce a unitary sensation at cortical levels higher than area SI. Table 4.1 summarizes the essential psychophysical characteristics of each channel.

Recently, the four-channel model has been applied to the vibrotactile sensitivity of the fingertip, based on tests of frequency selectivity, temporal summation, and spatial summation (Gescheider et al., 2002). The only difference between tests on the two body sites is that the amount of spatial summation, as determined by the reduced effect of contactor size, is relatively lower on the fingertip. The model has also been applied to the vibrotactile sensations that arise in response to the stimulation of hairy skin with vibratory stimuli, as previously discussed in chapter 3.

It is important to note the following caveat concerning the multichannel models for glabrous and hairy skin proposed by Bolanowski and his colleagues. These researchers make the general claim that their model mediates the "mechanical aspects of touch." While making a noteworthy contribution to our understanding of vibrotactile sensations, we emphasize that this model does not address the many other types of significant mechanical sensations that result from statically stimulating the skin or from applying spatially varying surfaces perpendicularly onto or tangentially across the skin.

As previously reported with mechanical taps, Craig (1974) also showed that when vibrotactile intensity discrimination is evaluated using a standard stimulus of 160 Hz presented in 200-ms bursts, the Weber fraction remains relatively constant as a function of increasing sensation level (in dB, relative to absolute threshold), except at very low levels of intensity.

The perceived magnitude of a 60-Hz vibration delivered to the fingertip grows as a power function of the vibrotactile intensity with an exponent of 0.95 (S. S. Stevens, 1959). The exponent of the function is influenced by many of the same factors that

affect the absolute threshold (e.g., site of stimulation, contactor area, age). The effect of frequency on the value of the exponent has proved to be controversial. For example, both S. S. Stevens (1975; also see figure 4.6) and Marks (1979) have shown that frequency influences the exponent, whereas Verrillo and Capraro (1975) have shown no such effect. Methodological differences may be responsible for such discrepancies.

The hand is also susceptible to vibrotactile "aftereffects." For example, when the skin is stimulated by an intense vibrotactile adapting stimulus of a given frequency for 1–2 min, the threshold for a neutral test stimulus (vibration of the same frequency) is lower than when the adapting stimulus has not been presented, demonstrating what is known as a negative aftereffect. (Sometimes, however, observers report that the test stimulus feels more intense following exposure to the adapting stimulus, that is, a positive aftereffect.) Aftereffects have been demonstrated at both threshold (Verrillo & Gescheider, 1977) and suprathreshold levels (Lederman, Loomis, & Williams, 1982) and have proved to offer a powerful, noninvasive methodology for testing the channel model of vibrotactile sensation.

Spatial Acuity

The spatial resolution of the skin has been measured in several different ways. The most traditional test, known as the two-point touch threshold, measures the smallest gap between two points contacting the skin that are experienced as two separate tactile sensations. Normal threshold values tend to be about 2–4 mm on the fingertips and about 10–11 mm on the palm (Weinstein, 1968). Researchers have long recognized that there are a number of parameters that may affect spatial acuity, including applied force, the areal extent of the points contacting the skin, and the criteria used to judge the number of sensations experienced. Figure 4.7 shows the mean two-point thresholds for different regions on the hand averaged over the left and right sides for females (panel A) and for males (panel B) (Weinstein, 1968). The most consistent difference between the palm and the five digits is that the digits resolve spatial detail more precisely than the palm does.

Johnson and Phillips (1981; also see Craig, 1999) have questioned the validity of using the two-point touch threshold as a measure of spatial acuity. They

Table 4.1. Four-channel model of mechanoreception

Psychophysical characteristics of channels	P channel[a,b,c,d,e]	NP I channel[a,c,d,f]	NP II channel[a,c,d,f]	NP III channel[a]
Sensation quality	Vibration	Flutter	Unknown	Pressure
Frequencies over which threshold defined				
Small contactor (0.008 cm^2)	None	3–100 Hz	80–500 Hz	0.4–3 Hz
Large contactor (2.9 cm^2)	35–500 Hz	3–35 Hz	None	0.4–3 Hz
Shape of frequency response function	U-shape	Flat (slight dip at 30Hz)	U-shape	Flat
Slope of "tuning curve"	−12 dB/oct	−5.0 dB/oct	−5.5 dB/oct	Unknown
Best frequency (bf) varies with temperature	Yes	No	Yes	No
Low temperature (15–25°C)	bf: 150–250 Hz	—	bf: 100–150 Hz	—
High temperature (30–40°C)	bf: 250–400 Hz	—	bf: 225–275 Hz	—
Temperature sensitivity	Yes	Slight	Yes	Slight
Temporal summation	Yes	No	No[g]	Indeterminate
Spatial summation	Yes	No	Unknown	No
Putative mechanoreceptive afferent type	FAII (PC)	FAI (RA)	SAII	SAI
Psychophysical procedures				
To lower threshold				
Best frequency	250–300 Hz	25–40 Hz	150–400 Hz	0.4–1 Hz
Surround presence	No	Yes	Unknown	Yes
Contactor size	Large (>2.9 cm^2)	Any	Very small (<0.02 cm^2)	Any
Skin temperature	> 32°C	15–40°C	> 32°C	15–40°C
Stimulus duration[g]	>500 ms	> 1 cycle	> 500 ms	> one cycle
To elevate threshold				
Frequency of masker/adapter	250–300 Hz	25–40 Hz	150–400 Hz	0.4–10 Hz
Surround presence	Yes	No	Unknown	No
Contactor size	Very small (<0.02 cm^2)	Any	Unknown	Any
Skin temperature	Cold (15°C)	Any	Cold (15°C)	Any
Stimulus duration[g]	<200 ms	Any	<200 ms	Any

The information in this table is limited to the palm of the hand and is based on the studies referred to by letter. Adapted from table 2.2 in Cholewiak & Collins (1991) and from Greenspan & Bolanowski (1996).

Abbreviations: P, Pacinian; NP, non-Pacinian.

[a]Bolanowski et al. (1988).
[b]Bolanowski & Verrillo (1982).
[c]Capraro, Verrillo & Zwislocki (1979).
[d]Gescheider et al. (1985).
[e]Verrillo (1968).
[f]Verrillo & Bolanowski (1986).
[g]Gescheider et al. (1985) described an effect of stimulus duration on threshold; however, a subsequent study (Gescheider, Hoffman, Harrison, Travis, & Bolanowski, 1994) found the effect to be due to the noise masker rather than to temporal integration.

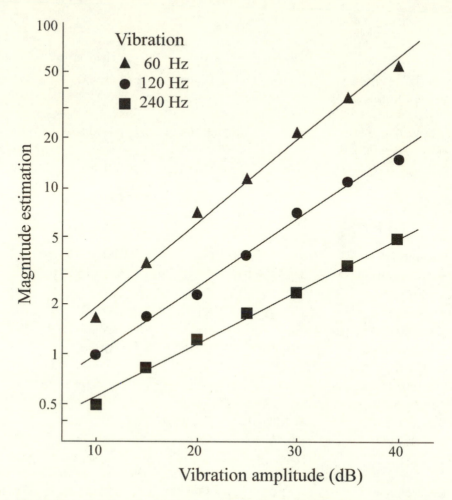

Figure 4.6. Magnitude estimation of vibration intensity for three vibration frequencies (60, 120, and 240 Hz). The vibrating stimulus was a thin metal rod (3.175 mm in diameter) that was held between the thumb and the fingers. Each point is the geometric mean of two judgments made by each of a group of observers. There were 12 observers for 120 Hz and 14 observers for the other two frequencies. The straight lines in log-log coordinates indicate power functions. The exponents are 1.0, 0.81, and 0.62. The coordinate scales give relative values only. Reprinted from S. S. Stevens, 1975, with permission of John Wiley & Sons Inc.

note that observers can use nonspatial (i.e., intensive) cues to resolve two points when they are spatially separated by distances considerably below the traditional two-point touch threshold. In its place, they propose a two-alternative forced-choice task in which observers are required to judge the orientation of a grating (horizontal versus vertical) for a number of spatial frequencies. (Spatial frequency is the inverse of spatial period, which consists of the sum of the widths of 1 element + 1 interelement gap.) With this method, observers are capable of resolving considerably finer gaps on the fingertip (~1 mm). Figure 4.8 shows the discrimination thresholds for grating

orientation on the index, middle, and ring fingers of each hand (Vega-Bermudez & Johnson, 2001).

Another measure of spatial acuity is point localization, which involves presenting one stimulus followed a short time later by a second stimulus that may or may not be in the same location. The observer must decide whether the two points are at the same or different locations. Errors of point localization on the hand are typically smaller than those for two-point touch, ranging from 1 to 2 mm on the fingertips for both men and women (figure 4.7); however, the two measures are typically highly correlated (e.g., $r = .92$ in Weinstein, 1968).

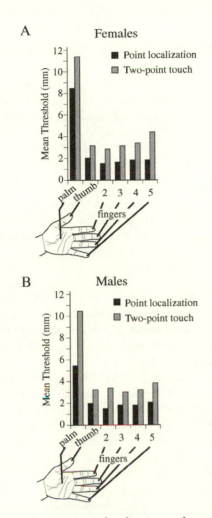

A   Females

B   Males

Figure 4.7. Mean point localization and mean two-point touch thresholds for the palm, thumb pad, and fingerpads of the index (digit 2), middle (digit 3), ring (digit 4), and little (digit 5) fingers for females (A) and males (B). Revised from Weinstein, 1968.

Localization of thermal sensations is known to be very poor compared to localization of tactile stimuli (Taus, Stevens, & Marks, 1975). It is very difficult to differentiate spatially two thermal stimuli owing to the fact that thermal stimulation summates spatially (Cain, 1973).

Functional interactions between the thermal and tactile modalities are known to occur. For example, Green (1977) demonstrated a fascinating thermal illusion in which observers were required to feel three thermal stimulators with the middle three fingers of one hand. The outer two were warm (or cool), while the middle one was thermally neutral.

Observers clearly experienced warm (or cool) sensations on all three fingertips. Thus, the thermal sensations experienced directly at the outer fingers were also "referred" or misdirected spatially to the middle finger, the site of the thermally neutral stimulus. A simple demonstration of the illusion may be produced by warming coins on a radiator or cooling them in a refrigerator. The illusion is depicted in figure 4.9. To summarize, when there was a spatial discrepancy between tactile and thermal patterns on the skin, the more spatially acute tactile sense dominated the less spatially acute thermal sense.

Finally, we consider the relative localization accuracy of electrocutaneous sensations in the hand. Higashiyama and Hayashi (1993) assessed the extent to which observers could localize such stimuli on the palmar surface of the middle three fingers using two types of electrode configuration. In the unifocal configuration, the anodal electrode was placed on the instep of the right foot, whereas the cathodal electrodes were placed at the targeted sites of study on the hand. In the concentric configuration, the distance between the cathodal and anodal electrodes was considerably smaller, being concentrically arranged in inner and outer disks, respectively. The site of stimulation varied across and within the middle three fingers of the right hand. Observers were required to select the apparent location of each pulse from a specified set of body sites on the fingers and wrist. For the unifocal configuration, localization was poor when the stimulating sites were on the middle finger (i.e., three phalanges) and relatively high for stimulating sites on the tips of either the index or ring finger; however, localization was relatively good across the fingers. In contrast, localization was very good for all stimulating sites when the concentric configuration was used.

A second study explored within-finger localization more thoroughly by using three stimulating sites within each of the index, ring, and middle fingers with both types of electrode configuration. As before, with the unifocal configuration, the errors were large. Localization was very poor on the proximal phalanx of all three fingers followed by the distal phalanx, whereas the middle phalanges showed little or no error. Thus, there was no clear proximal to distal gradient. In addition, localization errors on the proximal and distal phalanges were both biased toward the middle phalanx. In contrast, with the concentric configuration, localization was either very precise or biased a little toward the palm. The authors speculated that

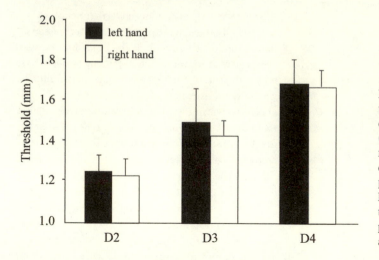

Figure 4.8. Mean grating-orientation discrimination thresholds for digits 2 (index), 3 (middle), and 4 (ring) of each hand. Error bars represent the SEM difference from each observer's own overall mean performance. Reprinted from Vega-Bermudez & Johnson, 2001, with the permission of ANN Enterprise/Lippincott Williams & Wilkins and the authors.

the poorer localization obtained with unifocal electrodes on the fingers may be due to greater spatial confusion caused by the fact that such electrodes stimulate both restricted superficial nerves and nerves that lie in deeper tissue, in contrast to concentric electrodes on the fingers, which stimulate only the more confined superficial nerves.

Temporal Acuity

Investigators have also measured how precisely people can resolve fine temporal differences in tactile stimulation. Various psychophysical methods have been used to address this issue. For example, the threshold for judging the successiveness of mechanical pulses (i.e., for perceiving two as opposed to one

pulse) is approximately 5 ms for the skin (Gescheider, 1974), whereas the temporal acuity for electrocutaneous pulses is about 50 ms, and therefore much inferior to mechanical touch (Higashiyama & Tashiro, 1988; Rosner, 1961).

How do the results describing the sensory resolving capacity of the hand compare to that of the eye or the ear? The visual system is clearly the most spatially acute sense, being able to resolve spatial differences between two points as fine as 1 min of arc under foveal viewing (Howard, 1982). In contrast, the auditory system can only localize a 1000 Hz target presented at varying angular distances from the median plane of the head to within about 1° (Mills, 1958). A comparable measure of tactile acuity is the two-point touch threshold. For the tongue, the most

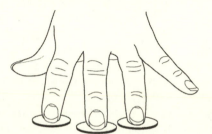

Physical Stimulation	warm	neutral	warm
	cool	neutral	cool
Perceptual Experience	warm	warm	warm
	cool	cool	cool

Figure 4.9. A thermal illusion may be easily generated, as described in the text. The observer is required to touch three coins with the middle three fingers, one set being physically warm/neutral/warm and another being physically cool/neutral/cool. The corresponding perceptual experience is that the three coins are all warm or all cool, respectively. Thus, the outer thermal sensations are referred to the thermally neutral inner coin (Green, 1977).

spatially acute locus on the body, the two-point threshold has a value of about 1 mm; the comparable value for the fingertip, the most spatially acute site on the hand, is 2–3 mm. (Based on the more precise grating-orientation task that has been proposed more recently, however, the spatial acuity of the fingertip is considerably finer than 2–3 mm, that is, around 1 mm.) However, as angular measures are not applicable to a contact sense, it is difficult to compare this value directly with the angular values for either audition or vision. With respect to temporal resolving capacity, the visual system can determine the successiveness of two pulses separated by about 25 ms, whereas the auditory system requires only 0.01 ms (Sherrick & Cholewiak, 1986). With a corresponding value of 5 ms, mechanical touch lies in an intermediate range between vision and audition, with audition being best and vision the worst. Numerousness, that is, the ability of the observer to count accurately a series of sensory events (visual, auditory, or tactile) presented within a given time period produces a similar temporal ordering of the senses (Lechelt, 1975). Such results are likely important when considering issues pertaining to the nature of multisensory integration (see Calvert, Stein, & Spence, 2004) and the design of multimodal interfaces for teleoperation and virtual environment systems (see chapter 10).

Tactile Motion

Studies have shown that stimulus movement across the skin can be detected by a passive observer in terms of its presence, location, direction, and speed (Bender, Stacy, & Cohen, 1982). For example, psychophysical studies that used brushes, edges, and probes to produce real continuous movement on the human hand have shown that discriminating the direction of stimulation depends on both the distance moved and the stimulus velocity (Essick, Franzen, & Whitsel, 1988; Norsell & Olausson, 1992).

In addition to the sensation of tactile motion that results from real movement across the skin, an illusory sensation of motion may also be produced, in this case, by delivering a series of pulses to adjacent locations in quick temporal succession (Kirman, 1974; Sherrick & Rogers, 1966). Smooth apparent motion, or beta motion, is detected at various sites (e.g., the back) when two tactile stimuli (optimally,

periodic such as 150 Hz) are delivered to two different sites on the skin. The stimulus appears to move between the two contact sites when the interval between stimulus onset falls within a certain range. The interval that is optimal for beta motion is a nonmonotonic function of the duration of the first stimulus, as in vision (Lakatos & Shepard, 1997; Sherrick & Rogers, 1966). For detailed comparison of the conditions under which tactile and visual apparent motion occur, the reader should consult Sherrick and Cholewiak (1986).

Gardner and Sklar (1994) have produced apparent motion on the fingers using a reading device for the blind known as the Optacon (figure 4.10) to assess the contribution of spatial and temporal properties of a moving tactile stimulus to the perceived direction of motion (for further details about the Optacon, see Bliss, Katcher, Rogers, & Shepard, 1970). Gardner, and Sklar (1994) were specifically interested in identifying the neural mechanisms that underlie the previously documented effects of distance and speed on perceived direction of motion (i.e., spacing of activated mechanoreceptor populations or number and overlap of peripheral receptive fields). They concluded that discrimination of movement direction is not just based on the localization of points on the skin. It also requires the complex central integration of a coherent spatiotemporal sequence of impulses.

Vision researchers have long known of highly robust aftereffects involving the visual perception of motion (e.g., the waterfall aftereffect). Visually adapting observers to the movement of a pattern in one direction results in their subsequently perceiving a stationary pattern as moving in the opposite direction. Do we find a similar illusory phenomenon on the skin? The results have been mixed. Hollins and Favorov (1994) confirmed the presence of a strong tactile motion aftereffect, which increases in magnitude and vividness with the duration of the adapting stimulus. In one study, observers placed their hands on a cylinder that was rotating at a rate of 280 mm/s, to which were attached raised, square-wave stripes of low spatial frequency covered with smooth adhesive tape. They experienced a vivid and reliable negative motion aftereffect that was very similar to what has been previously documented with vision. Hollins and Favorov explained the phenomenon using a model of cortical dynamics proposed by Whitsel and his colleagues (Whitsel, Favorov, Kelly, & Tommerdahl, 1991; Whitsel, Favorov, Tommerdahl,

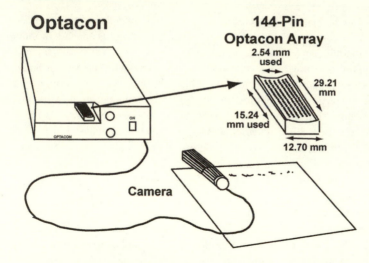

Optacon

144-Pin Optacon Array

2.54 mm used

29.21 mm

15.24 mm used

12.70 mm

Camera

Figure 4.10. The Optacon, a direct print-to-tactile pattern converter. Images from a camera (bottom) are displayed as point-for-point vibrating patterns on the 6-column × 24-row tactile array shown on the right. Reprinted from Cholewiak & Collins, 2000, with the permission of the Psychonomic Society.

Diamond, Juliano, & Kelly, 1989). In brief, the adapting stimulus tunes the responses of directionally sensitive somatosensory cortical neurons. When the adapting stimulus is replaced with a stationary stimulus, which in the unadapted state is presumed to elicit equal activity in neurons selectively sensitive to all directions of motion, the activity-weighted mean response of the adapted cortical network is displaced away from the direction of the movement of the adapting stimulus. It has been proposed that the observer interprets this as a shift in perceived movement in a direction opposite to that of the adapting stimulus. Unfortunately, researchers have not always been able to produce the tactile aftereffect reliably, as noted both by Hollins and Favorov (1994) and more recently by Lerner and Craig (2002).

Space–Time Interactions in Somatosensory Processing

Research has shown strong space–time interdependence with regard to visual, auditory, and also somatosensory processing. For example, the tau effect refers to the fact that the apparent distance between three equally spaced stimuli that are presented successively on the forearm depends upon the intervening temporal interval (Helson & King, 1931). More specifically, if the time between the first and second stimuli is shorter (longer) than the time between the second and third stimuli, people judge the corresponding distance between the first and second stimuli to be shorter (longer) than the physically equal

distance between the second and third stimuli. This effect has been demonstrated in all three modalities. The converse effect, known as kappa, refers to the fact that the apparent temporal interval between two successively presented stimuli depends upon their spatial separation. Specifically, the temporal interval is perceived to increase with increasing spatial extent. Although this perceptual phenomenon has been confirmed with respect to audition (Cohen, Hansel, & Sylvester, 1954) and vision (Cohen, Hansel, & Sylvester, 1953), it has not been demonstrated in the tactile domain (Yoblick & Salvendy, 1970). Moreover, none of the work to date has selected the hand as a site of stimulation. The occurrence of tactile tau and kappa should be explicitly evaluated with respect to the hand. Any successful theory of tactile space perception must be able to account for such space-time interactions.

A different but striking space-time interaction known as the saltation effect involves the illusory displacement of tactile stimuli (Geldard & Sherrick, 1972). By way of a simple demonstration, consider the placement of three identical contactors spaced about 10 cm apart on the forearm. A sequence of 15 very brief taps is initiated starting with the first contactor and ending with the third one, with each of the three in turn receiving 5 taps with no additional temporal gap between the stimulation of successive contactors. When this demonstration was actually presented to observers, much to the investigators' surprise, rather than perceiving three spatially separated bursts under the contactors, observers perceived a "slow, sweeping movement punctuated by taps" (Geldard, 1975). This intriguing perceptual phenomenon became affectionately known as "the rabbit" because initial observers reported that it felt like a rabbit hopping up their arms. This was picked up by a Norwegian newspaper cartoonist, who published one conception of the perceptual illusion, as shown in figure 4.11 (top panel).

When the conditions needed to produce the rabbit are reduced to their simplest form, two taps considerably separated in time (e.g., 800 ms) are presented to one skin site. The first tap serves to confirm the true location of the second tap, which may be considered the "attractee." The perceived position of the attractee changes with the temporal interval between it and a third tap, which is presented at a second location a short time later. When the time between the second and third taps is short (e.g., around 20 ms), the second tap appears to occur spa-

tially near the actual site of the third tap; however, as the time interval increases, the second tap will appear to occur toward its true location marked by the first tap and precisely at this locus when the interval is about 250 ms. The illusion can also be created by presenting multiple taps at a single site, followed in time by a tap to a second site. Observers report that they experience a sequence of taps that occurs at relatively equal distances between the two actual sites of stimulation (Geldard, 1975; Tan, Lim, & Traylor, 2000). Although initially observed with touch, such saltatory effects have also been documented using both vision and audition (see Geldard, 1975). The earlier work usually focused on body sites other than the hand. However, Geldard and Sherrick (1983) did confirm the occurrence of cutaneous saltation within both the fingerpad (index) and the palm. While the saltatory areas are similar

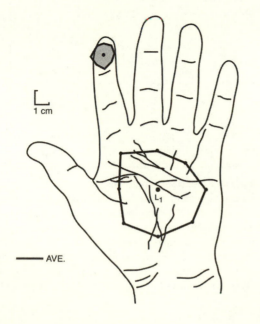

Figure 4.11. (Top panel) One conception of the "rabbit," supplied by a Norwegian newspaper cartoonist. Reprinted from Geldard, 1975, with the permission of Lawrence Erlbaum Associates. (Bottom panel) Area within which saltation occurs on the volar side of the hand (palm versus index finger). Reprinted from Geldard & Sherrick, 1972, with the permission of the Psychonomic Society.

in shape with the long axis of the oval parallel to the axis of the finger or palm as shown in figure 4.11 (bottom panel), the size is considerably smaller for the fingertip than for the palm. The authors note that the relative order corresponds to that for receptive field size in the two hand areas.

More recently, Cholewiak and Collins (2000) compared two display modes for drawing a dotted line on the index finger using an Optacon (figure 4.10). The veridical display mode produced bursts of vibration (230 pulses/s), creating a linear series of seven vibrating sites in sequence, with each site defined by a 2×2 arrangement of adjacent pins. The saltatory mode activated only three sites (7.6 mm apart), with three vibratory bursts at location 1, then three at location 4, and then one burst at location 7. Figure 4.12 presents visual representations of the contactor sites, stimulus patterns for both modes,

and the sensations that were experienced in both modes. Observers were not told how many sites would be activated on any trial. The design ensured corresponding temporal parameters for burst and interburst duration in both presentation modes. Observers judged perceived length, straightness, spatial distribution, and smoothness of the lines under different conditions relating to burst and interburst duration. The results indicated that effective lines were produced by both modes of presentation and that the two sets were perceptually equivalent.

Studies of real and apparent motion, motion aftereffects, and space-time interactions have important consequences for understanding how we normally process motion, space, and form via the skin. In addition, they are relevant to the design of tactile displays for communication and teleoperator and virtual environment applications, as discussed in chapter 10.

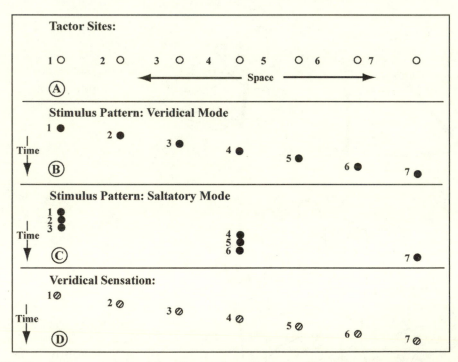

Figure 4.12. Visual representations of the vibrotactile spatiotemporal patterns. Several contactor sites (A) were aligned along the finger using an Optacon, with a computer replacing the camera (shown in Figure 4.10) to define the spatial layout and temporal parameters to be displayed and only a 2-column × 14-row subset of pins being used. In the veridical presentation mode, bursts of vibration were presented sequentially along all seven sites (B). In the saltatory mode, the seven bursts occurred at only three sites (C). In both cases, the interburst interval was constant over the series of bursts. In a typical veridical presentation (B), one feels a series of stimuli marching along the skin over time (D). Given the appropriate timing parameters, saltatory presentations (C) would also feel like this series (D). Reprinted from Cholewiak & Collins, 2000, with the permission of the Psychonomic Society.

Properties of Surfaces and Objects

We have previously noted that when an external agent stimulates the stationary hand with a stimulus (passive touch), observers tend to focus on their own internal sensations. In contrast, when they actively explore the stimulus (active touch), their perceptual experiences tend to be directed toward the external object and its properties (Gibson, 1962). With passive touch, the initial sensory inputs are provided solely by the peripheral populations of mechanoreceptive and thermal units, and they are subsequently integrated at higher levels of the nervous system (i.e., tactile sensing). Nevertheless, it is still possible to use tactile sensing to extract useful, albeit limited, information about the external world.

Klatzky and Lederman (1993) have suggested that tangible object properties may be hierarchically organized into two major categories: (a) material (e.g., texture, compliance, and thermal properties), and (b) geometry (e.g., size and shape), which occurs at both micro (fingertip) and macro (larger than fingertip) levels. Weight reflects the contributions of both material and geometric features since it is partly determined by object mass, which in turn is the product of object density (a material feature) and object volume (a geometric feature).

Stationary Versus Moving Stimuli

With no relative motion between the fingertips and the stimulus surface or object, the discharge rates of SA I units are relatively low (Johnson & Lamb, 1981). Moreover, without movement, the sensory experience of surface texture is degraded (Katz, 1925/1989). Subsequent to these studies, Hollins, Bensmaïa, and Risner (1998) showed more specifically that when relative motion is eliminated, discrimination of relatively smooth surfaces is impaired; in addition, the exponent of the psychophysical roughness function is reduced. In contrast, the absence of relative motion makes coarse surfaces feel smoother, but leaves the corresponding exponent relatively unchanged (Hollins et al., 1998). We will discuss the additional implications of these results for theories of texture perception under the next section, "Material Properties."

With no movement, it is still possible to distinguish large intensive changes in material properties relatively quickly. Lederman and Klatzky (1997) required observers to decide as quickly and accurately as possible whether or not a target (e.g., a very rough surface) was present among varying numbers of distractor items (e.g., very smooth surfaces) presented simultaneously to various combinations of one to six fingers selected from the middle three fingers of both hands. Performing this search task, observers could decide relatively quickly and with limited hand movements whether a surface was rough or smooth on some trials, hard or soft on others. (Differentiating between warm and cold was slower, presumably because of the time required for heat transfer.) In contrast, it took relatively more time, sometimes considerably more, to discriminate large geometric differences; for example, whether a raised dot lay to the left or right of a central indentation, the relative planar orientation of a raised line, or the relative three-dimensional orientation of a ramp. When observers made their binary perceptual judgments of a geometric property, they also chose to move their hands actively over the stationary display following initial contact (active haptic sensing).

Material Properties

Surface Texture

An early critical source of information about the tactual perception of surface texture is *The World of Touch* by David Katz (1925/1989). This book provides a wealth of original, varied, and highly informative observations on texture perception by touch and is based on both careful phenomenological observations and experimental results.

It is important to understand that the perception of texture is multidimensional. Hollins, Faldowski, Rao, and Young (1993) recognized this point by requiring observers to arrange a set of 17 stimulus surfaces into from three to seven groups according to their perceived dissimilarity. Surfaces were moved under an observer's stationary index finger. The dissimilarity judgments were then used to construct a perceptual space for texture using multidimensional scaling procedures. The data were fit reasonably well by a three-dimensional solution, as shown in figure 4.13. Rating judgments of the objects in terms of five adjective scales were obtained from the same observers and used to interpret the three-dimensional solution. Roughness/smoothness and hardness/softness proved to be reliable orthogonal dimensions. The

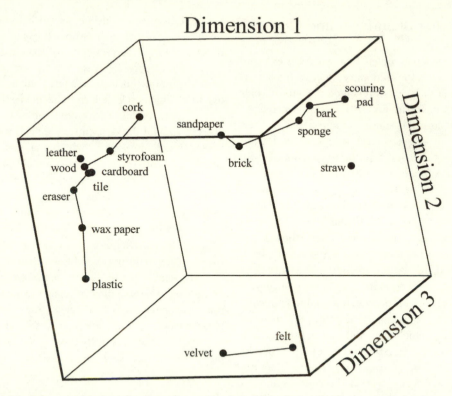

Figure 4.13. Cube representing the three-dimensional scaling solution for the perceptual texture space. The dimensions of this perceptual space are described in the text. Reprinted from Hollins, Faldowski, Rao, & Young, 1993, with the permission of the Psychonomic Society.

third dimension could not be clearly interpreted. In a subsequent study, a sticky/slippery dimension was also observed, although only for some observers (Hollins, Bensmaïa, Karlof, & Young, 2000). Hollins et al. (2000) concluded that the sticky/slippery dimension is less perceptually salient than either the rough/smooth or hard/soft dimensions, which were reliably observed in all participants.

People are remarkably good at detecting and discriminating surface textures. For example, a texture composed of either periodically ordered bars or dots only a fraction of a micron high (i.e., 0.06 μm and 0.16 μm, respectively) is detectable with 75% accuracy when moved laterally across the fingerpad (LaMotte & Srinivasan, 1991; cf. 6 μm for a single dot of 50 μm diameter and 1 μm for dots > 500 μm diameter). Moreover, with two-dimensional raised-dot patterns, people can discriminate differences in spatial period (that is, the sum of one element width and one interelement width) as small as 2% with 75% accuracy whether active or passive touch is used (Lamb, 1983).

Surface Roughness

Much of the research on tactile texture perception has focused specifically on the highly prominent perceptual dimension of roughness. Note that we have chosen to discuss this topic most extensively under the category of tactile sensing (cf. active haptic sensing) inasmuch as research indicates that the perception of roughness is similar under both conditions provided there is relative motion between the skin and the surface (Lamb, 1983; Lederman, 1981, 1983; Verrillo, Bolanowski, & McGlone, 1999). This suggests that cutaneous (as opposed to haptic, which includes both cutaneous and kinesthetic) inputs primarily determine perceived roughness.

In addition to considering the effect of mode of touch (tactile versus active haptic sensing), researchers have also evaluated the contribution of a number of parameters that pertain to surface microgeometry (e.g., dot width, interelement spacing, spatial period, ratio of element width to gap width) and to the manner in which the surfaces contact the skin (force, relative speed). The magnitude-estimation procedure has often been used to examine the rate at which the magnitude of perceived roughness grows as a function of the physical parameters of the textured surfaces and presentation parameters. Many different types of surface textures have been evaluated, including, for example, one-dimensional gratings (Lederman, 1974), two-dimensional dot patterns (Connor & Johnson, 1992), and abrasive surfaces (Hollins & Risner, 2000; Stevens & Harris, 1962).

Early work with incised one-dimensional rectangular gratings (figure 4.14A) suggested that the critical determinants of perceived roughness are the interelement spacing of incised rectangular gratings (G) and applied force (F), as shown in figure 4.14B. Perceived roughness grows as some function of $G*F^{\frac{1}{4}}$ (Taylor & Lederman, 1975). When not experimentally confounded with groove width, however, spatial period has no effect on perceived roughness (Lederman, 1983). In addition, the speed of relative motion demonstrates little effect with active exploration of one-dimensional gratings (Lederman, 1983) or with passive touch of two-dimensional raised-dot patterns (Meftah, Belingard, & Chapman, 2000). In another study, adapting the fingertip selectively to either a 40 dB_{SL} low-frequency (20 Hz) or 40 dB_{SL} high-frequency (250 Hz) vibration was initially shown to reduce the perceived magnitudes of a set of vibrations varying in intensity, provided the stimulus frequency was the same as that used during the adaptation period. Nevertheless, the same selective-adaptation procedures subsequently failed to influence the perceived roughness magnitudes of stimulus gratings (Lederman, Loomis, & Williams, 1982). On the basis of these three sets of converging results, vibration (and, therefore, temporal factors) appears to play no role in tactile roughness perception. Based on these psychophysical results, Taylor and Lederman (1975) proposed a quasi-static model of roughness perception in terms of the mechanical parameters of skin deformation. The model suggests that human roughness perception is encoded intensively, based on the volumetric deviation of skin deformed from its resting position during contact with the gratings. Note that while a spatial interpretation of those results is also possible, predictions based on the use of spatial versus intensive codes could not be disambiguated by this early work with one-dimensional gratings. To do so at the time would have required the technological capacity to produce sets of gratings that equated overall skin deformation while varying the spatial pattern. Spatial coding would predict perceptual equivalence in contrast to intensive coding, which would predict perceptual differentiation. For current purposes, we will not differentiate between the two forms of coding and therefore describe the model as spatial-intensive.

Human psychophysical experiments with two-dimensional raised-dot patterns suggest that observers also use a spatial-intensive (as opposed to a temporal) code to judge the magnitude of perceived roughness of dot patterns with centers separated by 1.3–6.2 mm (Blake, Hsiao, & Johnson, 1997; Blake, Johnson, & Hsiao, 1997; Connor, Hsiao, Phillips, & Johnson, 1990; Connor & Johnson, 1992). Complementary neurophysiological work with monkeys strongly implicates peripheral SA I tactile units (as opposed to FA I or FA II) in the encoding of stimulus roughness. Johnson and his colleagues propose that perceived roughness is represented in terms of a spatial-intensive neural code in which the spatial dot patterns are encoded as local spatial variations in the firing rates of adjacent SA I tactile units with receptive field centers separated by about 2 mm on the fingertip (i.e., by two primary afferent spacings in the skin of humans and monkeys) (Darian-Smith & Kenins, 1980; Johansson & Vallbo, 1979). According to this neural model, local spatial variation is computed in cortical area S-I (i.e., area 3b) and then summated in secondary somatic cortex (S-II) to produce an intensive code for perceived roughness (Hsiao, Johnson, & Twombly, 1993). Subsequent work by this group proposed that activity in SA I units can also account for human roughness perception of two-dimensional dot patterns with interelement spacings finer than SA I receptor spacings (Yoshioka, Gibb, Dorsch, Hsiao, & Johnson, 2001). More specifically, they have shown that there is substantial variation in the neural response (i.e., the stochastic variation in the spike rates of SA I units) despite the fact that the spatial variation in the stimulus pattern is much finer than the innervation density of the SA I units. Collectively, the three studies suggest a single-code theory of roughness perception.

Note that there currently exists some ambiguity about the shape of the psychophysical function for the perceived roughness magnitude of two-dimensional

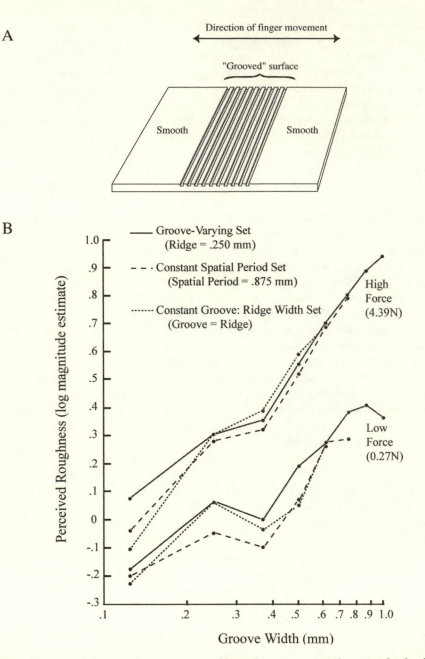

Figure 4.14. (A) Schematic of the stimulus gratings used by Lederman, 1983. The outer thirds of the plate are smooth; the inner third has parallel linear rectangular grooves cut along the width of the plate. Reprinted from Lederman & Taylor, 1972, with the permission of the Psychonomic Society. (B) Perceived roughness (log magnitude estimates) as a function of groove width (log scale) and fingertip force. The data for the groove-varying, constant spatial period, and constant groove-to-ridge sets are plotted separately for the 0.27 and 4.39 N force conditions. Reprinted from Lederman, 1983, with permission of the Canadian Psychological Association.

patterns. Connor et al. (1990) obtained an inverted-U shape when perceived roughness was plotted as a function of center-to-center dot spacing or spatial period (ranging from 1.3 to 6.2 mm, peaking at about 3.2 mm); however, Meftah et al. (2000) have documented a linear function over spatial periods ranging from 1.5 to 8.5 mm. The discrepancy is shown in figure 4.15 (panels A and B, respectively). Moreover, Lederman and her colleagues have consistently noted in the laboratory that when using their bare fingers, observers are reluctant to make judgments of roughness of two-dimensional raised-dot patterns with spatial periods more than about 3.5 mm. The reasons for such striking discrepancies need to be resolved if we are to understand fully the basic psychophysics of roughness perception and to apply the results of this basic research successfully to domains that involve the perception of surface roughness by touch.

Up to this point in the discussion, we have considered single-code models of roughness perception mediated by the population of SA I units. However, Hollins and his colleagues (e.g., Hollins et al., 1998; Hollins, Bensmaïa, & Washburn, 2001; Hollins & Risner, 2000) have proposed a duplex model of tactile roughness perception that differentiates between the perceptions of fine versus coarse textures. More specifically, small vibrations underlie the perception of very fine surfaces that are too fine to be resolved by any mechanoreceptor population, whereas a single spatial-intensive code underlies the perception of more coarse surfaces, as proposed earlier by Lederman, Johnson, and their respective colleagues. Results obtained by Hollins et al. (2001) reveal that adapting both RA and PC channels with a high-intensity adapting stimulus of 100 Hz produces poorer discrimination of very fine textures compared to an unadapted control condition. In contrast, there is no difference between the two conditions when coarse textures are discriminated. Such results support the proposal that fine texture perception is indeed mediated by vibration, whereas coarse texture perception is not. An experiment by Hollins et al. (2001) confirmed that when the RA and PC channels are selectively adapted to high-intensity 10 Hz and 250 Hz vibrations, respectively, fine texture discrimination is abolished when the 250 Hz, but not the 10 Hz, adapting stimulus is used, suggesting that perception of very fine textures is mediated by the PC channel. Bensmaïa and Hollins

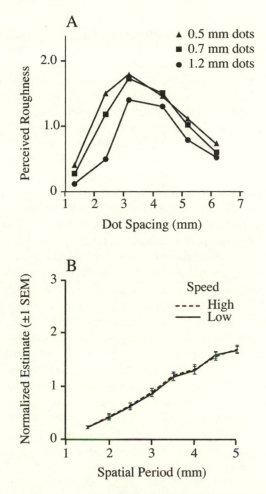

Figure 4.15. (A) The normalized roughness magnitude is plotted against dot spacing for three different dot-diameter sets. Reprinted from Connor, Hsiao, Phillips, & Johnson, 1990, with permission of the Society for Neuroscience. (B) Mean normalized roughness estimates as a function of longitudinal spatial period at low and high scanning speeds (51 and 96 mm/s). Each surface (transverse SP = 2 mm) was presented 6 times at each speed with 96 trials/observer. Data are pooled from eight observers. Reprinted from Meftah, Belingard, & Chapman, 2000, with permission of Springer-Verlag.

(2003) offered additional support for the proposal that perception and discrimination of very fine surface textures may be mediated by the PC channel and in addition that perceived roughness varies as a function of the vibratory intensity scaled to the spectral sensitivity of the PC system.

Cascio and Sathian (2001) and Smith, Chapman, Deslandes, Langlais, and Thibodeau (2002) have proposed additional roles for temporal cutaneous inputs in the perception of coarse roughness magnitude. In the former study, observers were required to provide magnitude estimates of the perceived roughness of periodic gratings, which varied in ridge or groove width, when scanned across the finger at different controlled forces and speeds. Along with the gratings' spatial frequency, variations in speed would therefore contribute to temporal frequency cues. A second task in the Cascio and Sathian study involved roughness discrimination in which temporal frequency was either held constant or exaggerated. In keeping with previous results, those of the current study confirmed a strong effect of groove width; however, a relatively small effect for temporal frequency for ridge-varying gratings was also found, thereby implying a small role for temporal coding in the perception of coarse roughness as well.

Finally, we should address the role of friction in judgments of roughness magnitude. An early study by Taylor and Lederman (1975) found no effects for a set of incised metal gratings varying in groove width from 0.125 to 1 mm (constant ridge width of 0.25 mm) when the surfaces were either dry or lubricated with detergent. However, a subsequent investigation by Smith et al. (2002) used raised columns of two-dimensional patterns of truncated cones with longitudinal spatial periods ranging in the transverse direction from 1.5 to 8.5 mm (dot diameter of 0.6 mm along the top, 1.0 mm along the base; corresponding interelement spacing values ranged from 0.9 to 7.9 mm and from 0.5 to 7.5 mm). Both normal and tangential forces were recorded throughout the active stroking interval. In the first experiment, perceived roughness correlated well with the mean normalized root mean square of the tangential force rate. In a second experiment, the changes in tangential force were dissociated from the surface geometry and the mean kinetic friction by requiring observers to estimate the roughness of dry versus lubricated surfaces. The results revealed that the mean normalized root mean square of the tangential force rate changed by approximately 21% and the subjective estimates of roughness by about 16%. The collective results from the two experiments suggest that the root mean square of the rate of change in tangential force may also contribute to perceived roughness magnitude.

Compliance

In judging object compliance, both cutaneous and kinesthetic cues are potentially available. People apparently use cutaneous inputs to the fingertip during both tactile and active haptic sensing to judge the softness of deformable elastic objects made of silicone rubber. In a psychophysical study that involved both rank-order and discrimination tasks, Srinivasan and LaMotte (1995) presented such objects to the observer's passive fingerpad so that the sensory information would be restricted to cutaneous inputs. Cutaneous information alone proved sufficient for discriminating the rubber surfaces. Presumably, because the contact force was constant, the spatial distribution of pressure within the contact region on the fingertip could be used to differentiate objects in terms of their relative compliance. Additional supporting data were obtained in another experiment in which observers used active haptic sensing while the fingertip was anesthetized, a procedure used to eliminate the cutaneous inputs. In keeping with the earlier passive-touch experiment, observers with the finger anesthetized were unable to discriminate even very large differences in compliance. In contrast, cutaneous information was insufficient on its own for discriminating the softness of rigid surfaces produced in the form of outer hollow cylinders surrounding inner cylinders containing springs. With such objects, both cutaneous and kinesthetic information (i.e., active haptic sensing) are clearly necessary.

Based on the measured peripheral responses of SA I and FA I units to the application of rubber specimens varying in compliance, Srinivasan and LaMotte (1996) concluded that tactile discrimination of softness is likely to be based on the response of SA I units, specifically their discharge rates.

Thermal Properties

Familiar materials (i.e., aluminum, glass, rubber, polyacrylate, and wood) can only be identified with variable success using static contact. Ino, Shimizu, Odagawa, Sato, Takahashi, Izumi, and Ifukube (1993) showed that while observers identified aluminum and wood with 91% and 82% accuracy, respectively, their performance was poor when attempting to recognize the remaining three materials. Overall, mean response times ranged from 1 to 2 s. We assume that observers used thermal sensing, inasmuch as neither texture nor compliance cues would have been useful because manual exploratory motions were prohibited. Ino et al. suggested that observers used the rapid changes in skin temperature, which were reported to occur within approximately 0.5 s. This figure varies markedly from that reported by Ho and Jones (in press), who found that the temperature of the finger-tip does not stabilize until about 5 s after contact and that the decreases in skin temperature on contact with the various materials are much smaller than those reported by Ino et al., as shown in figure 4.16. In this study, observers were required to choose the colder of two materials presented to the tips of the index fingers. The results indicated that observers were able to discriminate between materials using thermal cues when the differences in thermal properties (as defined by the contact coefficient, which is a function of the density, thermal conductivity, and specific heat of the material) were large, but could not discriminate between materials such as copper and bronze, this result being consistent with the predicted heat fluxes out of the skin.

In a further experiment, Ho and Jones (in press) found that the ability to localize thermal changes when three fingers on the same hand are stimulated is poor and depends on the thermal properties of the target material. In this experiment, observers were required to identify which of three materials presented to the fingertips differed from the other two. This task was considerably more difficult than the first discrimination experiment, as the overall correct response rate averaged 57%, as compared to 90% in experiment 1. It would be intriguing to determine whether the underlying thermal cue for material variations is referred to adjacent fingers, as

Green (1977) reported. Although the question is intriguing, the Ho and Jones experiment used a different experimental task (i.e., localization) and smaller thermal differences across the fingers, making it somewhat difficult to make comparisons between the two studies.

These studies indicate that thermal cues can be used to recognize objects by touch, but as the resting temperature of the skin is typically higher than the ambient temperature of objects encountered in the environment, the thermal conductance and heat capacity of different materials are used to assist in identifying what the object is made from. The absolute temperature is not informative because at room temperature it is usually the same for each object palpated. The coding of object temperature is therefore primarily determined by the activity of cold receptors, which signal the cooling of the skin.

Weight

Weight is perceived less effectively when the object rests on a stationary hand that is supported by a rigid surface (i.e., tactile sensing) than when it is lifted (i.e., active haptic sensing). For example, when Brodie and Ross (1984) required their observers to judge weight via tactile sensing, the Weber fraction (i.e., difference threshold/intensity of the standard) averaged 0.13 for a 54 g standard. This value was 1.46 times larger than that obtained via active haptic sensing. When tactile sensing alone is used, weight judgments are based solely upon cutaneous inputs. Under such circumstances, neither kinesthetic feedback nor corollary discharges from the efferent commands are available.

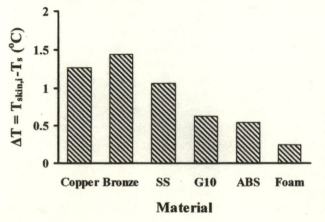

Figure 4.16. Change in skin temperature on the distal pad of the index finger after 5 seconds of contact with the materials indicated. SS = stainless steel; G10 = glass-reinforced epoxy; ABS = plastic. From Ho & Jones, in press, with the permission of the Psychonomic Society.

In 1846, Weber first noted a very powerful illusion involving weight perception (as cited by J. C. Stevens in 1979): a cold coin feels heavier than a warm coin of the same mass when applied to the forehead. Stevens (1979) further explored such thermal-weight interactions by requiring observers to estimate the heaviness of weights of varying temperature (neutral, warm, and cold) applied to different skin sites (palm, forearm, abdomen, thigh, upper arm, and back). The cold weights felt substantially heavier than the neutral weights at all body sites. Warm weights also tended to intensify heaviness sensations, but the results were only statistically significant on the palm, forearm, and upper arm. These results may be interpreted based on an earlier proposal offered by J. C. Stevens and Green (1978), namely, that thermal intensification of perceived heaviness is due to the failure of mechanoreceptors to respond selectively to mechanical stimulation. Several classes of mechanoreceptors are temperature sensitive, and hence their responses will change with variations in temperature. The results are more difficult to explain in terms of an additive cognitive model, which proposes that sensations of temperature, which are to be ignored, summate with weight to yield a stronger impression of weight. Unlike cold, which consistently magnifies weight perception, warmth intensifies perceived weight only in some regions of the body. Hence, the validity of such a cognitive interpretation seems less likely.

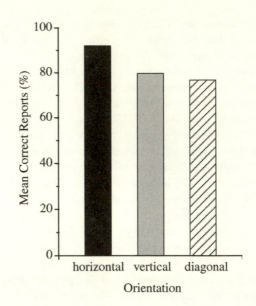

Figure 4.17. The percentage of correct reports of the stimulus pair containing the comparison orientation shown as a function of the deviation of the comparison orientation from horizontal, vertical, and diagonal (average of left 45° and right 45°) standard orientations. Adapted from Lechelt, 1988, with the permission of the author.

effect, this well-known anisotropy of perceptual space has also been demonstrated using active haptic sensing and visual perception (Appelle, 1972).

Material

As noted above, in principle, objects that vary in material may be identified and discriminated on the bases of their texture, compliance, and thermal conductivity, either singly or in some combination depending on the relative effectiveness of the cues available at the time.

Geometric Properties

Orientation

When line orientations are presented to the distal portion of the fingertip, observers can reliably discriminate changes in orientation better when the standard stimulus is oriented horizontally or vertically as opposed to obliquely (e.g., at 45°), as shown in figure 4.17 (Lechelt, 1988). Known as the oblique

Curvature

Curvature is the rate of change in the angle of the line tangent to a curve as the tangent point moves along the curve. For a circle or sphere, curvature is inversely related to the radius of curvature. For ease of interpretation, the following discussion describes stimuli primarily in terms of their radius of curvature with corresponding curvature values added in parentheses. Based on tactile sensing via the finger, observers can discriminate a decrease in the radius of curvature (or increase in curvature) of spherically curved segments of approximately 13% and 18% for corresponding standards of 3.5 mm (0.29 mm⁻¹) and 6.5 mm (0.15 mm⁻¹) (Goodwin & Wheat, 1992). A forced-choice paradigm was used in this work in which observers were required to judge whether a pair of stimuli were the same or different. The observers could also discriminate a convex spherical surface with a radius of curvature of 6.95 mm (0.14 mm^{-1}) from one of 6.33 mm (0.16 mm^{-1}) and a curvature of 3.48 mm (0.29 mm^{-1}) from one of 3.13 mm (0.32 mm^{-1}). These correspond to detecting a curvature change of ~10%. Applied contact forces of 0.20, 0.39, and 0.59 N all generated monotonically increasing estimates of the amount of curvature as a function of the decreasing radius of curvature. The radius of curvature extended across a fairly natural set of values from 1.44 mm (0.69 mm^{-1}) to a perfectly flat surface (Goodwin, John, & Marceglia, 1991). Overall, force has a variable effect on perceived curvature, as compared to the highly consistent effect due to changes in physical curvature. Finally, we note that curvature discrimination of strips presented statically to the hand is best when placed along as opposed to perpendicular to the fingers (Pont, Kappers, & Koenderink, 1998). When presented in the former orientation, observers can discriminate the curvature of a stimulus with a radius of curvature of 2500 mm (curvature = 0.0004 mm^{-1}) from one that is flat.

In experiments that examine microgeometry, information pertaining to an object's shape and size is contained in the population responses of SA I and FA I mechanoreceptors (LaMotte & Srinivasan, 1987a, 1987b; Srinivasan & LaMotte, 1987). In this series of studies, different shapes (e.g., sinusoidal steps, cylindrical bars) were applied both statically and via passive stroking to the stationary fingertips of monkeys. In contrast to intensive coding, the spatial discharge patterns of the SA I mechanoreceptor population best coded the geometric parameters, regardless of changes in the manner of skin-surface contact (e.g., force, orientation, and direction of object motion relative to the skin). Such neural-response invariance is of course essential for perceptual shape constancy.

Size

The well-known visual distortion effect known as the horizontal-vertical illusion has been documented with passive motion of a line stimulus across the forearm (Wong, Ho, & Ho, 1974), as well as with active touch (Day & Wong, 1971; Heller & Joyner, 1993; Millar & Al-Attar, 2000): The length of a vertical line is overestimated relative to that of a horizontal line, as depicted in figure 4.18. The form in which the vertical line bisects the horizontal line to form an

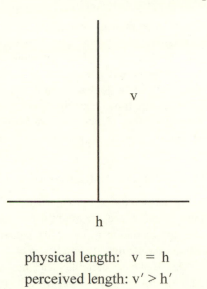

physical length: v = h
perceived length: v' > h'

Figure 4.18. Tactile and active-haptic versions of the well-known visual horizontal-vertical illusion. In the tactile-sensing mode, a line is passively moved vertically or horizontally across the skin on the forearm. In the active-haptic-sensing mode, the observer actively explores equal line lengths oriented horizontally and vertically. Regardless of sensing mode, when the line is displayed vertically, it feels longer than when it is displayed horizontally.

inverted T tends to produce a stronger effect than when the two lines meet at right angles to form an L figure (Heller & Joyner, 1993). Wong et al. (1974) have argued that the passively induced illusion operates independently of the active haptic variant, specifically suggesting that the latter is determined by the direction of exploratory hand movements, that is, radial as opposed to tangential with respect to the body (Day & Wong, 1971; for further discussion, see chapter 5). Only Heller and Joyner's study has specifically examined the effect of stimulus size on the magnitude of the illusion. The illusion was not obtained when small (i.e., hand-sized) inverted-T patterns were explored. Such studies with hand-sized stimuli may further inform our current understanding of the roles of relative speed of motion and duration on the perception of length via active haptic and tactile sensing (see Wong, 1977). For example, failure to observe an illusion with small stimuli could be due to the minimization of such differences, which can serve as potential sources of information about object length (Hollins & Goble, 1988), or to the use of finger, as opposed to arm, movements. To date, we know of no studies that have specifically tested for the horizontal-vertical illusion by passively drawing patterns on the hand, but it would be interesting to know whether there is an illusion, and if so, how it changes relative to the orientation of the hand (for the perceptual effects of hand orientation, see the next section, "Raised Two-Dimensional Patterns").

Raised Two-Dimensional Patterns

Observers use both cutaneous feedback from the fingers and kinesthetic feedback from the arm to maximize the discrimination of large two-dimensional shapes that vary in angle (Voisin, Lamarre, & Chapman, 2002). However, given the clear involvement of the kinesthetic system as well, we will consider this study more fully in the corresponding section in chapter 5.

The remaining discussion relates almost exclusively to the tactile recognition of fingertip-sized patterns, which are processed solely by the cutaneous system. Provided that tactile sensing involves relative motion between the skin and such stimuli, observers are capable of relatively quickly recognizing geometric patterns, such as Braille, geometric forms, and Roman letters (Craig, 1981; Loomis, 1981a). Johnson and his associates (Johnson & Phillips, 1981; Phillips, Johansson, & Johnson, 1990) have suggested that SA I tactile units serve as the primary neural peripheral mechanism in such tasks because of their relatively small receptive fields, although they do not rule out a subsidiary role for the FA I units with their slightly larger receptive fields. The work of this group of researchers has led them to argue that pattern recognition via tactile sensing is solely limited peripherally by the spacing of the SA I units.

Most of the research on tactile pattern recognition has focused on fundamental issues pertaining to the nature of the underlying processes and spatial representations, as opposed to issues pertaining to character legibility. For example, Loomis (1981b, 1982) demonstrated strong similarity between tactile (fingertip) and visual legibility of character sets differing in size and typography (e.g., Roman letters, Braille characters, etc.) when visual acuity was matched to tactile acuity via optical blurring (figure 4.19). During the practice phase, observers initially learned the names of the characters belonging to a

given set; in a subsequent recognition phase, they alternated between visual and tactile modes of examination. With the exception of a few systematic differences, Loomis subsequently (1982) noted marked similarities between the confusion matrices for letters and Braille characters. The procedure was similar to that used in his 1981 study. Collectively, Loomis's results further confirm that tactile pattern recognition is spatially limited only by the spatial resolution of the cutaneous system, as determined by the spacing of the peripheral SA I mechanoreceptors (see also Johnson & Phillips, 1981; Phillips, Johansson, & Johnson, 1990). Loomis's empirical studies culminated in a quantitative process model of pattern recognition by touch or blurred vision, which was designed to explain differences in legibility across character sets and to predict errors in the confusion matrix for an arbitrary set of patterns (Loomis, 1990). The model proposes that pattern recognition is the result of a two-stage process. In stage 1, an internal representation of the character is formed. The stimulus is processed through a linear, low-pass spatial filter, followed by nonlinear compression of stimulus intensity. In stage 2, the internal representation is matched to each stored representation contained in the character set. Response selection is based on Luce's (1963) unbiased choice model. Loomis's bottom-up model is quite successful in predicting legibility differences among character sets that vary in the type of character, size, and number of characters within the set. It also accounts for the similarity between Braille and Roman letter recognition. It is not as effective at predicting specific details of the confusion matrices.

Tactile pattern recognition on the finger is also limited by the temporal filtering characteristics of the cutaneous system. Craig has investigated the contribution of a number of factors to tactile pattern recognition. He used the tactile display from the Optacon (figure 4.10) to generate alphabetic patterns from an array of vibrating pins (6 × 24) with a computer substituting for the camera. For example, Craig (1980) initially showed that display time, mode of pattern generation (static versus scanned), and the temporal relationship between the presentation of the target pattern and a masking stimulus all influence performance. More specifically, observers were able to recognize letters well above chance when patterns were statically presented for durations as short as 4 ms. Surprisingly, the static mode

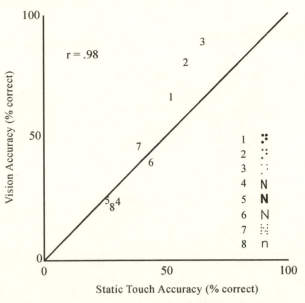

Figure 4.19. Scatter plot of visual and static touch accuracy with respect to eight different letter character sets. Reprinted from Loomis, 1981a, with permission of the Psychonomic Society.

proved to be superior to the scanned mode (i.e., Times-Square presentation) for all durations below 200 ms. Performance did not change meaningfully beyond 50 ms in the static mode nor beyond 400 ms in the scanned mode. These results highlight the need to consider both the mode of presentation and the stimulus duration with respect to single-letter recognition performance. Performance in the static mode was also considerably enhanced when the amplitude of the signal was increased, although this was not the case in the scanned mode.

As letters are not normally read in isolation, Craig also examined the extent to which patterns occurring relatively close in time to the target letter might interfere with letter recognition. Static and scanned modes were similarly affected by the relative timing between the letter stimulus and a masker created by activating all of the pins in the top 18 rows across all 6 columns. The end of the masker was separated in time from the onset of the letter (interstimulus interval) by a range extending from 8 ms (no masking stimulus present) through 0 ms (forward masking) and from 0 through to 250 ms (backward masking). Backward masking, in which the masker follows the letter, proved more deleterious than forward masking, in which the masker precedes the target letter (figure 4.20).

Craig (1982a) subsequently identified the type of masker as being important as well. He compared performance with pattern versus energy masks under several conditions (forward versus backward masking at several levels of intensity). Craig created the pattern masks by dividing the display space into equal segments, then randomly assigning one letter fragment (e.g., vertical line, curved line, etc.) to each segment. In contrast, an energy mask consisted of turning on all pins in the vicinity of the target letter. The overall results indicated that pattern maskers interfered with tactile letter recognition more than energy masks. Craig (1982b) also confirmed the importance of temporal integration (i.e., the time course over which temporally separated patterns can be integrated), using several different procedures. Collectively, the results suggest that the cutaneous system can totally integrate the pattern over a period lasting less than 10 ms and that the accuracy of tactile letter recognition asymptotes at about 50 ms. For more detailed interpretation of these results, the reader should consult Craig (1983a, 1983b) and Loomis and Lederman (1986).

Thus far, we have focused on how the spatial and temporal processing characteristics of the cutaneous system restrict tactile pattern perception and recognition. Another critical consideration is the spatial direction of the stimulated skin surface relative to the body and to the external environment. Some of the relevant research has involved drawing bidimensionally asymmetric patterns (e.g., the number 2, the letter L) on different body surfaces (e.g., forehead, back of head, palm) of a stationary observer. For example,

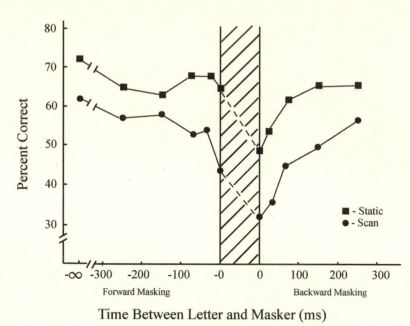

Figure 4.20. Percentage of correct letter recognition as a function of the time between the letter and a masking stimulus. Forward masking represents conditions in which the masking stimulus precedes the letter, whereas backward masking represents conditions in which the masking stimulus follows the letter. The duration of both target letter and masker was 26 ms. The point-∞ represents letter recognition with no masking stimulus present. Note that forward-masking times (negative ISI values) were measured from the onset of the masker to the onset of the target. Backward masking (positive ISI values) was measured from the offset of the target letter to the onset of the masker. The separation between the two zeros is only intended to separate the forward- and backward-masking effects in the graph. Reprinted from Craig, 1980, with permission of the American Psychological Association.

Corcoran (1977) showed that when a 2 is drawn on the back of the head, the observer's percept corresponds to the normally oriented pattern; however, when it is drawn on the forehead, the perceiver consistently reports experiencing an upright 2 that is left-right reversed. The same differences in pattern perception occur when the corresponding patterns are drawn on the back of a backward-facing hand and the forward-facing palm, respectively. However, when traced on the backward-facing palm with the elbow bent, the percept corresponds to a normally oriented 2; in contrast, when the palm is rotated away from the body with the thumb pointing upward, a reversed 2 is perceived (figure 4.21). Note that in this second example, the percept changes despite an unchanging pattern of cutaneous stimulation. Thus, the observer must be processing the cutaneous information within some allocentric (i.e., external) frame of reference. Based on a number

of informal observations, Corcoran proposed that observers view these patterns as if they were drawn on a transparent body via a "disembodied eye" positioned behind and a little above the observer.

Other researchers have obtained similar results (e.g., Krech & Crutchfield, 1958; Natsoulas, 1966). To maintain continuity, rather than reserving discussion of the related haptic work until chapter 5, we turn now to a study by Oldfield and Phillips (1983), which extends Corcoran's work to include haptic sensing. Observers were required to change the orientation of their fingertip (palm up versus palm down) or of their whole body (vertical versus horizontal) as a bilaterally asymmetric letter was pressed into their fingertip in the normal or mirror-reversed orientation. Overall, the results confirmed and extended Corcoran's by showing that, despite unchanging cutaneous stimulation, the perceived orientation of the letter depended on the spatial rela-

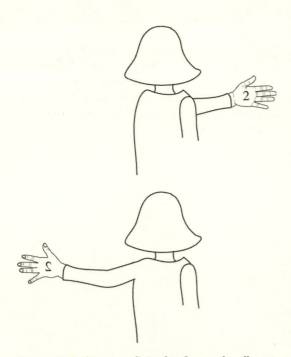

Figure 4.21. (Top panel) In this figure, the elbow is bent and the palm faces the head. The perception of the numeral 2 corresponds to the position. (Bottom panel) The palm is turned away from the body with the thumb up. The perception is reversed in this position, as shown by the reversed numeral 2. In both conditions, the actual cutaneous stimulation is the same. Adapted from Corcoran, 1977.

tions among the stimulated skin surface, the letter, and the observer. Their experiments further revealed that with respect to three-dimensional haptic space, observers make judgments with respect to the "above"/"below" axis in allocentric terms, with respect to the "near"/"far" axis in egocentric terms (i.e., with reference to the observer), and with respect to the left/right axis in terms of a combination of allocentric and egocentric factors. Oldfield and Phillips argued further that when haptically judging form, observers could take hand position into account with respect to the three-dimensional spatial framework that was selected. The data do not support the possibility that, at least in this task, perceivers converted their haptically derived representations into visually mediated representations.

Parsons and Shimojo (1987) performed a more comprehensive analysis of the perception of bidimensionally asymmetric patterns (normal versus mirror-reversed letters and numbers) traced on a large number of body sites. The results are complex and thus difficult to summarize succinctly. Observers were instructed to identify each pattern and its orientation based on initial impressions. The investigators began by showing that perception was influenced by the specific surface that was stimulated (e.g., forehead, tongue, hand). A second experiment clarified those results by showing that perception is affected not by the specific body site per se, but rather by the position (e.g., palm, back of hand) and orientation (e.g., forward, downward, etc.) of the stimulated body surface. A third experiment revealed that stimuli traced on the front and back of the head are judged with respect to local frames of reference located behind those surfaces: the frame tops match the top of the head, and the frame fronts are oriented opposite to the front of the head. In contrast, stimuli traced on surfaces other than the head are judged neither strictly within a local or whole-body frame of reference nor with respect to the frames used for the front and back of the head. Rather, the frames that are chosen appear to rely in a complex manner on the whole body and are selected on the basis of the position and orientation of the stimulated surface. Parsons and Shimojo described a set of five perceptual rules that highlight the adoption of three privileged types of reference frames: local frames for the front and back of the head, central frame(s) related to the upper body, and specific frames for the hand when it is in front of the chest or head and when it is above the head.

As previously mentioned, the generality of such perceptual demonstrations involving pattern perception is highlighted by the fact that they have also been confirmed using haptic sensing (Oldfield & Phillips, 1983; see also Loomis & Lederman, 1986, on larger two-dimensional patterns). Parsons and Shimojo (1987) have persuasively argued that the research on spatial frames of reference is very important because it may reveal how the tactile system represents cutaneous information about body parts, surfaces, and surrounding space for purposes of perceiving, thinking about, and acting on objects in the environment (for further developments, see Sekiyama, 1991). The results obtained to date highlight the considerable complexity with which the tactile system processes, represents, and interprets spatial patterns applied to the skin. In addition to being constrained by the spatial and temporal filtering characteristics of the tactile system, the observer is also influenced in a complex manner by the direction of the pattern that is applied relative to the body and to the external environment.

In this chapter, we have focused on sensory performance of the hand based on tactile sensing, one of four categories used to represent human hand function on the sensorimotor continuum (figure 1.1). This mode of touch uses sensory information from mechano- and thermoreceptors in the skin. In chapter 5, we consider active haptic sensing, which combines such information with the additional sensory inputs that arise during active use of our muscles, tendons, and joints.

5

Active Haptic Sensing

The Blind Men and the Elephant

It was six men of Indostan
To learning much inclined,
Who went to see the Elephant
(Though all of them were blind),
That each by observation
Might satisfy his mind.

The First approached the Elephant,
And happening to fall
Against his broad and sturdy side,
At once began to bawl:
God bless me! But the Elephant
Is very like a wall!
. . . .
The Sixth no sooner had begun
About the beast to grope,
Than, seizing on the swinging tail
That fell within his scope,
I see, quoth he, the Elephant
Is very like a rope!

And so these men of Indostan
Disputed loud and long,
Each in his own opinion
Exceeding stiff and strong,
Though each was partly in the right,
And all were in the wrong!

JOHN GODFREY SAXE, 1936

When we actively explore and manipulate objects with our hands, a wealth of sensory inputs is made available for further processing. Activation of mechanoreceptors embedded in the palmar skin of the hand (chapter 3) may provide valuable sensory information about objects and surfaces in the external world. In addition, because the fingers and palms are often flexed and extended during manual exploration and manipulation, the dorsal skin becomes stretched, therefore stimulating the SA II units. Such movement-related skin-strain cues may be used to assess the geometric properties of objects (Edin & Johansson, 1995). Mechanoreceptors in the muscles, tendons, and joints of the hand provide additional information in the form of kinesthetic inputs.

Many of the tasks addressed in the preceding chapter on tactile sensing (chapter 4) are relevant to our current discussion because active haptic sensing also involves the stimulation of cutaneous receptors. In the current chapter we will also discuss a number of additional tasks that specifically rely on the use of combined tactile and kinesthetic inputs.

It is important to emphasize that active haptic sensing per se constitutes a legitimate form of multi-modal perception inasmuch as this form of sensing uses both tactile and kinesthetic sensory inputs. It is critical that we understand how the haptic system functions in concert with other modalities, such as vision and audition; however, the details of such a complex topic lie well beyond the scope of this book. The interested reader should consult Welch and Warren (1986) and Calvert et al. (2004). Here, we provide a relatively limited discussion of multisen-sory processing with respect to selected topics for which there exists a substantial body of research.

Role of Hand Movements

It has been proposed that active manual exploration plays a critical role in the normal development of somatosensory perception in mammals (Bushnell & Boudreau, 1991; Nicolelis, de Oliveira, Lin, & Chapin, 1996). However, for fingertip-sized stimuli (e.g., tex-tures, Braille characters), provided there is relative motion between skin and surface, it matters relatively little whether the hand moves over a stationary sur-face or the surface moves across the stationary hand (Grunwald, 1965; Lamb, 1983; Lederman, 1981; Vega-Bermudez, Johnson, & Hsiao, 1991).

People normally choose to execute different stereotypical hand-movement patterns when they manually explore an object to learn about a specific property. Lederman and Klatzky (1987) have docu-mented a number of such movement patterns, or exploratory procedures (EPs). Their observers were initially presented with a standard object that could be described in terms of different attributes, such as texture, weight, or shape. On any trial, they were instructed to explore the standard object to learn about a designated attribute (i.e., texture, hardness, thermal properties, weight, coarse shape, fine shape, volume, motion of a part, or object function). Obser-vers were then presented in turn with three compari-son objects, each of which also varied in terms of a number of physical attributes. They were asked to choose the comparison object that best matched the

standard in terms of the previously named attribute. The hand movements were videotaped and subse-quently analyzed. Observers systematically chose to apply a specific EP, depending on the attribute they were instructed to explore. For example, they typi-cally moved their hands in a repetitive, back-and-forth manner across the surface (lateral motion EP) when extracting information about surface texture. In contrast, when judging the weight of an object, they lifted the object away from a supporting sur-face. Figure 5.1 shows the six EPs that have received the most research attention, together with the object property or properties for which each has been shown to be optimal. Two additional EPs initially doc-umented include the part-motion test (i.e., testing for motion of an object part) and the function test (i.e., testing for an object's function, as suggested by its structure). In a second experiment, the same match-to-sample task was used, only this time observers were constrained to perform each of the six EPs shown in figure 5.1 in conjunction with each of seven property-matching tasks. Response accuracy was used to order the EPs with respect to each prop-erty, with "ties" being broken in terms of associated response times. The results highlight the relative dif-ferences in EP precision. That is, for each property, one EP is optimal and, in the case of fine shape, nec-essary, as opposed to being merely sufficient or insuf-ficient to perform the task (i.e., chance level). It is noteworthy that the EP that observers voluntarily associated with a specific property in the first experi-ment proved to be the most precise (i.e., optimal) way to learn about that attribute. EPs also varied in terms of their generality (sometimes called breadth of sufficiency), that is, in terms of the total number of object attributes that each EP could extract with above-chance (sufficient) performance (Lederman & Klatzky, 1987). For example, lateral motion and pres-sure EPs are relatively less general because they pro-vide only three different types of object information (i.e., texture, hardness, and thermal properties). In contrast, both enclosure and contour following EPs are more general in that they provide some informa-tion about almost all of the attributes that have been studied. Finally, the EPs differ in terms of mean dura-tion or execution time. The relative precision, gener-ality, and duration of the six EPs are presented in figure 5.2. These factors serve to differentiate the EPs in terms of their relative costs and benefits.

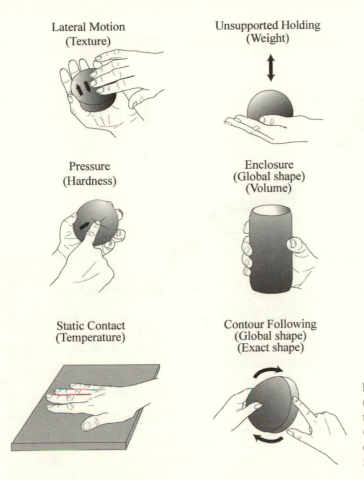

Lateral Motion
(Texture)

Unsupported Holding
(Weight)

Pressure
(Hardness)

Enclosure
(Global shape)
(Volume)

Static Contact
(Temperature)

Contour Following
(Global shape)
(Exact shape)

Figure 5.1. Exploratory procedures (EPs) and the associated properties that each EP is optimal at providing (in parentheses). Adapted from Lederman & Klatzky, 1987, with the permission of the authors.

Differences in the relative precision and speed with which the EPs are typically performed reveal a critical characteristic of haptic object processing: Material dimensions are generally processed more effectively than are geometric dimensions (Klatzky & Lederman, 2003a, 2003c). The material/geometric distinction applies as well when people are temporally constrained to processing inputs from an initial haptic "glance," which can be as brief as 200 ms. Across a set of haptic search tasks that targeted different object dimensions, Lederman and Klatzky (1997) showed that observers could process coarse differences in dimensions (e.g., rough/smooth, hard/soft, warm/cool) and the absence/presence of an edge relatively quickly and in parallel across the fingers of both hands, regardless of the number of fingers stimulated. In contrast, they processed geometric dimensions (e.g., relative position, relative orientation) more slowly and in sequence across the

stimulated fingers of both hands. Lederman and Klatzky concluded that following an initial brief contact, information about material properties and edges is available for further haptic processing earlier than is geometric information.

The differences in relative EP efficiency (as measured by accuracy and speed) led to the prediction that variations along material dimensions that are perceptually equivalent to those along geometric dimensions will be more cognitively salient to the haptic observer. This was confirmed in studies by Klatzky, Lederman, and Reed (1987) and Lederman, Summers, and Klatzky (1996). The Klatzky et al. study used custom-designed multiattribute stimuli produced by factorially crossing each of three values on two material (texture, compliance) and two geometric (shape, size) dimensions. Observers were required to sort the object set into piles such that objects within a pile were perceptually more similar

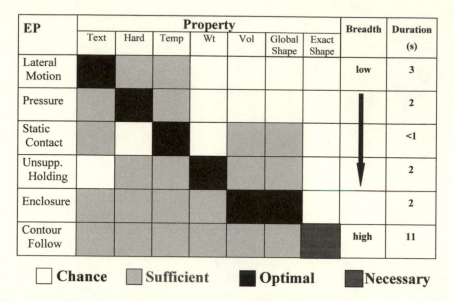

EP	Property							Breadth	Duration (s)
	Text	Hard	Temp	Wt	Vol	Global Shape	Exact Shape		
Lateral Motion	Optimal	Sufficient	Sufficient					low	3
Pressure	Sufficient	Optimal	Sufficient						2
Static Contact	Sufficient		Optimal						<1
Unsupp. Holding		Sufficient	Sufficient	Optimal					2
Enclosure		Sufficient	Sufficient	Sufficient	Optimal	Optimal			2
Contour Follow	Sufficient		Sufficient	Sufficient	Sufficient		Necessary	high	11

☐ Chance   ▨ Sufficient   ■ Optimal   ▰ Necessary

Figure 5.2. Costs and benefits of exploratory procedures (EPs) in terms of their relative precision (chance, sufficient, optimal, or necessary), breadth of sufficiency or generality as defined in the text, and average duration. Text = texture; Hard = hardness; Temp = temperature; Wt = weight; Vol = volume. Reprinted from Klatzky & Lederman, in press, with permission of Lawrence Erlbaum Associates.

to each other than to objects in any other pile. Physical values on each dimension were chosen on the basis of prior psychophysical testing to ensure that the objects could be differentiated within any of the four dimensions in perceptually equivalent ways. Different groups of observers were given different modality-biasing instructions, designed to favor visual versus haptic processing or vice versa. Based on previous findings regarding how efficiently each modality processes material versus geometric inputs, it was predicted and confirmed that when biased toward processing the objects visually, observers would weight the geometric properties, most notably shape, more strongly than the material properties in their similarity judgments. Conversely, when they were biased toward processing the objects haptically, they would weight the variation of material properties (texture and compliance) more strongly than the variation of geometric properties. Similar results were obtained when the objects differed in material (i.e., thermal and weight cues for wood versus aluminum), shape, and size (Lederman et al., 1996), as shown in figure 5.3. Collectively, the results confirm that haptic observers choose to form object representations that are based more on material than on geo-

metric properties when the perceptual differences within each dimension are perceptually equivalent across dimensions.

The functional significance of EP movement patterns is highlighted by disturbances in the hand movements used to interact with objects (tactile apraxia). This condition represents a specific dysfunction in the parietal lobe that is distinct from simple sensory or motor dysfunctions; rather, it is closely tied to stereognostic functioning (Binkofski, Kunesch, Classen, Seitz, & Freund, 2001) or object recognition, which constitutes the topic of the next section.

Recognition of Common Objects

People are extremely good at recognizing common objects using touch alone (Klatzky, Lederman, & Metzger, 1985). Blindfolded observers were able to recognize 100 common objects with a very high degree of success (regardless of the accuracy criterion adopted, i.e., stringent or lax) and very quickly (i.e., typically within only 2–3 seconds). These results are presented in figure 5.4. Lederman and Klatzky (1987) suggested that such excellent haptic performance can be ex-

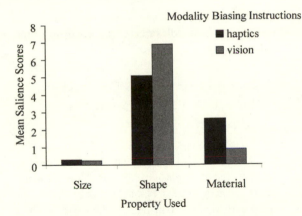

Figure 5.3. Mean cognitive salience (out of 8; based on subject sorts) as a function of object property for two different types of modality-biasing instructions. Adapted from Lederman, Summers, & Klatzky, 1996, with the permission of the authors.

plained by the fact that common objects possess multiple attributes that may serve as converging diagnostic cues to identity. Material cues can be efficiently extracted by executing the appropriate exploratory procedure(s); in addition, geometric properties of common objects fully vary in all three spatial dimensions and may be quickly, albeit coarsely, apprehended by means of a single grasp. Perhaps surprisingly, Kilgour and Lederman (2002) have since shown that

participants are also able to recognize live faces and face masks by hand. A match-to-sample task was used in which people chose one of three comparison faces that matched a standard face previously explored. The fact that recognition was higher with live faces (~80%) than with clay face masks of the same live faces (~58%) shows once again that material properties provide valuable information during haptic object recognition.

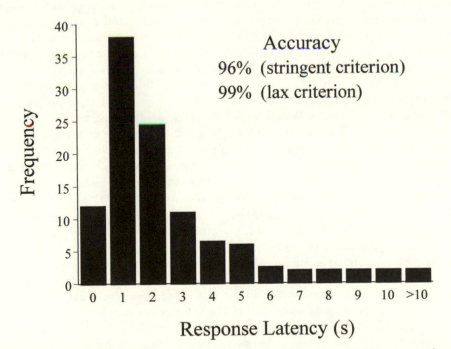

Figure 5.4. Frequency histogram showing the number of common objects (out of 100) recognized correctly as a function of response latency (in seconds). Overall mean accuracy values based on stringent or lax criteria are reported at the top right. Adapted from Klatzky, Lederman, & Metzger, 1985, with permission of the authors.

When performing a constrained object classification task at either the basic (e.g., "Is this abrasive surface sandpaper?") or subordinate (e.g., "Is this sandpaper rough sandpaper?") level of classification (Rosch, 1978), observers typically execute a two-stage sequence of manual exploration. Stage 1 consists of some combination of two EPs that are both relatively fast and general, that is, the grasp (enclosure EP) and lift (unsupported holding EP) (Lederman & Klatzky, 1990). Together, these broadly sufficient EPs provide coarse information about many different object dimensions. Stage 1 is performed regardless of which property is most diagnostic of the named object class. Manual exploration during stage 2 follows stage 1 and is optional. When it is performed, observers implement the EP(s) that provide(s) the most precise information about the property which is most diagnostic of the named class of objects.

One important issue with respect to object recognition concerns whether or not haptically derived object representations are view independent. Newell, Ernst, Tjan, and Bulthoff (2001) addressed this issue by initially allowing observers to explore visually or haptically unfamiliar three-dimensional objects constructed from Lego blocks that were attached to a table. Next, they were tested with an enlarged set of objects that included both new and old ones. Using either the same or another modality, observers were asked to indicate whether or not each object had been previously examined. Objects were presented in the same orientation during initial learning and subsequent test phases (0°), or they were rotated 180° about the X, Y, or Z axes during the test phase. Unimodal performance worsened while cross-modal performance improved when objects were rotated 180° about both the X (horizontal) and Y (vertical) axes between learning and test phases. Apparently, people recognize objects visually best from the front; in contrast, they haptically recognize objects best from the back, indicating that haptic and visual object representations are both view specific, but in strikingly different ways. Such a result further implies the need for some type of translation process. The cross-modal data also indicate that such intersensory processing can be relatively efficient.

A second important issue pertains to the effect on haptic object recognition of constraining manual exploration. As this topic relates most directly to end effectors, we will reserve that discussion until chapter 8.

Recognition of Two-Dimensional Raised-Outline Drawings of Common Objects

Other research has focused on the recognition of two-dimensional raised-outline drawings of common objects by sighted, blindfolded observers and by those with visual impairments. When exploration is free, performance by adults and children with such tangible displays is considerably poorer than with real common objects, although it has been well above chance in some cases (D'Angiulli, Kennedy, & Heller, 1988; Lederman, Klatzky, Chataway, & Summers, 1990; Magee & Kennedy, 1980). The relatively poor performance typically observed is not surprising given that these displays are usually limited to raised two-dimensional contours (i.e., there is no change in three-dimensional shape and little, if any, change in material properties). High demands on spatiotemporal integration and memory impose considerable constraints on processing raised-line pictures effectively (Loomis et al., 1991). We note that when manual exploration is guided by the experimenter, performance improves somewhat. Presumably by eliminating the need to establish and maintain contact with the contours in the display, observers are then freer to focus their attention on perceiving the nature of the two-dimensional pattern (Richardson & Wuillemin, 1981).

There is evidence that blind people with some previous vision form a visual image of the haptically derived information and use it for purposes of recognition (Heller, 1989; Lederman et al., 1990), however, visual experience and mediation by visual, as opposed to other forms of spatial, imagery are not prerequisites for haptic picture identification (Heller, 1989). Indeed, in the study by D'Angiulli et al. (1988), raised-line drawings of common objects that were frequently identified by congenitally blind children were similarly identified by blindfolded, sighted children. Studies with groups that vary in terms of the amount of visual experience help us to understand the role of vision in haptic object recognition. The interested reader should consult Heller (2000).

Material Properties

Texture

We have previously noted that, provided there is relative motion between the skin and surface and that it is equated with respect to speed and applied force, roughness perception is equivalent whether the surface is moved over the stationary finger (passive tactile sensing) or the observer moves the finger over the stationary surface (active tactile sensing) (Lamb, 1983; Lederman, 1981). This implies that cutaneous mechanoreceptors provide the primary source of information for roughness perception. The reader should therefore consult the corresponding section in chapter 4 for a comprehensive discussion of texture perception by touch.

Compliance

Unlike the relatively extensive study of tactile texture perception, there has been little investigation of important material properties related to compliance or other attributes, such as elasticity and viscosity. In 1937, David Katz published a fascinating article on the psychological judgments of different physical properties of dough by expert bakers. Katz argued that such a psychophysical study was important inasmuch as many of the bakers' complaints regarding the quality of a flour were based on their subjective impressions when making the dough. Katz proposed that if one goes beyond the subjective terminology adopted by the bakers, the experts are in fact evaluating four primary physical aspects of dough, namely, stickiness, elasticity, firmness or toughness, and extensibility (i.e., how much a batch of dough may be stretched before it breaks). He summarized the overall results of the various coarse psychophysical tests he conducted using a series of doughs that were made by varying the amounts of two different flours mixed together to produce a just noticeable change in the dough's properties. In brief, Katz noted that the bakers were most sensitive to changes in stickiness, a perceptual property that was described as very unpleasant. In addition, he concluded that the bakers were relatively insensitive to variations in elasticity and viscosity. Katz's conclusions are intriguing and worth extending beyond the application domain from which they emerged.

Future investigations will be able to employ the more precise methods of experimentation and stimulus control that are currently available.

When observers judge the stiffness (inverse of compliance) of springs enclosed in cylinders with rigid surfaces, a Weber fraction of about 17% is obtained. (The Weber fraction is equal to the difference threshold expressed as a percentage of the standard intensity.) Skin anesthesia and joint anesthesia have no effect on stiffness thresholds (Roland & Ladegaard-Pedersen, 1977), suggesting that cutaneous inputs do not play a role. Other studies have obtained Weber fractions with active touch that range from 8% to 22% depending on the cues available in the particular task. When terminal force cues are present, observers can use them to perform the task without actually having to discriminate on the basis of compliance (Tan, Durlach, Beauregard, & Srinivasan, 1995). However, discrimination based explicitly on compliance requires the use of both force and displacement. In the Tan et al. (1995) study, subjects grasped two plates between their thumbs and index fingers, squeezing them together along a linear track. Additional results support the fact that on their own, cutaneous cues are not sufficient for differentiating the stiffness of compliant objects with planar rigid surfaces; rather, both tactile and kinesthetic cues are necessary (Srinivasan & LaMotte, 1995). The stimuli consisted of telescoping hollow cylinders with the inner cylinder supported by four springs. In marked contrast, cutaneous information alone proved sufficient in additional experiments in which subjects assessed the compliance of rubber specimens with deformable surfaces (see the corresponding section in chapter 4). Depending on the particular experiment, subjects were required to rank the rubber specimens in terms of increasing softness or to discriminate between stimulus pairs in terms of their relative softness using a two-interval, two-alternative, forced-choice paradigm.

With respect to the rate of growth of sensation, the psychophysical functions describing the perceived magnitude of hardness and softness as a function of physical compliance (force/indentation) are best described by power functions, with exponents showing equal and opposite rates of growth for the two percepts (i.e., 0.8 and -0.8, respectively) (Harper & Stevens, 1964), as shown in figure 5.5.

Clearly, much psychophysical research remains to be done to more fully describe and explain the full

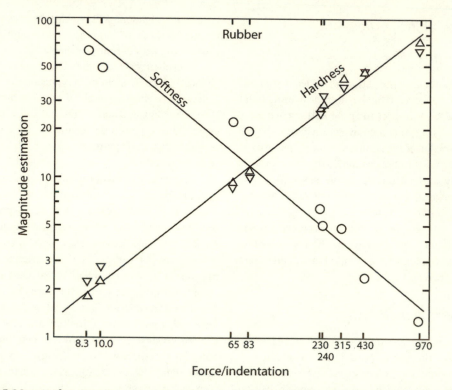

Figure 5.5. Magnitude estimations (geometric means) of subjective hardness and softness. The index of physical hardness, defined by force/indentation, is shown on the abscissa. The inverted triangles show the results of a repeat experiment. A total of 10 observers took part in each experiment. The specimens ranged from a fairly soft sponge to a fairly hard block of rubber. Reprinted from Harper & Stevens, 1964, with permission of Lawrence Erlbaum Associates.

range of percepts that relate, both singly and in various combinations, to the physical properties of materials that have been discussed in this section.

Thermal Properties

We know of no research that has directly compared thermal sensing under tactile sensing versus active haptic sensing with the locus and size of contact areas equated. However, there is no reason to expect any differences.

Material Composition

Although experimental details are somewhat sparse, Katz (1925/1989) has addressed the perception and recognition of object material per se using active touch. In one experiment, he reported that observers could distinguish between 13 pairs of different kinds

of paper surfaces (e.g., writing papers, blotting papers, drawing papers, etc.) quite well. In other experiments, he used a wider range of materials and required observers to recognize each material (e.g., wood, metal, plastic, glass, fabric, etc.). Unfortunately, as the experiments are only informally described, it is difficult to draw firm conclusions. Nevertheless, Katz makes some valuable observations related to the manner of touching and to the differential availability of several sensory cues to material identification. For example, as noted subsequently (Lederman & Klatzky, 1987), tapping actively with the finger (pressure EP) provides good tactile cues to compliance (e.g., vibration, net force, spatially distributed force profile). Statically contacting the surfaces (static contact EP) provides only coarse information about the differences in the thermal characteristics of the different materials. Finally, stroking a surface (lateral motion EP) provides the best information about its surface texture.

Katz (1925/1989) also noted that recognition improves as the person presses harder, presumably because increasing force enhances variations in the texture, compliance, and thermal characteristics. Visual cues and touch-produced sounds are also available to identify and differentiate materials.

Other investigators have also evaluated material identification. For example, Wynn Parry and Salter (1976) required a group of 40 observers to identify 10 different stimulus materials by name. The materials were sheepskin, leather (suede), silk, canvas, rubber, plastic, wool, cotton, carpet, and sandpaper. The average recognition time decreased from 5 s for sheepskin to 2 s for sandpaper. Robertson and Jones (1994) required their participants to match a 10 × 10 mm square of material to one of three samples. The materials were the same as those used by Wynn Parry and Salter. The average scores were excellent, 9.8 out of 10 (SD = 0.58) for both the left and right hands.

Weight

Weight is best judged by actively lifting an object away from a supporting surface, thus making available information provided by cutaneous and kinesthetic receptors and by neural correlates of the descending motor commands (also known as efferent copies or corollary discharges; L. A. Jones, 1986). The difference threshold for weight discrimination by lifting objects is 8%. Dynamically hefting or jiggling them further improves performance to 6% (Brodie & Ross, 1985).

Researchers have demonstrated that observers use more than object mass to assess apparent weight. Haptic weight perception is strongly influenced by the object's size, which is a geometric property. This distortion of weight perception is known as the size-weight illusion: A large and small object of the same mass do not appear to be equally heavy; rather, the large object is judged to be lighter than the small one (Charpentier, 1891). The classic form of this illusion involves both visual and haptic inputs, that is, observers heft the object while looking at it. However, Ellis and Lederman (1993) have confirmed the existence of a purely haptic version of the illusion. Their observers were required to provide magnitude estimates of the heaviness of cubes that were equal in mass but different in volume under varying size-exploration conditions, as shown in fig-

ure 5.6A. The results confirmed that unimodal haptic cues are both sufficient and necessary to obtain a full-strength bimodal (vision and haptics) size-weight illusion. In contrast, unimodal visual cues are merely sufficient, and they produce only a moderate size-weight illusion. Kawai (2002a, 2002b) has since provided empirical evidence that the haptic size-weight illusion reflects a specific instance of heaviness (and lightness) discrimination. Normally, this involves integrating weight and density inputs, with density in turn determined by dividing the weight by the haptically perceived size. In the case of the size-weight illusion, however, perceived heaviness is distorted because the cues to size and weight conflict.

Haptic weight perception is also influenced by object material, although less than previously noted with object size. Objects having the same mass but different material coverings (e.g., aluminum versus wood versus styrofoam) do not appear to weigh the same (Ellis & Lederman, 1998). This perceptual phenomenon is known as the material-weight illusion. As in their earlier 1993 study, Ellis and Lederman required observers to use a magnitude-estimation procedure to evaluate the heaviness of equal-mass objects, which in this study varied in their surface material cues (aluminum, styrofoam, and wood). Two sets of constant-mass objects were used, one low (~60 g) and the other high (~360 g) mass. Once again, four material-exploration conditions were included: unimodal haptics (whole-hand grasp), unimodal vision, bimodal haptics and vision (traditional illusion), and a control condition that eliminated both haptic and material cues. For the low-mass objects in all but the control condition, the aluminum object was judged to be the lightest, followed in turn by the wood, then the styrofoam object (figure 5.6B). In keeping with the size-weight illusion, the magnitude of the illusion was equivalent for the traditional bimodal and haptics-only conditions, only moderate for the vision-only condition, and did not occur in the control condition. Thus, haptic cues are both sufficient and necessary, whereas visual cues are merely sufficient. The illusion did not occur with the high-mass objects. Ellis and Lederman proposed that the material-weight illusion was the result of using the cutaneous inputs to judge object material, which in turn influenced judgments of weight. As with the size-weight illusion, they attributed the material-weight illusion

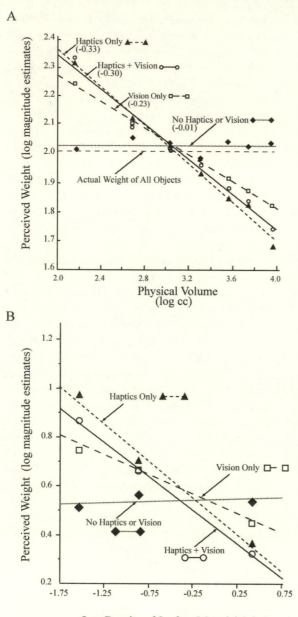

A

B

Figure 5.6. (A) Mean $\log_{10}$ magnitude estimates of weight as a function of $\log_{10}$ physical volume for each modality condition. Each data point represents the mean of 24 scores. The dotted line indicates the veridical heaviness estimate. Slopes are shown in parentheses. Object mass was constant at 350 g. A standard object (1093 cc) and modulus (100) were both used. Reprinted from Ellis & Lederman, 1993, with permission of Lawrence Erlbaum Associates and the authors. (B) Mean $\log_{10}$ magnitude estimates of weight as a function of the $\log_{10}$ physical density for each modality condition. Object mass was constant (58.5 g). The absolute magnitude estimation procedure was used to evaluate perceived weight magnitude. Each data point represents the mean of 16 scores. Reprinted from Ellis & Lederman, 1999, with permission of Lawrence Erlbaum Associates and the authors.

to the influence of sensory inputs, as opposed to higher-level cognitive expectations about the weight of objects made of different materials. Failure to obtain an illusion with the high-mass objects further supports this sensory interpretation. Ellis and Lederman speculated that the mass dependence of the material-weight illusion might be attributed to the effects produced when observers increased their grip force in order to stabilize the high-mass objects during the lift.

For an alternate theoretical approach to haptic weight perception based on the kinesthetic inputs generated during wielding, the reader should consider an experiment by Amazeen and Turvey (1996), which is discussed in the section "Attributes of Wielded Objects."

Geometric Properties

Curvature

Haptic curvature detection has been investigated across a broad range of spatial scales that presumably vary in their relevance for processing spatial information about fingertip-sized objects and displays that are larger than the hand (Davidson, 1972; I. E. Gordon & Morison, 1982; Henriques & Soechting, 2003; Louw, Kappers, & Koenderink, 2000). For example, Louw et al. (2000) used custom-designed shapes consisting of strip widths that ranged widely from 150 μm to 240 mm, with the upper surface being either convex or concave in Gaussian profile. A two-alternative, forced-choice method was employed. Observers were presented with one flat and one curved surface sequentially and asked to judge which one was more curved. Curvature thresholds (defined as 75% correct) varied from as low as 1 μm for the narrowest profile up to 8 mm for the widest profile, regardless of shape (convex or concave), as shown in figure 5.7. Above 1 mm width, the detection threshold increased as a power function of the spatial width of the Gaussian profile with an exponent of 1.3. Across the range of spatial scales, a number of different sensory cues are available: cutaneous inputs to detect the curvature of the narrowest profiles within the millimeter scale,

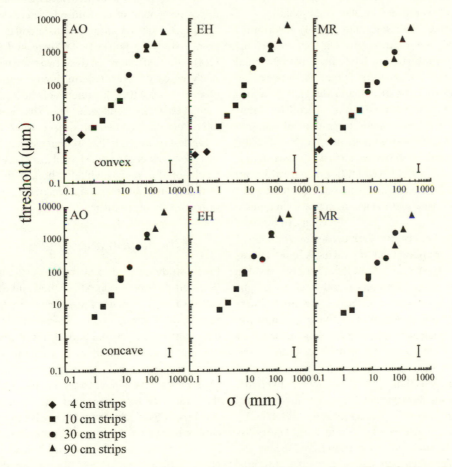

◆ 4 cm strips
■ 10 cm strips
● 30 cm strips
▲ 90 cm strips

Figure 5.7. Double logarithmic plots of discrimination thresholds versus the width of the Gaussian stimuli for three observers. Thresholds for convex (upper panels) and concave (lower panels) stimuli are plotted. The bars in the lower right corners indicate the size of mean errors. Note that the thresholds increase linearly with increasing width of the Gaussian stimuli on logarithmic scales. In addition, the thresholds for convex and concave stimuli are identical. The similarity between observers is also striking. Reprinted from Louw, Kappers, & Koenderink, 2000, with permission of Lawrence Erlbaum Associates and the authors.

kinesthetic inputs resulting from voluntary movements of the hand at larger spatial scales, and kinesthetic inputs from arm and shoulder muscles at still larger scales in the decimeter to meter range. The discrimination thresholds for the narrow-width stimuli were close to the 1 μm threshold for detecting the presence of a single asperity against a smooth background previously noted by Johansson and LaMotte (1983). Note that this value is an order of magnitude smaller than the diameter of the cutaneous mechanoreceptors (~10 μm), which lie approximately 0.8 mm below the skin surface at the interface between the epidermis and the dermis. As Louw et al. (2000) comment, this raises an interesting question concerning the nature of the mechanism that limits haptic resolution (cf. photons for visual thresholds and Brownian movement of the air at the eardrum for auditory thresholds).

Henriques and Soechting (2003) determined both absolute curvature detection and relative curvature discrimination of haptically explored virtual arcs varying in curvature. Stimuli were presented within the horizontal plane at different locations and orientations within the work space. The stimulus paths were sufficiently large that all required large multijoint arm movements. For the absolute curvature task, observers were asked to judge whether the stimulus path curved inward or outward from the body. No standard was presented. An adaptive double-staircase method was used to zero in quickly on threshold levels from both directions. In terms of the radius of curvature, the mean arc bias or point of subjective equality (PSE) was 1.8 m (0.56 m^{-1} curvature), which represents the outward or inward value of curvature that observers would judge as straight (i.e., without curvature). There were small differences in PSE that depended on the direction of the hand paths (sideways, front-back, diagonal left, and diagonal right). However, neither location in the work space nor distance from the observer affected curvature detection. The grand mean difference threshold for distinguishing between outward versus inward curvature was a radius of 0.90 m (1.11 m^{-1}), which corresponded to the hand being displaced 0.16 cm in either direction from the bias arc. In the experiment on haptic curvature discrimination, observers were required to decide which of two arcs was more curved. Stimulus pairs consisted of one straight fixed reference and one curved arc or two curved arcs, one of which was a fixed reference with a radius of curvature of 0.4 m (2.5 m^{-1}

curvature). A single adaptive-staircase procedure was used. The mean difference thresholds for the two curvature discrimination tasks were 0.44 m radius (2.26 m^{-1}) and 0.35 m radius (2.88 m^{-1}), respectively, both poorer than those obtained in the single-stimulus task in the previous experiment. The investigators emphasized the important contributions of cutaneous feedback, which is provided when the hand grasps and moves the manipulandum and of kinesthetic feedback derived from the movement of the joints and muscles of the hand and arm. These experiments provide valuable information about the precision with which people can haptically determine the geometric features of objects and their parts.

Until recently, researchers have assumed that the haptic perception of curvature is determined solely by surface geometry as opposed to lateral force. However, a study by Robles-De-La-Torre and Hayward (2001) showed that when these two cues, which normally co-vary, were made to oppose one another, observers based their judgments of the identity (i.e., bump versus hole) and location of shapes on lateral force cues, as opposed to surface geometry cues, regardless of whether the force cues signified a bump and geometric cues signified a hole, or vice versa. The results of this study highlight the fact that lateral force cues may well play a significant role in the haptic perception of curvature.

Orientation and Angle

In the study referred to above on curvature detection and discrimination, Henriques and Soechting (2003) also examined haptic sensitivity to the orientation (tilt) of straight paths. They defined tilt as the direction that the virtual path diverged from cardinal ones that were positioned parallel (purely sideways) or orthogonal (forward-backward) to the observer's frontal plane. Observers judged whether the path trajectory was tilted "in" (clockwise) or "out" (counter-clockwise), with respect to a specified cardinal direction. An adaptive double-staircase procedure was used to calculate thresholds. The mean absolute tilt bias was 4.57°, and it varied idiosyncratically across observers. The associated difference threshold was larger, 4.92°, but did not vary with position in the work space. By converting the bias and difference thresholds for haptic perception of tilt and curvature to corresponding hand-displacement values, Henriques and Soechting noted that haptic explorers were considerably less sensitive to tilt than to curvature.

Voisin, Benoit, and Chapman (2002) have also investigated the haptic perception of angle. Observers haptically discriminated angular differences between pairs of relatively large two-dimensional angles. They traced the pair of arms defining a two-dimensional angle with the index fingers of their outstretched arms. They were then required to decide which angle was larger, using a two-alternative, forced-choice procedure. The discrimination threshold for 75% correct performance was 5.2°. Voisin, Lamarre, and Chapman (2002) empirically confirmed that both cutaneous feedback from the fingers and kinesthetic feedback from the shoulders contributed to judgments in the previous task. Eliminating either source of sensory information raised angular thresholds, while eliminating both sources through passive touch with digital anesthesia prevented observers from performing the task at all.

As previously observed with tactile sensing (Lechelt, 1988), people are not equally sensitive to all line orientations (Appelle & Gravetter, 1985; Lechelt, Eliuk, & Tanne, 1976; Lederman & Taylor, 1969). For example, Appelle and Gravetter (1985) and Lechelt et al. (1976) have confirmed haptic versions of the oblique effect, in which observers are less accurate when discriminating lines that are oriented obliquely than those oriented either vertically or horizontally. Observers may use the more accurately perceived horizontal and vertical axes as perceptual anchors when haptically judging orientation (Lederman & Taylor, 1969). More recently, Gentaz and Hatwell (1995) have shown that the oblique effect occurs in both frontal and sagittal planes, whether the ipsilateral (same side) or contralateral (opposite side) hands are used during the exploration and response phases. It occurs in the horizontal plane as well, but only in the contralateral condition. As the orientation anisotropy occurs in all three spatial planes, Gentaz and Hatwell have suggested that the effect may depend on gravitational cues provided by the hand-arm system. When the active haptic mode is used, observers have access to information about orientation from cutaneous deformation patterns on the fingertip, from kinesthetic feedback resulting from manual exploration (particularly the contour-following EP), and possibly from corollary discharges.

Size

The exponent for the psychophysical function describing haptically perceived length as a function of physical length is a power function with an exponent close to 1, indicating that unlike most other sensory judgments, there is excellent correspondence between perceived and physical length (Teghtsoonian & Teghtsoonian, 1965). In this experiment, observers used a magnitude-estimation procedure to judge the length of rods held statically at their ends by their two index fingers. In another experiment, they were required to judge the apparent width of blocks held between their index finger and the thumb of the same hand (Teghtsoonian & Teghtsoonian, 1970). Haptically perceived width was an accelerating function of physical length. As the best-fitting power function had an exponent of 1.19, this means that doubling the block width increased the felt width by a factor of 2.3. The difference in the exponents for the two experiments may relate to the fact that the opening between the two digits was more limited anatomically in the 1970 study.

In some cases, the perception of size is systematically distorted. Gentaz and Hatwell (2004) suggest that the nature of the haptic exploratory movements can explain why some illusions occur both visually and haptically whereas others do not. We consider here two well-known visual illusions of size that have been evaluated haptically by a number of researchers. In the Müller-Lyer illusion (figure 5.8), a line enclosed by fins is overestimated relative to the same physical length enclosed by arrowheads. A haptic variant of this illusion has been confirmed as well (Casla, Blanco, & Travieso, 1999; Millar & Al-Attar, 2002; Wong, 1975). Collectively, the research on the Müller-Lyer illusion has shown that manipulation of

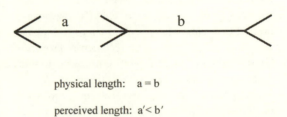

physical length: a = b

perceived length: a′ < b′

Figure 5.8. The Müller-Lyer illusion. Although the two shafts are physically equal, the shaft with the arrowheads (a) is perceived to be shorter than the shaft with the fins (b), both haptically and visually.

a number of different variables tends to produce similar perceptual effects on the haptic and visual forms of the illusion, indicating that the two modalities share similar processes. For example, Millar and Al-Attar (2002) propose that perception of the Müller-Lyer figure is distorted by disparate global shape and size cues as they relate to external (allocentric) and body-centered frames of reference. Susceptibility to the illusion was reduced when observers were instructed to use body-centered cues, but not when they were provided with an external, spatial reference in the form of an external frame. This seems reasonable inasmuch as body-centered cues are known to be important for spatial processing by the haptic system (Ballesteros, Millar, & Reales, 1998; Millar, 1985). A similar reduction has been documented when observers used body-centered cues with the visual form of the Müller-Lyer illusion.

Recall another powerful illusion of length distortion that has been demonstrated both visually and haptically: the horizontal–vertical illusion (figure 4.18). A vertical line is consistently overestimated relative to a horizontal line of the same length. However, the distortion may be broken down into two separate components: a bisection effect and a radial-tangential effect. The bisection effect consists of the bisected line being underestimated relative to the bisecting line, as demonstrated with T-shaped figures. Similar bisection effects have been documented both visually and haptically. However, the radial-tangential effect (Heller & Joyner, 1993; Marchetti & Lederman, 1983; Wong, 1977), an anisotropy of space, only occurs haptically. Wong (1977) proposed that radial movements toward and away from the body are overestimated relative to those that are tangential to the body. Fasse, Hogan, Kay, and Mussa-Ivaldi (2000) expanded the earlier investigations of haptically distorted space to include manual interaction with virtual objects. Observers were required to make judgments about the geometric properties of simulated objects, such as rectangular and triangular holes. The geometric properties included relative length, relative angle magnitude, and absolute object orientation. Observers demonstrated notable distortions in their haptic perception of all three properties; however, counter to prediction, their spatial percepts were geometrically inconsistent across those properties. These results clearly demonstrate the critical importance of exploratory movements in the haptic perception of size.

A different distortion of haptic space occurs when observers are required to reproduce the perceived Euclidean distance between two points. Euclidean estimates are progressively overestimated as the length of the complex curvilinear pathway connecting the end points increases, as shown in figure 5.9 (Lederman, Klatzky, & Barber, 1985). This is true whether or not an anchor is available. Once again, the haptic observers are influenced by their exploratory movements. In this case, they appear to take the duration of their exploratory movements into account when computing distance as-the-crow-flies.

Hollins and Goble (1988) have shown that observers typically underestimate the amplitude of a movement trajectory in which the arm movement velocity is relatively fast as opposed to slow. Such misperceptions might contribute to both the haptic radial-tangential effect and to the overestimation documented by Lederman et al. (1985).

Shape of Two-Dimensional and Three-Dimensional Forms Larger Than the Fingertip

Relatively little is known about how people haptically perceive the shape of objects larger than the fingertip. We do know that the shapes of relatively large planar and three-dimensional objects made of a homogeneous material are often poorly perceived relative to their material properties. In keeping with the critical material-geometric distinction previously highlighted, observers are less accurate when required to recognize haptically solid three-dimensional forms that vary only in their geometric properties (Klatzky, Loomis, Lederman, Wake, & Fujita, 1993; Norman, Norman, Clayton, Lianekhammy, & Zielke, 2004) than when the forms vary in both their material and geometric attributes (Klatzky et al., 1985).

Such research confirms that purposive manual exploration critically influences how objects and their properties are haptically processed and cognitively represented. Consider the EPs that are optimal or necessary for providing relatively precise spatial details about objects. Although an enclosure EP is relatively quick to execute, it can offer only relatively coarse details about shape from palmar contact. Moreover, if the fingertips are used, additional spatial integration across the sparsely distributed contact points is required. In contrast, the contour-following

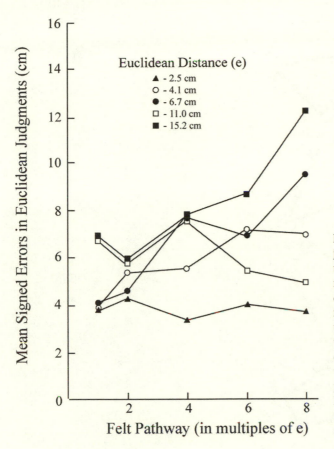

Figure 5.9. Mean signed errors in Euclidean judgments as a function of pathway distance, for each Euclidean distance. The mode of exploration involved the anchor condition, in which the observer's left index finger was placed on the start position and the right index finger slid across it to trace the raised pathway (as in the no-anchor condition). Similar results were obtained whether or not an anchor was used. When pathway distance = 1*e, the pathway is the Euclidean line (e). Reprinted from Lederman, Klatzky, & Barber, 1985, with permission of Lawrence Erlbaum Associates.

EP, which is necessary as well as sufficient for extracting fine spatial details, is by definition highly sequential and therefore relatively slow to perform. As a result, the observer is forced to integrate information over time. Observers typically have relatively more difficulty identifying common objects solely on the basis of their geometric differences when the tactile field of view and amount of sensory information are both reduced (Klatzky et al., 1993). More specifically, performance declines in terms of speed and/or accuracy as haptic exploration is progressively restricted from using an unconstrained five-finger search to a splinted five-finger search to a splinted single-finger search. We consider the consequences of constraining manual exploration in greater detail in chapter 8.

Observers may use their two hands strategically to overcome some of the constraints inherent in haptically processing relatively large shapes. For example, they tend to explore irregularly shaped planar objects as symmetrically as possible, sometimes slowing the leading hand to permit the second hand to catch up after it is delayed by some more complex contour (Klatzky, Lederman, & Balakrishnan, 1991; Lederman, Klatzky, & Balakrishnan, 1991). Such bimanual exploration enhances the detection of symmetry in raised-line shapes that are considerably larger than the fingertip (Ballesteros, Millar, & Reales, 1998).

Henriques, Flanders, and Soechting (2004) have explored the haptic perception of virtual planar outline shapes produced with a robot manipulandum whose force field was altered to simulate a planar closed quadrilateral boundary. Observers were required to reproduce the felt shapes haptically using the same manipulandum with the force field turned off or by tracing the shape on a touch screen in the frontoparallel plane with their eyes open. Regardless of the response mode, observers reproduced the shapes remarkably well, as shown in figure 5.10. This is reminiscent of an earlier study by Richardson and Wuillemin (1981), which showed that haptic performance is better when the experimenter guides the observer's hand movements (passive haptic sensing)

Manipulandum   Touch Screen

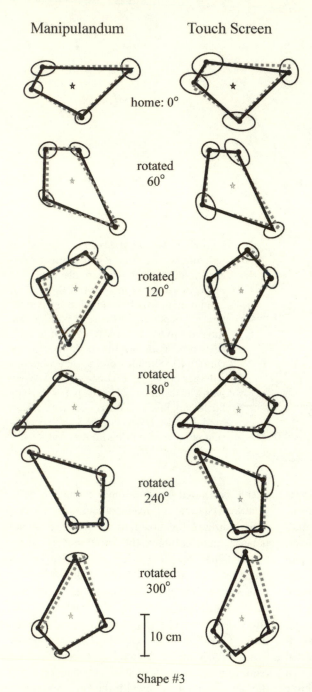

home: 0°

rotated
60°

rotated
120°

rotated
180°

rotated
240°

rotated
300°

10 cm

Shape #3

Figure 5.10. Drawings made with a manipu-landum (left column) and on a touch screen (right column), centered and scaled to match the traced quadrilaterals. Solid contours are the reproductions (averaged across trials and observers), whereas dashed contours are the veridical shapes for all six rotated forms of shape 3. Ellipses represent 68% confidence intervals for segment end points. Stars mark the centers of the traced and reproduced shapes. Reprinted from Henriques, Flanders, & Soechting, 2004, with permission of the American Physiological Society.

than when the observer is allowed to trace freely the contours of the two-dimensional displays (active haptic sensing). Exploratory movements were guided in the Henriques et al. study by the force feedback used to create the boundaries of virtual two-dimensional shapes.

The fact that the error patterns tended to be fairly similar in the two response modes (i.e., with the manipulandum or on the screen) further suggests that the distortion reflected the influence of percep-tual, as opposed to motor, processes. Short segments tended to be overestimated, while long ones were underestimated. The same pattern was documented with acute and obtuse angles, respectively. Reproductions tended to be more regular overall than the actual stimulus shapes; however, the distortions in the reproduced shapes could be interpreted in terms of haptic distortions that we have previously discussed in the section "Size" (the radial-tangential and oblique effects). Interestingly, the accuracy in judging shape was about the same as when judging single lines or angles, implying that there is no net error introduced when several line segments are integrated into an overall shape. Henriques et al. (2004) further speculate that the greater geometric consistencies observed with their study, as opposed to the puzzling inconsistencies found previously by Fasse et al. (2000), might be the result of the additional constraints imposed by presenting closed shapes in their study.

The strongly sequential nature of manual exploration may bias the manner in which the geometric structure of objects that are larger than the fingertip is processed. For example, Lakatos and Marks (1999) showed that when judging the relative similarity of pairs of geometric objects, haptic observers tend to focus more on local shape features than on global shape at the start, particularly when the time allowed for exploration is relatively brief. When rating the similarity of objects that are fairly similar in their global shape but not with respect to their local features, observers tend to decrease their emphasis on the local features as exploration time is increased, as shown in figure 5.11. However, for objects with dissimilar global shapes and no distinctive local features, similarity judgments are unaffected when exploration time is constrained. Lakatos and Marks (1999) further suggest that salient small-scale features may initially draw the observer's attention away from the global structure until a time when successive hand movements serve to perform a temporal low-pass spatial filtering, thereby diminishing the importance of the local features and, conversely, enhancing the significance of the global structure. As they remark: "The notion of touch as a modality that weights features differentially over time based

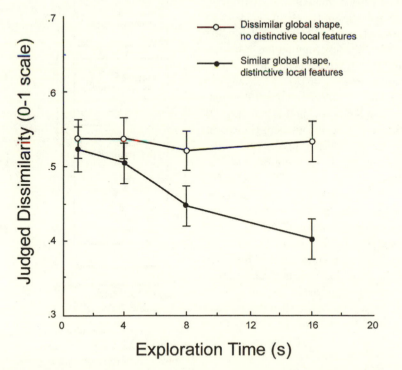

Figure 5.11. Judged dissimilarity ratings and standard errors plotted as a function of exploration time for two stimulus sets: pairs of objects with differing global features and no distinguishing local features (open circles) and pairs of objects sharing similar global shape but possessing distinctive local features (filled circles). Reprinted from Lakatos & Marks, 1999, with permission of the Psychonomic Society.

on a continuously evolving impression of local and global shape is one that merits further attention" (p. 908). The issue of global versus local haptic processing is further explored from a developmental point of view in the section "Object Recognition and Classification" in chapter 9.

Two-Dimensional Contact Locations and Their Layout Within Manipulatory Space

Regions that lie within easy reach of a person's hands (manipulatory or peripersonal space) often contain objects and manually accessible surfaces that can be explored without vision through sparse contact with one or two fingers. Such limited information can prove useful for learning about the parameters of the spatial layout (defined with respect to some spatial reference system) and for properly guiding one's hands to objects during reaching and manipulation.

When observers are required to set a test bar parallel to a reference bar within a planar (horizontal) work space, they demonstrate systematic biases in their perceptions of peripersonal space depending on the location and the task (unimanual or bimanual parallel bar setting, two-bar collinear setting, or pointing a test bar toward a marker; for details, see Kappers, 1999; Kappers & Koenderink, 1999). Zuidhoek, Kappers, van der Lubbe, and Postma (2003) interpreted the occurrence of consistent biases over many locations in these different tasks as indicating that the same egocentric (body-centered) and allocentric (centered in external space) frames of reference were used by the haptic system. To the extent that this is so, they predicted that providing a 10-s delay between the presentations of reference and test bars should reduce the distortion normally introduced by using an egocentric frame of reference when there is no delay. Their results confirmed that orientation errors were reduced in the delay condition, although this was limited to when the horizontal distance between the bars was 120, but not 60, cm. They interpreted their results as indicating a shift during the delay interval from using an egocentric to an allocentric frame of reference.

Klatzky and Lederman (2003b) have proposed that observers can adopt haptic spatial representations of manipulatory space at different levels. Figure

5.12 shows the levels of representation proposed, together with the successive computations required and the parameters represented at each level. Tasks were devised that demanded the use of different levels of representation. It was assumed that extracting parameters directly from a representation would produce smaller errors than when it was necessary to compute a higher level of representation so as to access the parameters required by the task. Observers were instructed to encode both the locations at which contact occurred and their spatial configuration. With respect to the latter, two metric properties of the configuration (each subject to translation or rotation) were considered: the distance (scale) and the angle between the contact positions relative to some

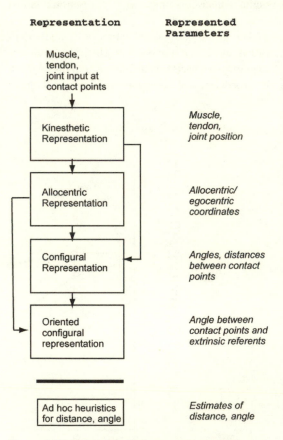

Figure 5.12. Proposed relations among spatial representations for kinesthetic contacts. The left column shows level of representation, with possible successive computations connected by arrows. The right column shows parameters directly conveyed by the representation at left. Reprinted from Klatzky & Lederman, 2003b, with permission of the Psychonomic Society.

reference direction. The observers' index fingers were moved passively by the experimenter to two locations in space for several seconds, then back to the start position. Observers were then required to estimate location, distance, and angle between the two locations in one of three ways: by replacing one or two fingers on the previously contacted points, by translating the two points that defined the configuration, or by directly estimating distance or angle. Based on their findings, Klatzky and Lederman concluded that their observers adopted a purely kinesthetic representation based on internal postural cues when asked to return their fingers to previously contacted positions. As indicated, such a representation preserves location information relatively well. However, metric parameters of the configuration (distance, angle) must be derived from this representation. The latter were mediated by an accurately oriented but more poorly scaled representation of the configuration of the contact locations that was not tied to a specific location in space. The fact that scale was not precisely represented may explain why people tend to adopt more ad hoc strategies for estimating distance (Lederman, Klatzky, & Barber, 1985).

We conclude this section by considering how our understanding of haptic space perception may be further informed by two intriguing perceptual distortions of space. The first is known as the Aristotle illusion (McKeon, 1941, as cited in Benedetti, 1985). When crossing one finger over an adjacent finger and then contacting a small ball with the two fingertips, people typically perceive two balls, not one. Benedetti (1985, 1986, 1988) used this illusion of external space in a series of papers that experimentally addressed how people process spatial information with crossed fingers. For example, in the 1985 study, observers were required to touch simultaneously a sharp point at the center of a disk and a ball at a fixed objective position directly to the side of the point. The third and fourth fingers were placed in three different positions: normal position, the third crossed over the fourth finger, and the third crossed under the fourth finger. Figure 5.13A shows the experimental paradigm for the third finger crossed over the fourth finger. The tactile inputs were processed very differently, depending on the relative orientation of the two fingers. Figure 5.13B depicts the spatial percepts corresponding to when the third finger was crossed under, versus over, the fourth finger. In a subsequent experiment, the fourth finger was fixed while the third finger was rotated to different

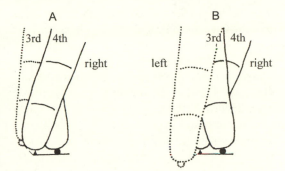

Figure 5.13. (A) Perceptual experience occurring when the third finger is crossed *over* the fourth. The ball is contacted by the third finger and perceived to be above the sharp point (contacted by the fourth finger) and in an uncrossed position. (B) Perceptual experience occurring when the third finger is crossed *under* the fourth The ball is again touched by the third finger, but now it is perceived to be below the sharp point (touched by the fourth finger) and in an uncrossed position (solid lines = actual position of fingers and stimuli; dotted lines = the perceived position of stimuli). Reprinted from Benedetti, 1985, with permission of the Psychonomic Society.

angular positions relative to the fourth finger. Collectively, the results confirmed that when the fingers are crossed, tactile spatial inputs are incorrectly processed as if the fingers were uncrossed. The results also indicated that the perceived angular position of the tactile stimuli varies linearly as a function of the angular position of the third finger up to the objective finger-crossing position. Beyond that range, the illusory spatial percepts remain unchanged; that is, they occur beyond the limits of the tactile spatial representation. Benedetti concluded that tactile stimuli are located within a body-centered frame of reference. Beyond the fingers' range of action, position is determined by the only source of kinesthetic information available, that is, by the limit of the fingers' range of action.

The Aristotle illusion provides cues about how people spatially interpret discrete points of contact on the hand in terms of the configuration of objects in external space (i.e., one or two objects?). Such knowledge is important to consider with respect to the accuracy with which hand movements (exploratory, grasp configuration) are controlled in peripersonal space. A haptic distortion that pertains to the felt size of one's own body parts is also relevant. For example, hand size is perceived to increase when cutaneous inputs from the thumb are reduced via anesthesia and when they are augmented by means of low-level electrical stimulation or painful cooling, although to a lesser extent (Gandevia & Phegan, 1999). Gandevia and Phegan showed that observers overestimated the two-dimensional area of the thumb by an amount that corresponded to doubling its volume. The illusion is reminiscent of our experience at the dentist's office when we perceive a cavity explored with the tongue to be larger after anesthesia (Anstis, 1964). The haptic distortion of thumb size was documented using both perceptual matching and sensorimotor drawing tasks. It would be informative to determine whether similarly altering the peripheral cutaneous inputs from the thumb or index finger would interfere with the formation of a proper grasp configuration as well.

Attributes of Wielded Objects

Turvey and his associates (e.g., Turvey, 1996) have adopted an entirely different conceptual approach to touch, focusing on what they call "dynamic touch" as a haptic subsystem. Dynamic touch occurs when an object is grasped and wielded in some way (e.g., raised, lowered, pushed, turned, transported). The primary focus of the research on dynamic touch is how people perceptually judge the attributes of objects wielded in the hand, including, for example, length, weight, width, shape of the object tip, and the orientation of hand and object relative to each other. Turvey and his colleagues have presented a substantial body of empirical evidence to support their argument that "the spatial capabilities of dynamic touch result from the sensitivity of the body's tissues to certain quantities of rotational dynamics about a fixed point that do not vary with changes in the rotational forces (torques) and motions" (1996, p. 1134). More specifically, Turvey describes resistance to wielding in the form of an inertia tensor, a 3×3 matrix whose elements represent the resistance to rotational acceleration about the axes of a three-dimensional coordinate system imposed on the object around the center of rotation (i.e., the wrist). The eigen values of the matrix correspond to the principal moments of inertia or the resistances to rotation about a nonarbitrary coordinate system that uses the primary axes of the object (i.e., those around which the mass is balanced). He further proposes that the perceived attributes directly map onto specific patterns of com-

ponents (eigen values and eigen vectors) of the inertia tensor, with the specific feature determining which component.

For example, in one experiment, Solomon, Turvey, and Burton (1989) required their observers to judge the length of rods ranging from 0.61 to 0.91 m by dynamically hefting them. They indicated their response by placing a board that could be seen at the distance that matched their perception of the location of the tip of the unseen rod. The investigators proposed that perceived length maps onto the moments of inertia (i.e., resistance to being turned). To test this, they altered the distribution of mass for a fixed point of rotation by attaching a weight at increasing distances from the hand, as shown in figure 5.14A. The moment of inertia increases the farther the rotational axis is from the center of mass. In another experiment, they changed the moment of inertia by changing the position of the point of rotation. Observers were required to grasp the rods at different positions along the rods, as shown in figure 5.14B, thus altering the distance of the rotational axis from the center of mass. In support of the moment-of-inertia hypothesis, perceived length mapped onto the moment of inertia in a 1:1 fashion. This relation is shown for the experiment with the added metal ring in figure 5.14C.

For another example of how this approach has been applied to direct haptic perception by wielding, consider a study by Amazeen and Turvey (1996). They proposed that haptic weight perception, both veridical and illusory (i.e., the size-weight illusion), is based on the resistance to rotation imposed on the limbs as one holds or wields an object. They argued that such information derives directly from the inertia tensor rather than from some cognitive inference pertaining to heaviness that is based on integrating mass and volume percepts. To obtain evidence for their theoretical stance on the haptic size-weight illusion, they created tensor objects in order to manipulate directly the pattern of eigen values associated with traditional size-weight objects, while holding both mass and volume constant. The results showed that haptically perceived heaviness changes in ways predicted by the specific pattern of resistance to rotational acceleration, despite the fact that neither mass nor volume changed.

The study of dynamic touch does not relate to touch as traditionally defined by tactile or haptic sensing. Rather, dynamic touch involves kinesthetic sens-

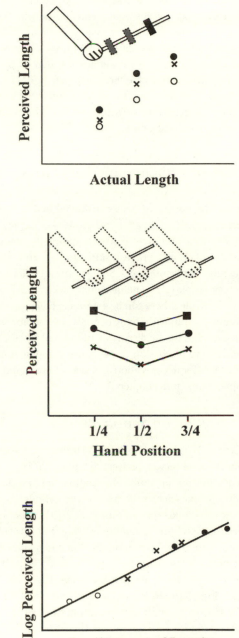

Figure 5.14. The hypothesis that perceived length is a function of rotational inertia can be tested by altering the mass distribution of rods that are of the same length. With an attached metal ring at different distances along a rod from its handheld end, perceived length by wielding should increase with the ring's distance. The hypothesis can also be tested by having participants grasp a rod at different points along its length and wield it. Perceived length should be the least for a grasp in the middle; for grasps at equal distances from the near and far ends, perceived lengths should be the same. These predictions have been supported. (a) The effects of three different ring positions and (b) the effects of three different hand positions on the perception of rods of three different lengths. Perceived length increases exponentially with moment of inertia. To obtain an estimate of the exponent, the nonlinear relation can be made linear by the use of logarithms. (c) The data from the experiment with the attached metal ring are plotted in logarithmic coordinates. The exponent (the slope) is approximately 0.33. Reprinted from Turvey, 1996, with permission of the American Psychological Association.

ing (as defined in Loomis & Lederman, 1986). Turvey and colleagues have noted that "deformations of muscles, tendons, and ligaments are inevitable accompaniments of manipulation with the consequence that the role of dynamic touch in the control of manipulatory activity may be both more continuous and fundamental than that of vision" (Turvey, Burton, Amazeen, Butwill, & Carello, 1998, p. 35). This statement pertains to action guided by dynamic touch and seems quite plausible. In such cases, the object serves as an extension of the hand and, therefore, as a tool for performing actions via teleoperation. Focusing on teleoperational action tasks would seem to offer a valuable research direction for the future. We would also suggest considering whether it is possible to apply this theoretical approach to hand-sized objects. To date, dynamic touch experiments have typically involved wielding relatively large stimulus objects, as opposed to the smaller ones that are more commonly manipulated (e.g., pens, pencils, forks).

Despite the value of studying dynamic touch with respect to motor function, the research program has tended to focus more on the use of dynamic touch for purposes of object perception. Observers in dynamic touch experiments have typically been required to make perceptual judgments about different object properties (e.g., length, weight, width, shape of the object tip, and orientation of hand and object relative to each other) by wielding. The case to be made for studying the conscious perception of objects and their properties by wielding seems most relevant to understanding human tool use and teleoperation with larger end effectors.

One final issue that has been frequently raised by opponents of dynamic touch and its theoretical underpinnings pertains to the absence of any attempt to relate the critical mathematical constructs in their theory to neurophysiology. There are no known biological sensors in muscles that respond to the inertia tensor or any of its components, whether singly or in some combination. How then does the observer/actor directly determine eigen vectors and eigen values as required by the theory? Any mapping between a percept and the inertia tensor cannot be direct because these parameters too must be calculated from the available receptor outputs.

Sensory Gating with Active Movements

Research has shown that voluntary or active movements of the limb will reduce the transmission of cutaneous signals through the dorsal column–medial lemniscal pathway during movement (Chapman, 1994; see the section "Sensory Cortical Projections" in chapter 3). Such movement-related gating also impairs perception. Motor activity reduces both the detection of near-threshold cutaneous stimuli (i.e., detection thresholds increase) and the perceived magnitude of suprathreshold vibrotactile stimuli. Discrimination thresholds remain unchanged, perhaps because the relative differences between the pairs of

stimuli are retained. If voluntary movements generally reduce perceived stimulus magnitude, then how can we explain the apparent equivalence in pattern recognition and perceived roughness magnitude using active versus passive modes of touch, as previously noted in this chapter? Chapman (1994) suggests that more efficient active control over the touching process may serve to counteract the effect of movement-related gating.

Sensing via Intermediate Links

Up to this point, we have considered active haptic perception with the bare hand. However, people also explore real and virtual worlds indirectly using a variety of tools, probes, skin covers, and more recently, haptic interfaces. Some of these devices can be very simple (e.g., a point-contact device, such as a pencil), while others can be extremely complex (e.g., a robotic hand) with many degrees of freedom, force sensors, and haptic display of the sensed forces. We will reserve discussion of the more complex haptic interfaces for teleoperation and virtual-environment applications until chapter 10.

Vibration Sensitivity

Brisben, Hsiao, and Johnson (1999) have shown that when observers actively grasp a cylindrical rod that contains an embedded motor that produces vibration parallel to the axis of the cylinder, some observers could sense vibrations at 200 Hz that were less than 0.01 μm in amplitude. Note that this value is not much greater than the thickness of a cell membrane. The usual U-shaped function (see chapter 4) is obtained when the average threshold for detecting vibration is plotted as a function of vibratory frequency, with a minimum between 100 and 150 Hz (figure 5.15). However, those thresholds are lower and the functions steeper than all but one previous study by Békésy (1939, as cited in Brisben et al., 1999). Additional psychophysical experiments showed that with the exception of contact force, stimulus location, contact area, direction of vibration, and contact shape all affect transmitted vibrotactile thresholds. Such factors, in addition to whether or not a surround was used, may help to explain why the thresholds were lower with the

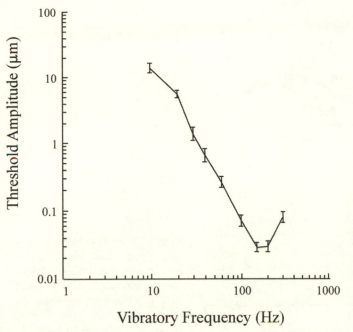

Figure 5.15. Detection threshold as a function of vibratory frequency while lightly grasping a cylinder (means based on 19 observers). The ordinate displays the mean threshold (half of the peak-to-peak excursion). The error bars represent standard errors of the mean. Reprinted from Brisben, Hsiao, & Johnson, 1999, with permission of the American Physiological Society.

vibrating tool than reported previously by studies that directly activated the skin perpendicularly.

Recognition of Common Objects

When observers haptically explore objects with a rigid probe or with a finger enclosed in a rigid sheath, their speed and accuracy are both notably impaired relative to a bare-finger control (Lederman & Klatzky, 2004). Lederman and Klatzky argue that intermediate interfaces such as these constrain manual exploration in several different ways that reduce or eliminate any of the following sources of information: cutaneous spatial cues, thermal cues, and kinesthetic spatial and temporal cues. These will be discussed in greater detail in chapter 8 and the implications of these results in chapter 10, where we deal with the relevance of this work to the design of haptic interfaces for teleoperator and virtual-environment systems.

Surface Roughness

The research to date has shown that roughness perception via a probe is only moderately impaired relative to performance with the bare finger. Lederman and Klatzky have adopted a psychophysical approach

to understanding the remote perception of roughness via rigid probes (Klatzky & Lederman, 2002; Lederman & Klatzky, 2001). The surfaces consist of spatially jittered, two-dimensional raised-dot patterns varying in interdot spacing. The shape of the psychophysical roughness functions (i.e., log perceived roughness magnitude as a function of log interdot spacing) obtained with a rigid probe frequently differs from the linear functions more typically obtained with the bare finger. The corresponding probe-based functions are consistently quadratic, as shown in figure 5.16 (but see Connor et al., 1990), and are based solely on the vibrations generated when the probe is moved across a surface. Klatzky, Lederman, Hamilton, Grindley, and Swendsen (2003) argue that the quadratic function is determined by the contact mechanics as the tip of the rigid probe rides along the tops of the elements; beyond some interelement spacing value, the tip drops down into and across the base of the intervening spaces and up and over the leading edges of the successive elements. The resulting psychophysical functions are quadratic in shape, the peak indicating the point along the interelement spacing continuum where the probe begins to fall between the elements (the drop point). The speed of motion and the size of the probe tip both affect peak position

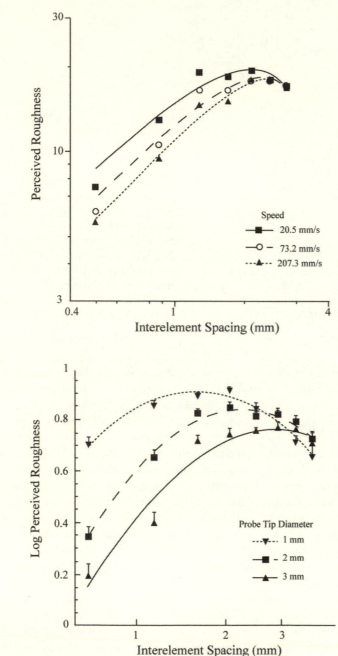

Figure 5.16. (A) Geometric means of normalized roughness magnitude estimates as a function of interelement spacing on double log scales for slow (20.5 mm/s), medium (73.2 mm/s), and fast (207.3 mm/s) speeds. Each data point is based on 20 observations, with each observation being the mean of two replications. Reprinted from Klatzky, Lederman, Hamilton, & Ramsay, 1999, with permission of the American Society of Mechanical Engineers. (B) Mean log roughness magnitude estimates by interelement spacing (measured as the average distances from each element to all other elements that lie adjacent diagonally, as well as horizontally and vertically; mm, log scale) for three probe sizes. Quadratic functions have been fit to the data. Each data point is based on 20 observations, with each observation being the mean of two replications. Bars indicate +1 SEM. Reprinted from Klatzky, Lederman, Hamilton, Grindley, & Swendsen, 2003, with permission of the Psychonomic Society.

(Klatzky et al., 2003; Lederman, Klatzky, Hamilton, & Ramsay, 1999), as shown in figure 5.16 (panels A and B, respectively). More specifically, as probe speed and probe diameter both increase, the peak shifts to a larger interelement spacing that is determined by a similarly increasing drop point. As we discuss in chapter 10, the psychophysical results with a probe may provide data that are relevant to rendering virtual textures for use in virtual environments.

Compliance

LaMotte (2000) has examined the remote perception of softness when a rigid stylus is used to tap or press down on specimens varying in compliance. Performance on a perceptual-ranking task proved to be equivalent whether direct (finger) or remote (stylus) sensing was used. Under both conditions, tapping proved to be more accurate than pressing down on the specimen. An additional two-interval, two-alternative, forced-choice task performed only with the stylus confirmed that discrimination was also more accurate for tapping than pressing. Finally, for both tapping and pressing, observers were more accurate when they actively initiated contact with the object than when they were passively contacted by the specimen. Presumably, the higher success in the active condition is the result of the additional kinesthetic cues that become available during active sensing.

Recognition and Discrimination of Object Material

Katz (1925/1989) suggested that although people do not typically use the hand to identify particular materials, such as wood, metal, cardboard, and porcelain, remotely it may be possible to do so based on the vibrotactile cues generated by contact between the material and an intermediary link (e.g., fingernail, rigid probe held in the hand). For example, he observed that, with their ears plugged, observers could successfully recognize an iron plate by striking the surface with a hammer very briefly (3–5 ms)., Katz also provided clear evidence that material discrimination was relatively unaffected when observers were required to differentiate fine paper differences through a layer of glue or adhesive tape or with a wooden rod. Under such circumstances, the primary information would be in the form of vibration differences, presumably generated by the variations in surface texture. As Katz further

noted, under normal circumstances, a perceiver would also have access to a wealth of auditory information generated by contact between the intermediate link and the surface. Such acoustic cues are particularly useful for recognizing and identifying specific materials. Cutaneous pressure and kinesthetic cues offer yet additional sources of sensory information about the relative compliance of various materials. Thermal cues, normally available for identifying materials through direct contact, are usually missing; however, researchers are now attempting to provide such information artificially in the form of thermal displays that simulate the changes in skin temperature that occur on contact with an object (Ho & Jones, 2004).

Shape and Size

People can be quite good at judging specific material properties (e.g., roughness, compliance) with a rigid probe or rigid finger sheath (e.g., Klatzky & Lederman, 1999; Lederman & Klatzky, 1999). Like direct touch, however, remote touch via a rigid link (e.g., probe, finger sheath) slows people down or renders them inaccurate when identifying common objects that are classified primarily by their geometric features (Lederman & Klatzky, 2004). The implications of these results will also be discussed in chapter 10.

In this chapter, we have considered the sensory performance of the hand when active haptic sensing is used. This topic is further addressed when we deal with end-effector constraints in chapter 8, hand function across the lifespan in chapter 9, and applied aspects of hand function in chapter 10. We turn now to the third category in our conceptual framework of human hand function (figure 1.1): prehension or grasping.

6

Prehension

*[T]he act of grasping an object includes more than the accurate direction of
the movement on to the target; the pressure of the grasp must neither
crush an egg nor allow it to slip from the fingers.*

PHILLIPS, 1986

Many tasks considered under the rubric of prehension involve reaching with the arm and hand to grasp and manipulate an object. The task objective is usually defined in terms of the action that will be performed with the object (e.g., drinking from a cup, writing with a pen). Although it is important to know the properties of the object, such as its shape or weight, in order to accomplish the task, this is usually considered secondary to the task's objective. This is best exemplified by the complaint of people with damage to the peripheral nerves innervating the sensory receptors in the hand, who complain not of the attendant sensory loss but of their motor deficiencies, which are evident in their clumsiness and difficulties in grasping and holding objects (Moberg, 1962).

Experimental studies of prehension have examined how the hand is transported and configured to grasp and manipulate objects. This has been studied in terms of the kinematics of hand and finger movements, the coordination of fingertip forces as the object is grasped, and the role of visual and somatosensory feedback in these processes. In general, studies that focus on reaching have emphasized the kinematic features of the hand and arm movements as the hand reaches to grasp an object and have been performed independently from those concerned with grasping. Studies of grasping generally start at the point of contact with the object and concentrate on analyzing how the forces produced by the fingers as an object is grasped and lifted are adapted to the properties of the object. These studies have also investigated the contribution of afferent feedback from cutaneous and muscle mechanoreceptors to the modulation of finger forces.

Reaching to Grasp

There are several classes of reaching movements that can be differentiated on the basis of the accuracy requirements of the movement and the configuration of the hand as the arm moves. Pointing, aiming, and reaching-to-grasp movements generally have similar kinematic features even though reaching-to-grasp movements involve changing the posture of the hand as the movement progresses so that it can grasp an object. In both pointing and reaching-to-grasp tasks, the duration of the movement increases monotonically with task difficulty, as defined in terms of movement distance and object size (Marteniuk, Jeannerod, Athenes, & Dugas, 1987). In this chapter on prehension, we will review the literature on reaching movements that involve moving the arm so that an object can be grasped. Pointing and aiming movements are treated as nonprehensile skilled movements and will be examined in chapter 7.

The act of reaching and grasping an object involves at least three distinct phases: moving the arm from its initial position to a location near the object (the reaching or transport phase), adjusting the posture of the hand as it approaches the object so that it can be grasped (the grasp phase), and finally the actual manipulation of the object (manipulation phase). This final phase will be covered in the section on grasping. The first two phases are coordinated in space and time as will be described below. The reaching or transport phase of the movement is typically considered in terms of the kinematics of the wrist's movements and so variables such as movement time, velocity profile, peak acceleration, and peak height are analyzed from data recorded from sensors (e.g., electromagnetic angular position sensors), passive reflective markers, or infrared emitting diodes (IREDs) mounted on the wrist and hand. The transport phase of reaching is usually characterized by a bell-shaped velocity curve and a single acceleration and deceleration peak (Jeannerod, 1984). Figure 6.1 shows the trajectory of the hand as it reaches to grasp objects located at three different distances from the observer, the bell-shaped relation between hand velocity and movement duration during these movements, and the relation between grip size and movement duration. The velocity curve becomes more asymmetrical with a prolonged deceleration phase when the object being grasped becomes smaller or more fragile (Marteniuk, Leavitt, MacKenzie, & Athenes, 1990). In general, the trajectory of the wrist during reaching movements does not vary as a function of movement speed or load on the arm (Soechting & Lacquaniti, 1981). However, movements to more distant objects have a longer latency, accelerate to peak velocity more quickly while attaining a higher peak velocity, and take longer than movements to near objects (Jakobson & Goodale, 1991).

During a reach-to-grasp movement, the posture of the hand changes as it is shaped to conform to the dimensions and properties of the target object. A number of kinematic features have been used to characterize these changes in the configuration of the hand, with the objective of determining how the hand's posture changes as the arm moves toward the target object and how the arm and hand trajectories are coordinated in space and time. The movement of the hand is usually recorded from markers or sensors placed on the tips of the thumb and index finger and sometimes on the joints of other fingers, if a more comprehensive analysis of the grasping motion is required (e.g., Mason, Gomez, & Ebner, 2001). A motion-analysis system is then used to calculate the position in space of the body points on which the markers or IREDs are mounted (e.g., Gentilucci, Castiello, Corradini, Scarpa, Umilta, & Rizzolatti, 1991). In some studies, a glove with sensors embedded in its surface has been worn so that measurements of the metacarpophalangeal (MP) and proximal interphalangeal (PIP) joint angles of the four fingers can be made in addition to recording the MP and interphalangeal joint angle of the thumb (Santello & Soechting, 1998). With this type of measurement system, it is possible to determine the coupling or synergies between the fingers as the hand reaches for and grasps an object (Santello, Flanders, & Soechting, 2002; Santello & Soechting, 1998).

The main spatial feature of the grasp that is analyzed is the amplitude of the maximum grip aperture, which is correlated with the size of the target object and with the object's distance from the observer (Gentilucci et al., 1991; Jakobson & Goodale, 1991). The change in grip aperture and velocity as a function of movement time are shown in Figure 6.2 for objects varying in width from 2 to 5 cm and in distance from 20 to 40 cm. For each 10 mm increase in object size, Marteniuk et al. (1990) reported that the maximum grip aperture increased by 7.7 mm. The aperture of the hand is adjusted for changes in object size even when these are not consciously perceived. Gentilucci, Daprati, Toni, Chieffi, and Saetti (1995) modified the size of the object being grasped by creating a small discrepancy (5 mm) between the virtual, visually perceived size of the object (created by producing a virtual image of the object on a semireflecting mirror) and the real object that was grasped, without informing observers. The peak apertures of their grasps were modified to compensate for the change in object size, even though the observers were unaware that the object had changed in size. This indicates that tactile and proprioceptive information from the hand about the size of the object can be used

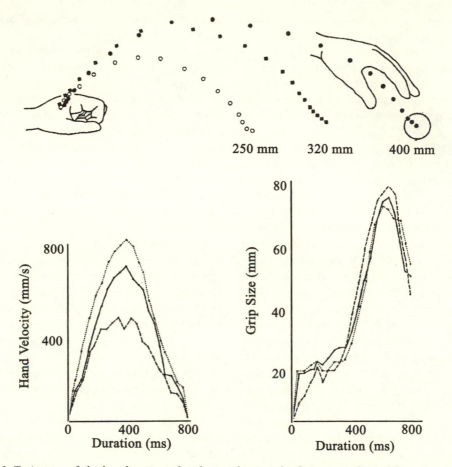

Figure 6.1. Trajectory of the hand as it reaches for an object at the distances indicated from the participant (upper figure). The dots represent the position of the wrist measured every 40 ms. The velocity of the hand and the grip size during these "transport" movements are shown in the lower two panels as a function of movement duration. The dashed line represents movements made to the object positioned 250 mm from the participant; the solid line represents movements made to the object at 320 mm; and the dotted line represents movements made to the object at 400 mm. These movements were performed without vision of the arm. Adapted and reprinted from Jeannerod, 1984, with permission of the Helen Dwight Reid Educational Foundation.

to modify the kinematics of reach-to-grasp movements, even when visual cues remain unchanged. A dissociation between visually perceived object properties (vision for perception) and the voluntary movements executed to interact with those objects (vision for action) has been noted in other contexts. For example, Aglioti, DeSouza, and Goodale (1995) showed that when a visual illusion (Titchener's circles) is used to change the apparent size of an object, it has no effect on the maximal aperture of the grip when reaching for it. However, when the perceived length of an object is modified using the Müller-Lyer illusion, there is a small effect on grip aperture

(Daprati & Gentilucci, 1997) and on the grasp position adopted for a balanced lift (Ellis, Flanagan, & Lederman, 1999). In the latter two studies, the effect of the visual illusion on grasp was much smaller than the visually perceived changes; the results are, therefore, still consistent with the idea of some dissociation between visual perception and visually guided action (see Milner & Goodale, 1995).

The point in time at which the aperture of the hand is at its maximum is readily identifiable and occurs during the deceleration phase of the reaching movement. Two temporal features of grasp formation that are frequently analyzed are the time to

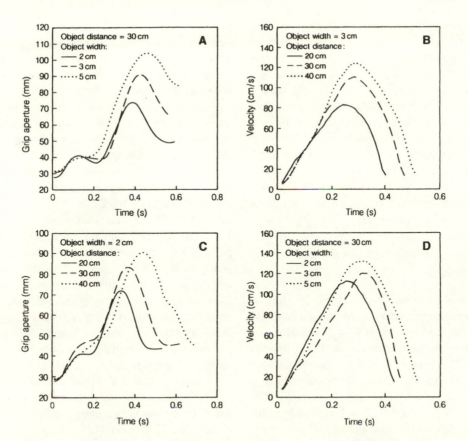

Figure 6.2. Representative data from a single participant demonstrating (A) scaling of maximum grip aperture to object size; (B) scaling of velocity and duration to object distance; (C) scaling of grip aperture to object distance; and (D) scaling of peak velocity and duration to object size. Reprinted from Jakobson & Goodale, 1991, with permission of Springer-Verlag and the authors.

maximum grip velocity and the time to maximum grip aperture. In two-digit and whole-hand grasps, the peak aperture occurs at approximately 60–70% of the total movement time, as illustrated in figure 6.2 (Chieffi & Gentilucci, 1993; Santello & Soechting, 1998). At this point, the correlation between the hand posture and its final configuration is not strong, but after the peak aperture is reached, the hand shape becomes more distinct and its form more closely resembles that of the target object. Multivariate statistical analyses of the changes in finger joint angles during the latter part of the reaching movement indicate that it is possible to distinguish between concave and convex target objects from the relative amount of flexion in the index and little fingers (together) as compared to the ring and middle fingers (Santello & Soechting, 1998). These analyses reveal that despite the large number of mechanical

degrees of freedom available during grasping, the number of independent degrees of freedom actively controlled is considerably smaller. These findings also indicate that there is a gradual evolution of hand shape as the reaching movement progresses and that the final hand posture is specified toward the end of the arm's trajectory.

The role of tactile input from the hand as the arm reaches to grasp an object has been investigated by eliminating tactile feedback from the distal phalanges and determining the effect on the kinematics of the movement, when the hand is obscured from view. When the fingertips are anesthetized, there is an increase in the time taken to reach maximum aperture, and the peak aperture is increased relative to normal, but the finger-closure time does not change (Gentilucci, Toni, Daprati, & Gangitano, 1997). It appears that a more conservative movement strategy is adopted in the absence of tactile feedback from the fingertips to cope with the uncertainty about finger position and the specific posture of the hand. In addition to these changes in the grasping component of the movement, there is a marked increase in the frequency with which the object slips between the fingers as it is grasped. The latter finding indicates the important role played by tactile afferents in signaling the initial contact with an object and is consistent with microneurographic studies of these units, which show a burst of activity in FA I and SA I units at the beginning of contact with an object (Westling & Johansson, 1987).

Visuomotor and Digit-channel Theories of Reaching

On the basis of a number of studies of reaching to grasp an object, Jeannerod (1984) proposed that the visuomotor mechanisms involved in reaching and grasping an object are independent, but temporally coupled. This became known as the "visuomotor channel hypothesis" and has been the subject of extensive study and debate (e.g., Jeannerod, 1988; Smeets & Brenner, 1999). Reaching was hypothesized to involve the transport channel, which extracts spatial information from the environment on the location of objects and on the distance of objects from the observer, and so deals with the extrinsic properties of objects. This channel then generates the motor commands that activate the muscles involved in moving the hand and arm to the object. The manipulation channel is involved in grasping, and it extracts information concerning the size and shape of the object (intrinsic properties) and then transforms that information into commands for the distal musculature, which grasps the object. Jeannerod (1988) observed that these two channels are frequently synchronized. He noted that as a person reaches to grasp an object, the time to maximum hand aperture, which is a function of object size, occurs at approximately 70% of the total movement time and always occurs at the same relative point in time, independent of movement duration (Jeannerod, 1984).

A number of studies have called into question the independence of the transport and manipulation channels as proposed by Jeannerod. The distance of the object from the observer and its size, shape, and fragility have been shown to affect both components of reaching (Santello & Soechting, 1998). The maxi-

mum aperture of the hand is scaled to the object's size, and its occurrence is correlated with the deceleration peak of the transport phase (Gentilucci, Chieffi, Scarpa, & Castiello, 1992). In addition, both the transport time and peak velocity of the reaching movement vary as functions of object size (Jakobson & Goodale, 1991; Marteniuk et al., 1987). If the size or position of the object is suddenly changed during the course of a reaching movement, adjustments are made to both the transport and grasping components within 100 ms, again showing a dependence of the two (Paulignan, MacKenzie, Marteniuk, & Jeannerod, 1991; Paulignan, Jeannerod, MacKenzie, & Marteniuk, 1991). Finally, if the arm is unexpectedly perturbed as it reaches for an object, there is a compensatory adjustment of both the wrist trajectory and the hand aperture. In this situation, it appears as if the spatial relation between the components of the reaching movement is being preserved (Haggard & Wing, 1991). These findings from a diverse set of experiments suggest that the control mechanisms underlying reaching and grasping are affected by the same task constraints and that temporal and kinematic coupling is a requirement for the accurate control of arm transport and manipulation.

Smeets and Brenner (1999) have proposed an alternative to the visuomotor channel hypothesis in which they suggest that there are separate channels for the finger and thumb during reaching and grasping. In this conceptualization, grasping is essentially reduced to an analysis of pointing with the thumb and finger and is modeled by a minimum-jerk (first derivative of acceleration) approach (described in chapter 7 for pointing movements). The "digit channel" hypothesis (Mon-Williams & McIntosh, 2000), as this model has been termed, postulates that the digits approach the object's surface perpendicularly and that grasping simply involves moving the fingers and thumb to the correct spatial locations. This hypothesis avoids one limitation of the visuomotor channel hypothesis, namely, determining what part of the hand is controlled during the transport phase of grasping, and can account for changes in hand posture with variations in the orientation of the target object. In the latter case, the intrinsic and extrinsic properties of the object remain unchanged, but the grip size and movement of the arm can change with the object's orientation (Smeets & Brenner, 1999). This model has also been shown to predict finger trajectories during grasping, the relation between the size of the target object and the aperture of the grip, and the increase in grip size with movement speed (Wallace, Weeks, & Kelso, 1990). It cannot, however, account for asymmetries in the velocity profile nor can it predict movement time when reaching and grasping with obstacles in the path (Mon-Williams & McIntosh, 2000).

Not all contact points on an object will lead to stable grasps, and so it seems reasonable that the points of contact of the hand on the object will be specified in the movement trajectory. Analyses of the final spatial positions of the fingers on an object indicate that they do vary with its size (Gentilucci, Caselli, & Secchi, 2003), center of mass (Lederman & Wing, 2003), and intended use (Napier, 1956). Gentilucci et al. (2003) reported that when observers used a tripod grasp to pick up spheres varying in diameter from 18 to 65 mm, the position of the thumb remained relatively constant along the sagittal, transverse, and vertical axes, although its relative position on the spheres changed with size. In contrast, the position of both the index and middle fingers changed along the sagittal and vertical axes with variation in object size. These results are consistent with the concept that grasping involves two "virtual fingers" formed by the thumb and one or more fingers that act in opposition to it, as proposed by Arbib, Iberall, and Lyons (1985). This concept is discussed more fully in chapter 8. The relative invariance of the thumb's position is also consistent with the hypothesis that it is the motion of the thumb, rather than the wrist, that is planned in the reaching movement (Haggard & Wing, 1997).

The placement of the fingers on an object is influenced by its perceived center of mass, which needs to be taken into consideration for asymmetrically shaped objects. If such an object is grasped at off-axis contact points, it will become unstable and begin to rotate in the hand. The influence of object symmetry on grasp-point selection was studied by Lederman and Wing (2003) using a range of planar objects that were either symmetric or asymmetric in shape. Observers were required to grasp and lift these objects using the thumb and middle finger, and the spatial location of the digits on the objects was recorded. They found that the grasp axes formed by the thumb and middle finger passed through or were close to the center of mass of the objects. The average error, which was defined as the perpendicular distance of the grasp axis from the center of mass, was only 3–4 mm. Furthermore, the error for grasp-

ing symmetric objects was 31% lower than that for asymmetric objects. A stable grasp was therefore achieved based on object geometry, including visual cues to symmetry.

Visual Feedback and Reaching

When reaching for an object, the central nervous system makes use of visual information about the location of the object and proprioceptive feedback regarding the position of the arm and hand, so that it can produce a pattern of muscle activity that moves the arm toward the object. The object's position must be computed in body-centered visual space, whereas the position of the moving arm with respect to that of the object is calculated from proprioceptive signals. If the position of the object suddenly changes during a reaching movement, corrections to the movement trajectory can be made within 100 ms, which is the minimum delay needed for visual and proprioceptive afferent signals to influence the movement (Paulignan, Jeannerod, et al., 1991). Visual analysis of the object's size, orientation, and shape assists in determining how the hand should be configured as it reaches for the object. Studies of eye-hand coordination during reaching movements have shown that observers visually fixate the object and landmarks involved in the task, rather than the moving hand (Johansson, Westling, Backström, & Flanagan, 2001).

There has been a considerable amount of research on the extent to which visual feedback is required to control reaching-to-grasp movements and how the movement trajectory and hand posture change under different visual-feedback conditions (e.g., Connolly & Goodale, 1999; Winges, Weber, & Santello, 2003). In these experiments, the performance of participants under full vision is compared to that under reduced visual-feedback conditions, namely, with only the object visible (in a normally illuminated room with a screen occluding view of the arm, or in darkness with a fluorescent target object), or with all visual feedback of the hand and object eliminated when the eyes are closed (Schettino, Adamovich, & Poizner, 2003). In some studies, visual feedback of the arm has been eliminated at different times during the reaching movement by using liquid crystal spectacles that can switch from a transparent to occluding state in 3 ms (e.g., Winges et al., 2003). In other studies, all visual feedback from the environment has been eliminated, and only the thumb and index finger are illuminated as the hand reaches toward the visible object (Churchill, Hopkins, Rönnqvist, & Vogt, 2000). When visual feedback is eliminated, there are predictable changes in the temporal and, to a lesser extent, spatial profiles of the reach-to-grasp movement. However, it is important to note that whether observers use visual cues to guide reaching movements can depend on the predictability of these cues. If visual feedback is randomized so that observers do not know in advance whether visual feedback will be continuously available, in some studies they have tended to adopt a default strategy in which they operate as if no visual cues are available, even when they are (Jakobson & Goodale, 1991).

When visual feedback of the arm is eliminated as people reach to grasp a visible object under normal illumination, the duration of the movement increases in comparison to movements made with full visual feedback, as shown in figure 6.3 (Connolly & Goodale, 1999). The maximum aperture of the hand and the relative time taken to reach maximum aperture remains unchanged, even though the arm is occluded from view (Connolly & Goodale, 1999). This suggests that the posture of the hand can be controlled without direct visual feedback, presumably on the basis of proprioceptive feedback from the hand and visual cues from the object. These results differ from those reported in an earlier study by Jakobson and Goodale (1991) in which there was an increase in the maximum grip aperture in the absence of visual feedback. In this study, the open-loop trials (i.e., no visual feedback) were run in complete darkness, and the increase in grip aperture may have been a

compensatory strategy adopted by observers as it permitted them to make more adjustments of the hand during the course of the movement. Churchill et al. (2000) also noted that there were differences in the grip aperture when the unseen hand reached in darkness for a luminous object.

Continuous vision of the object as the hand reaches toward it is not essential for the hand to be preshaped to the object's geometry. However, without continuous visual input, the reaching movements are longer, particularly during the final approach of the hand to the object as the fingers close (Winges et al., 2003). In these experiments, vision of the arm and object was occluded at different latencies (0–750 ms) from the start of the reaching movement. The hand path remained fairly consistent in the different visual conditions, and the kinematics of the movement (i.e., peak wrist velocity, time to peak velocity) did not change. The rate at which the hand was shaped to fit the geometry of the object was also not affected by reducing the time during which visual feedback was available. These findings indicate that the configuration of the hand as it reaches for an object is computed prior to movement onset and that it can be updated or modified on the basis of proprioceptive information from the fingers. If these proprioceptive signals are not available because the person has a peripheral neuropathy that affects afferent feedback from the arm and hand, then large directional and amplitude errors occur when reaching for an object without visual feedback (J. Gordon, Ghilardi, & Ghez, 1995).

When reaching movements are made using only haptic and proprioceptive cues, that is, when partici-

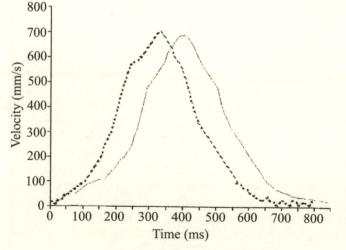

Figure 6.3. Velocity profiles of a person's reaching movements during open-loop conditions (gray line), when only the object but not the arm is visible, and closed-loop conditions (black line), when full visual feedback is available. Adapted and reprinted from Connolly & Goodale, 1999, with the permission of Springer-Verlag and the authors.

pants never see the object to be grasped but feel its size haptically and perceive its location on the basis of proprioceptive cues, the kinematics of the reaching movement are different from those performed with full visual feedback. Chieffi and Gentilucci (1993) found that participants were able to calculate the amplitude of the movement using only proprioceptive cues and so reached and grasped the objects correctly. Moreover, the peak velocity and time to peak velocity were similar in the visual and nonvisual conditions, even though the grasp time was significantly longer in the nonvisual condition (738 ms as compared to 616 ms). The trajectory of the wrist was more variable when only proprioceptive cues were available, and both the maximum finger aperture and the velocity of finger aperture were greater in the nonvisual condition. Figure 6.4 illustrates the rela-

tion between object size and maximum grip aperture when reaching for objects that are visually perceived and those that are only haptically perceived. These results show that although both the transport and grasp components of reaching movements are updated via visual feedback when it is available, considerable accuracy can still be achieved when only haptic and proprioceptive information is available.

In summary, studies of the kinematics of human reaching movements have demonstrated the close coupling between the arm's trajectory and the posture of the hand as it reaches to grasp an object. The spatial location of the object influences the arm's trajectory and total movement time, whereas the object's shape, size, and symmetry have all been shown to affect how the hand posture changes during the reaching movement. In the absence of visual feedback of the hand

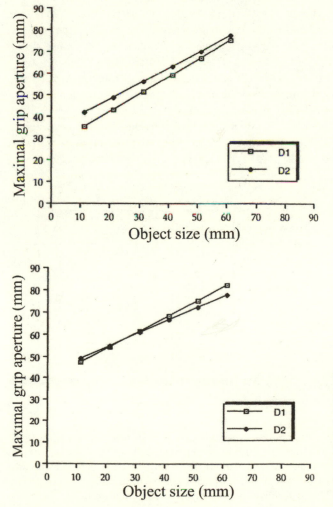

Figure 6.4. Relation between object size and maximal grip aperture when object size and location is perceived visually (upper figure) or only haptically (lower figure) at two distances, D1 (70 mm) and D2 (175 mm), from the observer. Mean regression lines are shown. Adapted and reprinted from Chieffi & Gentilucci, 1993, with the permission of Springer-Verlag and the authors.

and arm, there is an increase in the final deceleration phase of the movement, but the kinematics of the movement remain relatively similar to those produced with full visual feedback. Both tactile and proprioceptive feedback from the hand provide information that can be used to compensate for the loss of visual feedback, and on their own, they provide sufficient information to enable a person to reach and grasp an object.

Grasping

Research on grasping has received considerable attention since the 1980s, and the topics covered range from how the intrinsic properties of an object (e.g., its mass and surface texture) influence the grasping forces generated to how task constraints determine the choice of a particular hand configuration. Most manipulative tasks require the precise coordination of forces generated at the fingertips in order to hold the object in a stable grasp. This has been studied extensively using an experimental apparatus and procedure first developed by Johansson and his colleagues (Johansson & Westling, 1984). The protocol requires that a person grasp an object between the thumb and index finger and lift it from a supporting surface. The temporal sequence of this process is broken down into a number of phases as follows: the preload phase in which the grip is

formed; the loading phase in which there is a parallel increase in the grip force normal to the object surface and the load force tangential to the surface until the load force overcomes gravity; the transitional phase when the object is lifted to the desired position; the static phase in which the object is held stationary and the forces reach a steady state; the replacement phase in which the object is lowered to the resting position; and finally the unloading phase when the object is finally released from the hand (Johansson & Westling, 1984; Westling & Johansson, 1987).

A schematic drawing of the grip-force apparatus, together with the temporal sequence of the experimental protocol, is shown in figure 6.5. This apparatus and experimental protocol have been widely adopted by researchers interested in human motor control (e.g., Flanagan & Wing, 1995) and have been used to study fine manual control in young children (Forssberg, Eliasson, Kinoshita, Johansson, & Westling, 1991), the elderly (Cole, 1991), and people with peripheral nerve (Nowak & Hermsdörfer, 2003), cerebellar (Müller & Dichgans, 1994; Nowak, Hermsdörfer, Marquardt, & Fuchs, 2002), or cortical lesions (Hermsdörfer, Hagl, & Nowak, 2004; Robertson & Jones, 1994). A similar experimental apparatus was developed in the 1970s to study the effects of pyramidal tract lesions on precision grip control in monkeys (Hepp-Reymond et al., 1974; Hepp-Reymond & Wiesendanger, 1972).

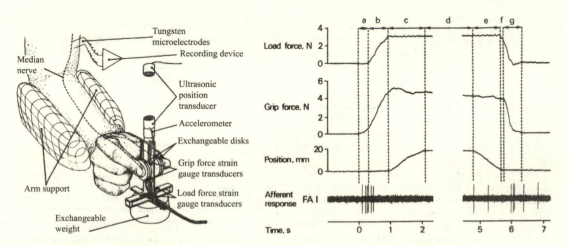

Figure 6.5. Schematic illustration of the grip force apparatus and the position of the arm during testing (left). Phases of the lifting trial identified in the analyses (right): (a) preload phase, (b) loading phase, (C) transitional phase, (d) static phase, (e) replacement phase, (f) delay, and (g) unloading phase. Load force and grip force are shown together with the position of the object and a typical response from an FA I afferent unit. Reprinted from Westling & Johansson, 1987, with permission of Springer-Verlag and the authors.

In these experiments, the test object usually has displacement (or acceleration) sensors or IREDs attached to its surface so that changes in its position can be monitored, and force transducers are mounted on the object to record the normal and tangential forces. The object itself may be passive (e.g., a box whose weight or shape may vary) or active (a servo-controlled motor), with the latter enabling perturbations to be delivered as the object is manipulated. The effects on grip forces of various object properties, such as the texture on the grasping surface, the shape and curvature of the object, and its weight, have been investigated. An additional focus of these studies has been on the role of cutaneous mechanoreceptors in the coordination of fingertip forces.

When an object is grasped with the thumb and index finger and lifted from a supporting surface, the rate at which the grip (normal) and load (tangential) forces increase is maintained at an approximately constant ratio, which is consistent with a coordinated pattern of muscle activation in the hand and arm muscles (Johansson & Westling, 1988). Once the object is supported in a stable position during the static phase, the normal force stabilizes and is approximately proportional to the weight of the object (see figure 6.6). These normal forces have to be large enough to prevent the object from slipping

between the fingers but not excessive, as this may cause damage to a fragile object or muscle fatigue, if the forces have to be maintained for some period of time. The ratio of the normal (grip) force to the tangential (load) force must exceed the inverse of the coefficient of friction for a grip to be stable and an object secure. The minimum force at which an object begins to slip between the fingers, known as the slip force, has been measured by asking people to extend the thumb and index finger very slowly until the object being held slips from the fingers. Using this procedure, Westling and Johansson (1984) found that the slip force was proportional to the load force and varied as a function of the friction between the skin and the object, as shown in figure 6.6. A comparison between the grip and slip forces revealed that there were large interindividual differences in grip force, although the corresponding slip forces were equivalent. They coined the term "safety margin" to describe this difference between the slip and grip forces and noted that this tended to be smaller in more dexterous people.

Grasp stability during manipulation is maintained by automatically increasing and decreasing the grip forces in parallel with the load force. If an object begins to slip between the fingers, there is an automatic increase in grip force, which occurs within

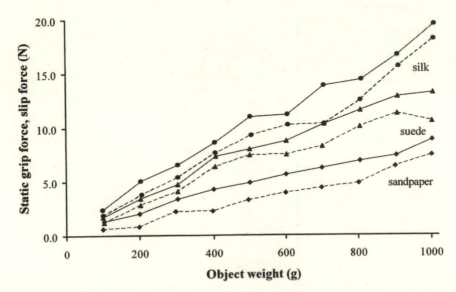

Figure 6.6. Relation between the weight of an object and grip and slip forces when the surface is covered in silk, suede, or sandpaper. Data are from a single observer. Safety margins are represented by the areas between the solid and dashed lines for each surface texture. Redrawn from Westling & Johansson, 1984, with permission of Springer-Verlag and the authors.

70 ms of the slip and results in a more stable grasp (Johansson & Westling, 1987). This delay represents about half the latency of a voluntarily initiated change in force that can be elicited by cutaneous stimulation of the fingers and twice the latency of the fastest spinal reflex in intrinsic hand muscles (Matthews, 1984). The latency is, however, comparable in magnitude to long-latency reflex responses and so is presumably supraspinally mediated. Microneurographic recordings indicate that these slips readily excite tactile units in the fingerpad, even though they are usually not perceived. The slip responses are detected by FA I, FA II, and SA I mechanoreceptors, with the primary response coming from FA I and SA I units (Johansson & Westling, 1987).

Effect of Friction on Grip Forces

The friction between the skin of the hand and the object being grasped depends on the material on the grasping surface, the amplitude of the grip force at low forces (< 1 N), the contact area, and the degree of hydration of the skin (Buchholz, Frederick, & Armstrong, 1988). The influence of friction on the

forces used to grasp objects has been evaluated using materials such as silk or sandpaper, which are mounted on the grasping surface. Westling and Johansson (1984) used Amonton's law to calculate the coefficient of friction for sandpaper (grade 320), suede, and silk in contact with skin and reported average values for 10 people of 1.21, 0.69, and 0.35, respectively. With these materials mounted on the grip-force apparatus, they measured the change in normal force as participants grasped and lifted the test object. They found that, as the friction between the skin and object decreased, higher normal forces were required to maintain grasp stability, as illustrated in figure 6.6. The main effect of the different surface materials was on the rate with which the normal force changed during the preloading and loading phases, whereas the time course of the change in tangential force was similar for all three surfaces (see figure 6.7). Force coordination therefore varied as a function of surface material. This adjustment in force with surface material appeared to be based on friction per se, rather than the textural properties of the object; when the skin was washed and made less adhesive, the normal force was

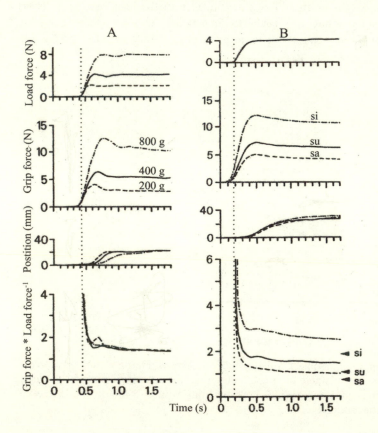

Figure 6.7. Coordination between grip and load forces when lifting objects of different weights (200, 400, and 800 g) (A) and with three different surface textures (silk [si], suede [su], and sandpaper [sa]) (B). Arrows indicate mean slip ratios for the three surfaces. The data in (A) are from a single participant (averaged across eight trials per condition) and in (B) are averaged across nine participants. Adapted and reprinted from Johansson & Westling, 1984, with permission of Springer-Verlag and the authors.

adapted to this change even though the surface material of the object remained unchanged (Johansson & Westling, 1984).

These results have been confirmed in a more recent study by Cadoret and Smith (1996), who showed that people rely on the friction of an object against the skin to optimize grip force, independent of whether the friction comes from macroscopic surface features in contact with the fingers or a coating (talc, sucrose, or water) between the object's surface and the skin. The friction between the object surface and the fingers not only influences the dynamics of prehensile force control as described above, but also affects the more proximal musculature in the arm, such as the elbow and shoulder muscles that are involved in moving the hand and object from one location to another (Saels, Thonnard, Detrembleur, & Smith, 1999).

When the surface material in contact with the digits changes, the finger forces are influenced by this new frictional condition within 100 ms of the initial contact. This initial adjustment in normal force is sometimes insufficient and a further secondary adjustment in force occurs later during the loading phase (Johansson & Westling, 1984). If the friction is now varied independently at the contact surfaces on the thumb and index finger by using different surface materials on the two grasping surfaces, the fingertip forces automatically adjust to the surface conditions at each digit. The tangential force at the digit in contact with a low-friction surface is significantly smaller than the force at the digit touching the high-friction surface, which causes the object to tilt slightly. This control strategy provides similar safety margins against slips at both digits (Edin, Westling, & Johansson, 1992). If the grasp configuration is now changed so that the object is held between the index fingers of the left and right hands, the same adjustment in tangential force with friction condition is seen. The ratio of the normal to tangential forces during the hold phase reflects the local friction conditions at each interface; the forces are relatively unaffected by conditions on the opposite finger; and appropriate safety margins against slips are maintained (Burstedt, Edin, & Johansson, 1997). These findings indicate that the forces produced by the digits can be individually controlled, even though this may entail activating overlapping muscle groups.

When the fingertips are anesthetized, thereby eliminating feedback from cutaneous mechanore-ceptors, the normal force no longer reflects the friction conditions at each digit, and large and variable safety margins occur at both digits. The normal forces are greater and so the safety margins are larger because forces are no longer optimally adapted to the weight of the object and to the skin-surface friction (Häger-Ross & Johansson, 1996; Johansson & Westling, 1984; Monzée, Lamarre, & Smith, 2003). Figure 6.8 illustrates these changes in grip force as a function of object weight and surface texture, when the objects are lifted normally and when the fingers are anesthetized. Under these conditions, the object often slips between the fingers and is frequently dropped as tactile afferents no longer signal the slip on the fingerpad (Augurelle, Smith, Lejeune, & Thonnard, 2003). The temporal coupling of grip and load forces is, however, preserved in the absence of cutaneous feedback, which suggests that the modulation of these forces is centrally mediated and less affected by tactile feedback from the fingers. These results demonstrate the critical role played by tactile receptors in the coordination of grasp forces to the friction requirements of the task (Edin et al., 1992). The adaptation of fingertip forces to changes in friction is most likely signaled by FA I afferents as their responses are strongly influenced by the material on the grasped surface: The more slippery the material, the stronger the initial response in these afferents (Johansson & Westling, 1987).

Grip- and Load-Force Coordination

The grip forces produced when grasping and lifting an object are modulated as a function of the weight of the object, as illustrated in figure 6.7. The duration of both the loading and unloading phases usually increases with heavier objects, as a longer period is now required before the grip force is sufficient to lift the object. The balance between the grip and load forces does not vary as a function of weight, whereas it does change under different friction conditions (Johansson & Westling, 1984). Weight-related information is not available until the object is actually lifted from the supporting surface, and so the grip and load forces must be programmed in anticipation of the movement, based on an estimate of the object's weight. With unfamiliar objects, people often infer the weight of the object by assuming a default density that is within the range of commonly encountered

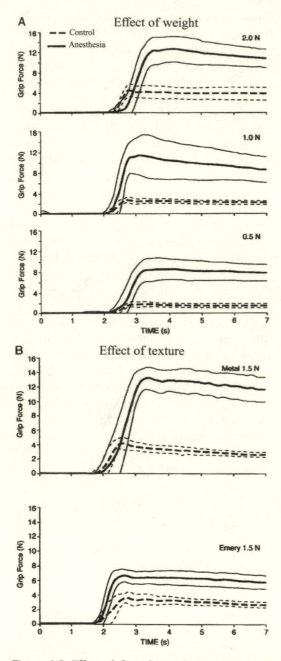

Figure 6.8. Effect of digital anesthesia on the grip forces used to resist three different motor forces of 0.5, 1, and 2 N (A) and on the grip forces used when the grasping surface was either metal or emery (B). The data are from a single participant and show the mean grip force +SD for 20 trials, with cutaneous sensation intact (dashed lines) and during digital anesthesia (solid lines). Reprinted from Monzée et al., 2003, with permission of the American Physiological Society.

densities (Gordon, Westling, Cole, & Johansson, 1993). On some occasions, there is a mismatch between the anticipated force requirements and those apparent in the task. This can occur, for example, when there is a discrepancy between the predicted and actual weight of the object. Under these conditions, the force output is adjusted within 100 ms of object contact. Presumably, feedback from muscle spindle receptors signals that the force produced is inappropriate for the task and that the object has not moved or moved too rapidly as it was grasped (Macefield & Johansson, 1996). After a single trial, the force rates are adapted to the object's weight, provided it remains constant (Johansson & Westling, 1988).

When the fingers have to restrain an active object, such as a servo-controlled motor, the grip forces are graded to the load amplitudes (resistive forces) in a similar manner to the adjustment to passive objects (Johansson, Riso, Häger, & Bäckström, 1992; Jones & Hunter, 1992). There is a proportional change in grip force with load force that serves to maintain an adequate safety margin that prevents slips, although the safety margins employed with an active object are slightly higher than those used for passive objects being supported against gravity at similar load forces (Johansson, Riso, et al., 1992). The higher safety margins may be due to the less predictable nature of the physical stimulus (i.e., motor), which can generate arbitrary forces at unanticipated rates rather rapidly. With the active object, the grip-force responses are adapted to the rate at which the load force changes, and the latency between the start of the loading phase and the onset of the grip response decreases with increasing load-force rate over the range of 2 N/s to 32 N/s (Johansson, Häger, & Riso, 1992). The latter finding suggests that the grip responses are elicited at some threshold load force, which will be reached more rapidly at higher load-force rates.

A variety of grasp configurations and tasks has been used to examine the coordination of grip and load forces, including applying tangential loads to the hand either by tilting the object (Goodwin, Jenmalm, & Johansson, 1998) or using an object with its center of mass located off the grip axis (Salimi, Hollender, Frazier, & Gordon, 2000). In addition, the influence of inertial and gravitational forces on grip-force control has been studied by having participants move their arms while an instrumented object is held in their hands under normal gravitational conditions (Flanagan & Wing, 1995, 1997) or during the periods of microgravity that occur with parabolic flight maneuvers (Hermsdörfer et al., 2000). Despite the presence of these destabilizing forces, the grip and load forces are modulated concurrently with fluctuations in the acceleration-dependent inertial loads (Flanagan & Wing, 1995), and the predictive coupling of grip and load forces occurs even during transitions between gravity levels (Hermsdörfer et al., 2000). In addition, the scaling of the grip forces is directly related to the minimum grip force required to maintain an adequate safety margin that prevents slips (Flanagan & Wing, 1995; Goodwin et al., 1998). When pushing and pulling forces are applied to the hand as it grasps an object, the grip forces produced are modulated with the applied load forces. However, this modulation is affected by the direction of the load force, as shown in figure 6.9, with higher grip forces being produced in response to pulling load forces at larger force amplitudes, particularly for more slippery surfaces (Jones & Hunter, 1992). The greater grip forces generated in response to the pull force suggest that the friction between the skin and the object varies as a function of the direction of load force, perhaps reflecting differences in the anisotropic extensibility of the skin (Wilkes, Brown, & Wildnauer, 1973).

Tactile mechanoreceptors in the fingertips respond to the changes in normal and tangential forces, with FA I and, to a lesser extent, SA I receptors primarily signaling the changes in force during the preloading and early loading phases of lifting (Macefield, Häger-Ross, & Johansson, 1996). Westling and Johansson (1987) proposed that the initial responses of the FA I and SA I units reflect the skin deformation changes that result from the increasing grip force. At low grip forces of around 1 N, the area of contact is approximately 70% of that at 5 N (Serina et al., 1997), as described in chapter 2 and shown in figure 2.10. This means that the contact area increases rapidly with relatively small changes in grip force. At force levels of around 0.5 N, it has been estimated that the receptive fields of approximately 350 tactile units would be stimulated, and about 66% of these would be FA I units. At 1 N, approximately 450 tactile units would be stimulated, most of them extremely sensitive to small skin deformations (Westling & Johansson, 1987). As the grip force increases further, there is little additional

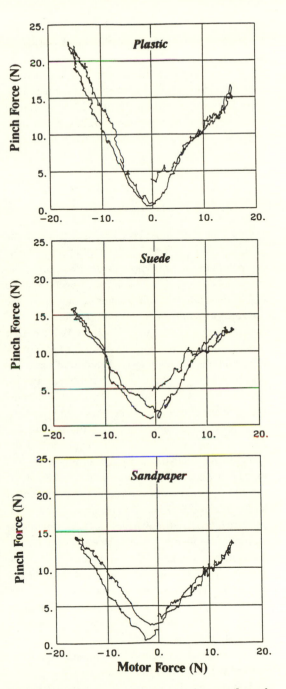

Figure 6.9. Relation between motor force and pinch force for three different surfaces for a single participant. Negative motor forces refer to a pulling force and positive motor forces to a pushing force. Reprinted from Jones and Hunter, 1992, with permission of the Helen Dwight Reid Educational Foundation.

change in the contact area, and so the FA I and SA I units would be less effectively stimulated. During the static phase, when the grip force remains constant, SA II units discharge and are strongly influenced by both skin indentation and lateral forces, and so they also probably contribute to the control of finger forces (Westling & Johansson, 1987). As described in chapter 3, it is these units that play a role in proprioception in the hand.

Object's Shape and Size and Grip Forces

Both tactile and visual cues are used to modulate grasping forces, but their influence occurs at different points in the grasping process. Visual information is used primarily to identify the properties of the object, determine the grasp requirements, and make anticipatory adjustments to the grip forces as the hand comes in contact with the object. Features that are identified visually include shape, surface curvature, and object weight as predicted from size (Gordon, Forssberg, Johansson, & Westling, 1991a; Gordon et al., 1993; Jenmalm, Dahlstedt, & Johansson, 2000). The shape of the object imposes constraints on force coordination, as the sum of the forces and moments applied must equal zero if the object is to be held in a stable grasp. When the surface angle on the object increases (i.e., the object is tapered upward), both the normal and tangential forces increase as a function of surface angle; the balance between these forces is, however, maintained independently of the surface angle, and so the safety margin remains constant (Jenmalm & Johansson, 1997). Even when tactile feedback from the fingertips is eliminated with digital anesthesia, the normal (grip) and tangential forces are still adjusted to the object's shape, which is visually perceived, although the grip forces are considerably larger when compared to those used for lifting with normal sensibility (Jenmalm & Johansson, 1997).

Some geometric features, such as object curvature, have little or no effect on the magnitude of the grasping forces, but do influence the safety margins that are used to prevent slips. Jenmalm, Goodwin, and Johansson (1998) reported that when people grasp objects with either concave or convex surfaces, the minimum grip forces required to prevent slips are influenced by surface curvature. Under these conditions, the slip ratio is not only a function of the fric-

tion at the digit-object interface but is also affected by the characteristics of the mechanical contact between the fingers and the object. They found that the safety margins adopted to prevent slips were higher for either concave or convex surfaces as compared to those used for objects with flat surfaces.

When people are asked to lift two objects of identical weight but varying in size, the smaller of the two objects is typically perceived to be heavier than the larger object, an effect known as the size-weight illusion (described in chapter 5). When these objects are lifted, however, participants quickly learn to scale the grip forces appropriately for the actual weight of each object, not its perceived weight, and so the grip forces during the holding phase of the lift are the same independent of object size (Flanagan & Beltzner, 2000; Gordon et al., 1991a). This occurs regardless of whether size is perceived visually (Gordon et al., 1991a) or haptically (Gordon, Forssberg, Johansson, & Westling, 1991b).

Object's Center of Mass and Grasping

In most studies on prehension, the object is grasped with two fingers, and the center of mass is usually midway between the fingers in the plane of the two contact surfaces. In this situation, the load force at each digit is equal to half the object's weight. If the center of mass does not lie on the grip axis, then torques tangential to the grip surfaces can develop and result in the object rotating in the hand. To prevent these rotational slips, the grip forces must be scaled appropriately at each digit. If the center of mass is located distal to the grip axis that connects the tips of the digits, the grip forces are scaled to the resulting load torque (Wing & Lederman, 1998). When the center of mass is parallel to the grip axis, that is, the center of mass is located laterally to the center of the object, the load forces must be partitioned asymmetrically at the thumb and index finger if the object is to be lifted without tilting. Using this type of manipulation, Salimi et al. (2000) found that within three to five lifts, participants were able to modulate the load forces developed at the two grasping surfaces so that the object was lifted with minimal tilt. This anticipatory control of the force was, however, specific to the effectors manipulating the object. When the object was rotated 180° in the same hand or translated to the other hand, the load forces were not appropriately scaled at each digit, and the object twisted in the hand. Within a few trials, however, participants were able to scale the forces correctly and presumably had developed a stable internal representation of the object's center of mass.

Anticipatory Models and Grasping

The anticipatory scaling of grip and load forces as the hand makes contact with an object has been proposed to result from an internal model of the task that includes both its dynamic and kinematic characteristics. Previous manipulatory experience is used to develop an internal representation of the relevant physical properties of objects and of the relation between these properties and the forces required to support the objects. This model then contributes to the preprogramming of forces and to the modulation of forces under different grasping conditions (e.g., when rotating the object in the hand). If visual cues about the object and its grasping surface are available prior to contact, they can be used to scale these grip forces and grasp kinematics (e.g., at which point to place the digits). If these cues are not available, shape and size can also be perceived using tactile and kinesthetic information (Jenmalm & Johansson, 1997). If the object is novel, the scaling of grip forces to its physical properties is established within a few trials of lifting and persists in memory for up to 24 hours (Gordon et al., 1993). However, if tactile feedback from the fingers is eliminated following local anesthesia, this internal model is unable to compensate for the lack of peripheral sensory information even though participants have had extensive experience lifting the object. Under these conditions, the grip forces remain consistently high (Augurelle et al., 2003; Monzée et al., 2003). This suggests that continuous or relatively frequent sensory information from the fingers is required for the internal model to function effectively.

The internal representations of some of the properties of objects that are grasped do not depend on the particular end effector used to acquire that information. Cues obtained using one hand regarding the texture (Johansson & Westling, 1984) or the weight (Gordon et al., 1993) of an object are available to the other hand during subsequent lifts. In contrast, when

an object's center of mass does not lie in the grip axis, information about the center of mass is not represented in a way that other end effectors can utilize when they make initial contact with the object (Salimi et al., 2000). This suggests that there may be different levels of representation of different object properties that may reflect the coordination required to execute the task. The instances in which there have not been transfers of these object properties (i.e., center of mass and different textures at each digit) are those in which there must be a partitioning of the force responses at the two digits in contact with the object, and so they cannot be controlled as a unit.

It has been suggested that there must be many internal models within the central nervous system that represent the properties of objects encountered in the environment (Witney, 2004), given the ease with which people can switch between objects in their daily manual activities. There may be generic object models that are initially accessed based on visual and tactile information that are updated once contact is made with the object. Davidson and Wolpert (2004) have shown that people can form internal models of two objects of varying weight, but identical appearance, and repeatedly switch between them as they grasp and lift one of the objects. Moreover, people can additively combine these models to preprogram the forces required to lift a single, combined (stacked) object.

In summary, the grip forces produced when lifting an object in the hand are precisely coordinated in space and time and are of sufficient magnitude to prevent the object from slipping between the fingers. The ratio between the grip and load forces is adapted to the friction at the finger-object interface and so maintains an adequate safety margin. Afferent information from cutaneous mechanoreceptors in the fingertips plays a critical role in the regulation of grip forces when manipulating objects. Their sensitivity to mechanical events at the finger-object interface, particularly to small slips between the object and the skin, permits the hand to maintain stable and efficient grasps. In the absence of this feedback, either acutely due to anesthesia (Monzée et al., 2003) or chronically because of peripheral or cortical lesions (Nowak & Hermsdörfer, 2003; Robertson & Jones, 1994), the precise control of prehensile force is lost, and the hand becomes clumsy and inefficient.

7

Non-prehensile Skilled Movements

To imagine that the dexterity of the musician, the artist or the craftsman lies in the special anatomical perfection of the human hand is a delusion that is apparent in the writings of most of those who have given attention to the subject. Such elevating studies can only be undertaken by those who mistakenly ascribe to the primitive human hand those aptitudes conferred on it by the specialized cerebral cortex.

WOOD-JONES, 1944

Skilled hand movements that are non-prehensile cover a diverse class of activities ranging from the gestures made as part of normal speech (gesticulation) or as a substitute for speech (sign language) to the movements involved in depressing the keys on a keyboard. With the exception of pointing and aiming movements, these activities typically involve all fingers and both hands (Tallis, 2003). The categories of skilled movements considered to be non-prehensile are shown in table 7.1, together with the types of manual activity subsumed under each class.

Gestures and Signs

Studies of gesture have focused on variables such as the temporal relation between gesticulatory movements and speech phrases, the repertoires of codified gestures (often referred to as quotable gestures) in different speaker communities, and the differences between cultural groups in their use of gestures (Kendon, 1993). When the use of speech is limited due to environmental conditions (e.g., the noisy work environment of a sawmill [Meissner & Philpott, 1975]) or restricted because of cultural or reli-

gious reasons (silence required during periods of mourning or in monastic religious orders), quotable gestures are often used in place of speech and have properties similar to those of spoken language.

The gesture systems that have arisen as substitutes for speech within environments in which there is no direct access to spoken languages, such as in deaf communities, are as elaborate as spoken languages and structurally analogous (Klima & Bellugi, 1979). These gestures are different from the gesticulations that are used to supplement or illustrate the spoken word, and so are commonly referred to as signs. Monastic sign languages (Umiker-Sebeok & Sebeok, 1987) and sign language systems that are developed as alternatives to speech by people who have full and normal access to speech, such as Australian aboriginal women in mourning, share many similarities to the sign languages used in deaf communities. Comparable to quotable gestures, signs have a specified form, and variations in the shape of the hand, movement of the hand, location of the hand in space, and palm orientation can all be used to convey morphology and syntax. For example, in American Sign Language (ASL), the nouns *apple* and *onion* are similar signs that are differentiated by the

Table 7.1. Categories of non-prehensile skilled movements and conditions under which they have been studied

Gestures	Pointing	Aiming	Keyboard skills	Bimanual music skills
Gesticulations	Single joint or multijoint	Reciprocal (repetitive) or discrete	Typing	Playing stringed instruments
Quotable gestures	With/without visual feedback	With/without visual feedback	Piano playing	Playing wind instruments
Sign language	Active or passive movements	Target contact with finger or stylus	Typing Morse code	Playing percussion instruments
Finger spelling	Target position encoding			

place of articulation, to the right of the chin or the right of the eye, respectively. In contrast, the words *please* and *sorry* differ only in hand shape and are articulated at the same location (in front of the chest) and with the same movement, as shown in figure 7.1 (Emmorey et al., 2003).

These three components of signs, namely, location, hand shape, and movement, are considered analogous to the phoneme in spoken language in that they occur below the level of the actual sign. In contrast to spoken languages, they occur simultaneously and not sequentially but, similar to spoken languages, they vary in different sign languages (Deuchar, 1999). For example, the cheek and the ear are distinct locations for signing in British Sign Language (BSL), but are not distinct in ASL, whereas in Chinese Sign Language, a location under the arm is used that is not found in ASL (Klima & Bellugi, 1979).

Sign languages used by the deaf have received widespread analysis by linguists since the 1970s. Unlike spoken languages, they have tended to develop in isolation from one another, and so even sign languages that are used in English-speaking countries, such as the United States (ASL) and the United Kingdom (BSL), are quite different from each other (Deuchar, 1999). For example, the BSL sign for *true* means *stop* in ASL. Even signs that are iconic, such as that for *tree*, which resemble their meanings in physical contours are different in American, Danish, and Chinese sign languages (Bellugi & Klima, 1976).

Uncommon or unfamiliar words are often spelled out by sign language interpreters using finger spelling. Studies of the hand movement sequences used when finger spelling indicate that the hand shape for a particular letter is influenced by the letters that precede and follow it. This has been interpreted as being analogous to the phenomenon of coarticulation in speech, in which certain phonemes are articulated differently depending on which phonemes follow (Jerde, Soechting, & Flanders, 2003). Consistent with the findings

SORRY

PLEASE

Figure 7.1. Signs in American Sign Language (ASL) that contrast in hand configuration but occur at the same location.

on typing and piano playing reviewed in this chapter, these results suggest that the organization of movement sequences is not strictly serial and that the context in which the hand movement is made influences its production.

Pointing and Aiming Movements

Another class of hand movements that falls within the rubric of non-prehensile manual activities is pointing and aiming movements. The distinction between these two types of movement is based on the end point of the movement trajectory. Pointing movements are made by the hand and arm toward a visually or proprioceptively defined target, but contact is not usually made with the target. When fingertip contact is made with the target, the accuracy of pointing movements improves (Rao & Gordon, 2001). Aiming movements do involve contact with a target and are therefore more spatially constrained in terms of endpoint accuracy. The contact with the target can be with the fingers or with a stylus held in the hand. When studied experimentally, aiming movements are usually produced repetitively toward one or more target positions. Both of these types of movement are considered to be non-prehensile because the objective of the task is to transport the hand to a defined spatial location. In contrast to reaching-to-grasp movements, the position and forces produced by the fingers and wrist do not have to be explicitly controlled in aiming and pointing movements. Nevertheless, there are many similarities in reaching and pointing movements in terms of the kinematics of the movements and the factors that influence movement time and accuracy.

Pointing Movements

Pointing movements have been studied from a variety of perspectives, ranging from analyses of the kinematics of the movements and the conditions under which these change (Atkeson & Hollerbach, 1985; Morasso, 1981) to studies of the errors made in pointing and what these suggest about the computational processes involved in planning movements (Soechting, Tillery, & Flanders, 1990). There has also been a considerable amount of research conducted on whether movement amplitude or final position is the controlled variable in pointing movements (e.g., Bock & Eckmiller, 1986; Feldman, 1974; Kelso, 1977) and on how kinesthetic cues are used to encode these movements (Adamovich, Berkinblit, Fookson, & Poizner, 1998).

The kinematics of pointing movements have been studied in detail, and the characteristic features of the velocity profiles have been described. For both single-joint and multijoint pointing movements in which the constraints on spatial accuracy are low, the velocity profile of the wrist's movement is bell-shaped and has a single peak, as shown in figure 7.2 (Morasso, 1981; Soechting, 1984). As the amplitude of the movement increases, so too does the peak velocity, which means that movement time remains essentially constant (Georgopoulos, 1986). If the accuracy requirements of the pointing task become stringent, the velocity profile becomes segmented, which reflects the fine positioning movements made by the hand as it approaches the target.

When the arm is moved in the sagittal plane toward a target, the path of the hand is usually slightly curved (Soechting & Lacquaniti, 1981), with the extent of curvature varying with move-

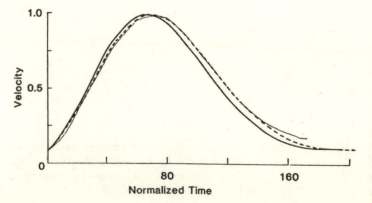

Figure 7.2. Velocity profiles for movements of the wrist to large (thin line) and small (dark line) targets. The dashed line shows the data for movements to the small target replotted on a time scale which has been expanded by a factor of 1.05. The data are from a single subject. Reprinted from Soechting, 1984, with permission of Springer-Verlag and the author.

ment direction and from person to person (Morasso, 1981). This curvature in the hand's trajectory is not affected by the speed of the movement nor by the addition of large loads (Lacquaniti & Soechting, 1982; Lacquaniti, Soechting, & Terzuolo, 1982). The curvature does, however, increase when movements are made in the absence of visual feedback of the limb and target (Sergio & Scott, 1998). It is also influenced by the visual experience of the individual, as blind people follow much straighter paths than do blindfolded sighted people (Miall & Haggard, 1995; Sergio & Scott, 1998). Two hypotheses have been proposed to account for the curvature of voluntary arm movements, one of which proposes that the trajectory is selected so that it optimizes control variables related to the dynamics of arm movement and minimizes the changes in torque producing the movement (Hollerbach & Atkeson, 1986; Uno, Kawato, & Suzuki, 1989). In this model, known as the minimum torque change model, the arm follows a trajectory planned in intrinsic (joint and muscle) coordinates. The alternate hypothesis, the minimum jerk model, argues that movements are planned in extrinsic coordinates and that the desired path is straight but deviations occur due to inaccuracy in the controller, visual misperception of straight lines, and unavoidable errors (Flash & Hogan, 1985). The differences in the movement trajectories of blindfolded sighted and blind participants (Miall & Haggard, 1995) indicate that visual experience does affect the spatial path of the hand during pointing movements and that trajectory planning is not limited to simply minimizing some dynamic function.

When pointing to a target, information about the location of the target is provided by the visual system whereas proprioceptors in muscles signal the initial positions of the hand and arm. The sensory signals that specify target location have to be transformed into motor commands, which are sent to motor neurons controlling arm muscles, and so there needs to be a common frame of reference for the visual (extrinsic) and proprioceptive (intrinsic) coordinate systems. Based on their analyses of pointing errors to remembered visual targets in three-dimensional space, Flanders and Soechting (1990) proposed that two separate channels are involved in this visual-proprioceptive transformation; first, arm elevation is computed from target distance and elevation, and second, arm yaw is calculated from target azimuth. They argue that the preferred coordinate system for

controlling the kinematics of the arm is centered on the shoulder and not the head, because most of the errors in pointing were with respect to the distance that the arm had to be moved and not the direction of the movement (Flanders, Helms Tillery, & Soechting, 1992). This hypothesis was based on the consistency of the errors when they were evaluated using a spherical coordinate system that was centered on the shoulder. In these experiments, participants moved their hand, which was resting at their side, to the remembered target location either in the dark or with the lights on. It is not clear to what extent the interpretation of these results can be generalized to other, more natural, pointing tasks in which the initial position of the hand varies and the target is visible. The movement errors that provide the basis for this model of sensorimotor transformations are not seen under normal conditions when both the hand and the target are visible. Under these conditions, it would appear that the position of the hand in space, rather than the orientation of the arm, is more critical for accurately locating the visually defined target position.

Pointing movements can also be made in the absence of any visual feedback, and in this context the contribution of proprioceptive cues to the encoding of movements has been studied, together with an analysis of how the conditions under which these cues become available influences pointing accuracy. In these experiments, participants moved their arm actively to a target location or the arm was guided passively by the experimenter or a robotic arm. The person was then required either to move the arm to the same location after it was returned to the original or to a different starting position or to move the opposite arm to the same location as the target limb. Paillard and Brouchon (1968) showed that pointing errors are significantly smaller (6 mm versus 22 mm) if people actively move their arm, as compared to having it passively moved by the experimenter (with the arm in a cradle). There is, however, no difference in the errors if the final position of the arm is actively or passively maintained by supporting the arm in a cradle. These findings indicate that the proprioceptive feedback available during active movements improves the encoding of the end point of the movement and that the difference between active and passive movements presumably reflects the enhanced muscle spindle afferent discharges associated with active movements (Prochazka, 1996).

More recently, Baud-Bovy and Viviani (2004) have used a robotic arm to position the hand and have shown that pointing accuracy is not affected by the kinematics of the initial locating movement nor the active pointing movement itself. The pattern of errors that they observed was, however, similar to those reported when pointing to visual targets, suggesting that the errors reflect the mechanisms by which space is represented internally and are not modality specific.

Aiming Movements

Goal-directed aiming movements have been studied for more than a century, ever since Woodworth (1899) first proposed that there are two components to these movements: a ballistic movement that covers most of the distance, which he termed the initial impulse, followed by a deceleration phase to the final goal position that is under visual feedback control, called current control. The experiments that Woodworth conducted required that participants make back-and-forth horizontal sliding movements with a pencil on paper that was attached to a drum rotating orthogonally to the pencil's movement. Of particular interest to Woodworth was the contribution of vision to current or feedback control. This was studied by measuring movement accuracy with the eyes open and closed during movements of varying speed. He determined that at movement times of approximately

450 ms, there was no difference between the errors obtained in the eyes-open and eyes-closed conditions. Subsequent studies of discrete aiming movements, not reciprocal movements as Woodworth used, have shown that the time required for the visual feedback loop to operate is considerably shorter than Woodworth's estimate and is probably between 100 and 200 ms (Keele & Posner, 1968, Zelaznik, Hawkins, & Kisselburgh, 1983). Even for very rapid movements, visual feedback of the arm and target enhances movement accuracy, particularly during the last 25% of the movement trajectory (Elliott, Helsen, & Chua, 2001).

Since the 1950s, reciprocal aiming movements have been extensively studied using a task in which a person holds a stylus in the hand and moves it rapidly between two target positions, as shown in figure 7.3. Typically, the size of the targets and the distance between them are varied, and measurements are made of movement time and sometimes the trajectory of the movements (Jagacinski, Repperger, Moran, Ward, & Glass, 1980; Langolf, Chaffin, & Foulke, 1976). In 1954, Fitts suggested that the degree of precision required to execute such a movement was a function of the number of alternative movements that could be made. If a movement is being made to a wide target, then a large number of movements are possible, whereas the range of movements is constrained if the target is very small. Similarly, the range of possible movements is affected by

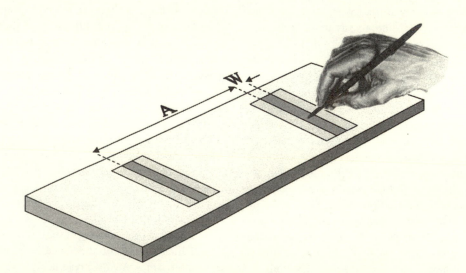

Figure 7.3. Schematic representation of the apparatus used to study rapid reciprocal aiming movements as initially described by Fitts, 1954. *A* = movement amplitude; *W* = target width.

the amplitude of the movement (i.e., distance). Fitts noted that the time taken to move between two targets depended on the movement amplitude and the width of the targets and proposed the following equation to predict movement time:

$$MT = a + b \log_2(2A/W),$$

where MT is movement time, a and b are constants that depend on the individual and the task, respectively, A is movement amplitude, and W is the target width.

This equation characterizing the speed-accuracy trade-off became known as Fitts' law and has been found to generalize to many types of hand and arm movements (Crossman & Goodeve, 1983; Fitts, 1964). The speed with which movements can be performed is a function of the index of difficulty (ID), which is defined as $\log_2(2A/W)$, and so movements with higher indices of difficulty take longer to perform, as shown in figure 7.4. The index weights movement amplitude and target width or tolerance equally, which means that doubling the distance and halving the target width are equivalent (see figure 7.4). This implies that the amount of information transmitted by the neuromuscular system is constant over time. In a typical reciprocal tapping task, the information transmission rate has been found to be 10 bits/s (Fitts, 1954). Fitts' law has been found to be a good predictor of movement time for a wide variety of precise manipulative motions, including microassembly and peg transfer tasks performed using a microscope (Langolf et al., 1976). However, it does not predict accurately the movement times in short movements (Klapp, 1975), nor in bimanual tasks in which the hands move to targets of very different widths (i.e., discrepant indices of difficulty). Under these conditions, the arm movements are synchronized so that the hands arrive at the targets simultaneously, as shown in figure 7.5 (Kelso, Southard, & Goodman, 1979).

Analyses of the movements made during reciprocal tapping tasks indicate that the increase in movement time with increases in ID occurs in the terminal phase of the movement, where feedback control is assumed to be important (Carlton, 1979). This is reflected in the velocity profile, which changes from being relatively symmetric when aiming at large targets to asymmetric for smaller targets. For the latter, more time is spent in the deceleration phase of the movement to make corrections as the hand approaches the target (MacKenzie, Marteniuk, Dugas, Liske, & Eickmeir, 1987).

Further support for the importance of visual feedback to movement accuracy in the terminal phase of goal-directed aiming movements comes from studies in which movements are made with and without vision. When visual feedback of the limb and the target is available, participants spend proportionally more of the overall movement time in the interval after the peak velocity than when moving in the absence of vision, even though early kinematic markers, such as peak velocity and time to peak velocity, are not affected by the presence or absence of visual feedback (Elliott, Carson, Goodman, & Chua, 1991). In addition, there are usually more discontinuities in the movement trajectory during the deceleration phase in the presence of vision than without (Chua & Elliott, 1993), although this does not always occur (Meyer, Abrams, Kornblum, Wright, & Smith, 1988). Errors are, however, always greater in the

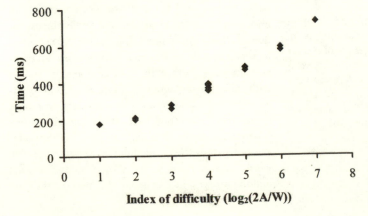

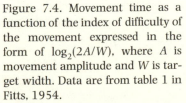

Figure 7.4. Movement time as a function of the index of difficulty of the movement expressed in the form of $\log_2(2A/W)$, where A is movement amplitude and W is target width. Data are from table 1 in Fitts, 1954.

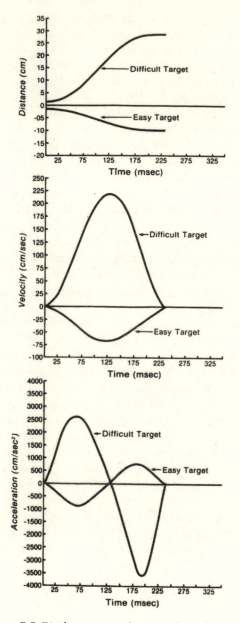

Figure 7.5. Displacement, velocity, and acceleration of the arms during two-handed movements of unequal difficulty, as defined in terms of target width. Note the almost perfect synchrony between the peaks in the velocity-time and acceleration-time curves. Reprinted from Kelso, Southard, & Goodman, 1979, with permission of the American Psychological Association and the authors.

no-vision conditions—often by up to 300% (Elliott et al., 1991)—and so it appears that with visual feedback the extra time after peak velocity is used to adjust the movement's trajectory and to reduce aiming error, as originally proposed by Woodworth (1899).

Keyboard Skills

Two non-prehensile skill areas that have been extensively studied are the keyboard skills of typing and piano playing (Rumelhart & Norman, 1982; Salthouse, 1986). These two tasks share some similarities in that they both involve reaching with the fingers to depress a sequence of keys, but they differ in several important dimensions. With piano playing, there is a rhythmic requirement to the timing of the key presses, whereas typing has no such restrictions, other than that the keys are struck in the correct order. In addition, the force with which the keys are struck is not usually regulated when typing, provided it exceeds the key switch actuation or "make" force. When playing the piano, the force applied to the key and the time that the key remains depressed are actively controlled as they influence the loudness and duration of the resulting tone.

Much of the research on these complex motor skills has focused on the nature of the internal representation of the task and the translation of this representation into action and so has considered both the cognitive and motor aspects of the skill (Gentner, 1987; C. Palmer, 1997; Rumelhart & Norman, 1982). For typing, a theoretical model comprising four components of the skill has been proposed: the input stage in which the text is converted into chunks; a parsing phase in which the chunks are decomposed into strings of characters; a process of translation in which the characters are converted into movement specifications; and finally the execution of these movements (Salthouse, 1986). It is assumed that these four processing phases can operate simultaneously with each component dealing with different segments of information. The influence of these various stages of processing on typing speed and errors has been studied in numerous experiments in which the characteristics of the material being typed (e.g., meaningful versus nonsense words, high-frequency versus low-frequency words) and the amount of preview of the material (i.e., the number of visible characters) have been varied (Gentner, 1987; Inhoff, 1991; Salthouse, 1984).

To understand these keyboard skills better, some researchers have simplified the task to eliminate the linguistic (typing words) or musical (piano playing) components and have required that participants tap a series of keys in a prescribed order or according to some temporally defined sequence (Povel & Collard, 1982). On the basis of the latency profiles of the participants' tapping responses on this type of task, models of the internal timekeeper have been proposed (e.g., Shaffer, 1981; Wing & Kristofferson, 1973).

Typing

The first typewriter to achieve commercial success was developed by Sholes and Glidden (Sholes, Glidden, & Soulé, 1868), and by 1874 it was being manufactured by the Remington Arms Company. It was a machine devised without empirical research and developed in response to the limitations of other machines, whose keys tended to jam at moderate typing speeds. It was originally designed for "hunt and peck" operation and not for touch typing; the performance objective in terms of speed was to make a machine that could print words at a speed equivalent to handwriting (Noyes, 1983). It achieved success while many other rivals failed, in part because it had an excellent system for product improvement, with as many as 30 models coming out over 4 years (Tenner, 2004). The QWERTY keyboard had become the international standard by 1905, and despite criticisms over the years as to its suboptimal layout and the excessive use of the weaker fingers, alternative keyboards have met with little commercial success (Dvorak, 1943; Noyes, 1983). It has been characterized as one of the most maladaptive human-machine interfaces ever invented (Salthouse, 1984).

Typing is a motor skill that requires explicit training to acquire expertise, that takes a relatively long time to become an expert, but that once acquired is resilient to loss (Salthouse, 1984). Touch typists are trained to type in a prescribed manner, with the "home position" for each finger defining a standard initial posture of the hands. Each letter of the keyboard is typed with a particular finger, which returns to its home position after the letter is typed. A keystroke to type a single letter therefore involves three components: moving the finger from its home position to a position over the target key, depressing the key with sufficient force to register the letter, and then returning to the home position. It has been estimated that over a 10-year period a typical professional typist types more than 25 million words at an average rate of 60 words/min (wpm), which corresponds to an average of 5 keystrokes/s. Very fast professional typists can maintain a typing speed of more than 9 keystrokes/s for up to an hour (Gentner, 1988).

When the performance of expert and novice typists is compared, the most obvious differences are that experts type faster, 60–120 wpm as compared to 13–25 wpm for novices after 4–8 weeks of typing classes, and have much lower error rates. The interval between successive keystrokes is shorter in experts as compared to novices (see figure 7.6), and the average speed of the finger movements of expert

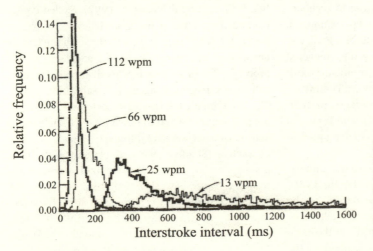

Figure 7.6. Distribution of interkey stroke intervals for typists of varying levels of expertise as indicated by the number of words typed per minute (wpm). Adapted and reprinted from Gentner, 1983, with permission of Elsevier.

typists is more than two times (353 versus 152 mm/s) that of novices (Gentner, 1988). The improvement in typing speed with experience mainly reflects the ability of skilled typists to overlap their finger movements so that two and occasionally three keystrokes are occurring concurrently, as successive letters are processed in parallel (Gentner, 1983, 1988). This continuous, fluid motion of the fingers in skilled typists stands in contrast to the awkward hunt-and-peck methods of the unskilled typist, in which each keystroke is made in sequence.

Piano Playing

The musical keyboard has existed for more than 500 years and has its beginnings in ancient organs, which were played with open hands or fists, but not with individual fingers. From these origins came the great medieval organs with keys instead of sliding horizontal bars, clavichords in which the keys actuated brass surfaces that made contact with metallic strings, and harpsichords in which the strings were plucked by quills held by wooden bars that were actuated by the keys (Tenner, 2004). However, it was not until the mid-18th century that keyboard technique underwent a major innovation when C. P. E. Bach proposed a fingering system with an emphasis on the use of the thumb as a pivot. Since that time, there have been numerous attempts to improve the design and construction of the piano, some of which were immensely successful, such as the cast-iron frame that could support much higher string tensions and was less sensitive to seasonal fluctuations in temperature and humidity. Other innovations that attempted to overcome deficiencies in keyboard layout met with much less success. For example, a keyboard was designed in the 1880s by Paul von Janko in which the keys were arranged in staggered tiers of whole tones (similar to the organ keyboard) so that the octave was now six keys wide. This meant that it was easier to play complex passages, particularly for people with smaller hands. The keyboard was criticized, however, for making challenging passages too easy (Tenner, 2004).

Some of the earliest research on the motoric aspects of piano playing was conducted in the 1930s by Seashore, who attached balsa-wood fins to the hammers inside the piano and used continually moving film to record the timing and intensity of the hammers' movements (Seashore, 1938). Analyses of the data he collected from concert pianists playing Beethoven and Chopin revealed that the pianists used considerable variation in tempo (rubato) in the timing of notes, bars, and phrases, and yet these variations were consistently reproduced in subsequent performances. More recent studies of piano playing have used a variety of tasks to study musical skill. These include performing unfamiliar music from notation (sight-reading), performing well-learned music from memory or notation, improvising, and playing by ear (C. Palmer, 1997). Performance on these various measures tends to be relatively highly correlated and improves with training (McPherson, 1995). Comparisons of the performances of expert and less-accomplished pianists indicate that expert performance is usually characterized by a superior ability to reproduce reliably the timing and loudness variations of the musical score in consecutive performances (C. Palmer, 1989). In addition, expert pianists are better than less-accomplished pianists at maintaining independent timing in each hand when playing (Shaffer, 1981).

Eye–Hand Span and Keyboard Skills

Typing and piano playing are often considered to be among the more complex forms of skilled serial action performed by human beings. Analyses of the sequencing of finger movements indicate, however, that they are not strictly serial processes as the time for each keystroke is not based on the previous keystroke. When typing, there is a considerable anticipatory component to performance, particularly when consecutive keystrokes are executed by different hands (Soechting & Flanders, 1992). Skilled typists commit themselves at the motoric level to typing a particular letter approximately three characters in advance of the current keystroke and can visually process up to eight characters in advance of the character being typed (see figure 7.7; Salthouse, 1984). The latter finding is based on the deterioration in typing performance when the amount of preview available on a display screen is limited to fewer than eight characters (Shaffer, 1976). Consistent with these findings, recordings of eye movements during transcription typing have shown that skilled typists are typically reading between three to five characters in advance of where they are typing (Butsch, 1932; Gentner, 1988; Inhoff, Briihl, Bohemier, & Wang, 1992). This difference between the text that is visu-

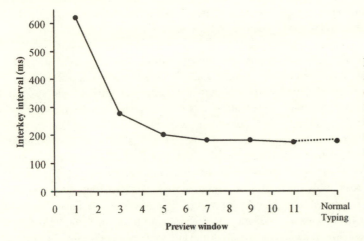

Figure 7.7. Median interkey interval as a function of the number of visible characters during transcription typing. The rightmost point represents typing from printed text. The material to be typed was displayed on a video monitor with the designated number of preview characters. The data are the mean results from 74 typists in the Salthouse, 1984, study. Redrawn from Salthouse, 1986, with the permission of the author.

ally fixated and the letter that is typed at a particular instant, known as the eye–hand span, increases with typing speed. It appears that, with practice, skilled typists learn to look ahead of the currently typed text and prepare for future keystrokes by positioning their fingers over the appropriate keys. This interpretation is supported by the finding that the greatest difference between novices and experts in typing speed occurs when consecutive keystrokes are executed by fingers on different hands (Gentner, 1983). When expert typists are prevented from previewing the text that is to be typed, their typing speed becomes similar to that of novice typists (Salthouse, 1984).

A similar span of preview is evident when pianists reproduce unfamiliar musical sequences. Sloboda (1974) reported that the eye–hand span for sight-reading music was between four and seven events beyond the point at which the musical notation disappeared from view. As was noted for typing, the eye–hand span in pianists was positively correlated with sight-reading ability, as defined by the number of mistakes made before the musical score disappeared from view (Sloboda, 1974). Using a different task, in which pianists were required to perform briefly displayed musical extracts, Bean (1938) also noted that experienced pianists had a greater "span of apprehension" than did novices; that is, they played more notes correctly. These studies on eye–hand span in pianists and typists indicate that skilled performers are able to look further ahead in the material being read than are novices and on the basis of this preview make preparatory finger movements.

An interesting analogue to the eye–hand span in keyboard skills is the ear–hand span reported for telegraphers receiving Morse code (also known as continuous wave, or CW). In their 1899 study of telegraphy, Bryan and Harter observed that expert telegraphers prefer to keep 6–12 words behind the transmitting instrument, which means that they have processed in the order of 230 clicks (dots and dashes) before typing the received message. In contrast, the beginner "must take each letter as it comes" (Bryan & Harter, 1899, p. 353) and learn not to guess ahead in order to avoid time-consuming errors. With practice, the beginner goes from waiting for letters to words, and after many years of experience the processing of these elements becomes almost automatic in the expert telegrapher (Bryan & Harter, 1899). For telegraphy, typing, and piano playing, expert performance is associated with being able to remember a longer sequence of letters or notes that are realized in finger movements. Independently of whether this sequence is represented as a visual or auditory code, the span is similar: between 6 and 12 elements.

Sensory Feedback and Errors in Performance

Typing errors are usually noted by a typist even when no visual feedback is available and are responded to quite rapidly as reflected in the increase in between-keystroke times immediately following an error (Salthouse, 1984). These responses suggest that kinesthetic feedback about movement trajectories is used to evaluate the success of the movement after its execution. There are four main categories of typing errors: substitutions, intrusions, omissions, and transpositions. Substitution errors often entail typing an adjacent key that is either horizontal or vertical to the target key or involve a mirror-image substitution in which the same finger on the opposite hand types the key (Logan, 1999). Analyses of substitution errors indicate that, in many cases, it is not that the trajectory of the finger is incorrect, but that the wrong finger is executing the keystroke, and so it is an error in finger assignment, not finger movement (Salthouse, 1986). Intrusion errors often involve very short interkey intervals in the vicinity of the error and appear to result from incorrectly positioning the finger above the target key so that two keys are struck by the same keystroke (Grudin, 1983). In contrast to these errors, omission errors are often followed by a much longer interkey interval, which is consistent with the finding that many of these errors occur because the key was not depressed with sufficient force and so did not register. Omission errors occur more frequently for keys that are difficult to reach or that are typed with the little fingers (Shaffer, 1975). The majority of transposition errors, which refer to errors in which two adjacent letters in a word are reversed, occurs across the hands (Shaffer, 1976). These can be accounted for by the anticipatory aspect of typing noted above, in which several fingers may move simultaneously toward their target keys.

Musical performance is evaluated in terms of the quality of the interpretation and the consistency of dynamic changes applied across successive performances of the same piece (Ericsson, Krampe, & Tesch-Römer, 1993; Sloboda, 1984) and not simply in terms of errors. Pianists tend not to correct musical errors in performance, as it is considered more important to preserve the rhythm of the piece rather than the exact melody (Shaffer, 1981). Nevertheless, error analyses indicate that pitch deletions (missing a note) are more likely to occur within phrases (a partition of the musical sequence), and perseverations (repeating a note) occur at phrase boundaries (C. Palmer & van de Sande, 1995). The particular notes played in error often preserve the harmony of the music in skilled pianists and can be corrected extremely rapidly (within 66 ms). The speed with which these errors are corrected is considerably shorter than the auditory (140–160 ms), visual (180–200 ms), and tactile (140–270 ms) reaction times (Leonard, 1959; Welford, 1980). This suggests that they are detected and a corrective response is initiated as they occur (Shaffer, 1981). The minimum response latency to tactile stimuli in the distal hand muscles in adults is 60 ms (Johansson & Westling, 1987).

When typing requires that two successive keystrokes are made by the same hand, the second keystroke is initiated within 20 ms of the first. If alternate hands are used, the movements overlap, and the second keystroke is initiated before the first key press, indicating that tactile information from the fingerpad is not used to cue subsequent movements (Soechting, Gordon, & Engel, 1996). If the fingertip of one digit is anesthetized so that observers no longer receive tactile input from the finger but can still move it, the timing of finger movements and the excursion of the fingertip are not affected (Gordon & Soechting, 1995; Rabin & Gordon, 2004). However, the loss of tactile information does affect movement accuracy and the detection of errors. There is an increase in misdirected movements, with the majority of errors involving striking an adjacent key. These errors are not usually recognized, and so the typical increase in the interval before the next keystroke normally seen after mistyping a letter is not evident when errors are made by the anesthetized finger (Gordon & Soechting, 1995). Figure 7.8 shows the intervals between keystrokes when typing an error normally and when the fingers are anesthetized. If tactile feedback is removed from all fingers by having participants type above the keyboard without contacting the keys, then there is a substantial decrease in typing speed and significant increase in the number of errors (Terzuolo & Viviani, 1980). These findings indicate that tactile afferent feedback from the fingertip is important for ensuring movement accuracy and for detecting errors during typing, but is not critical for maintaining typing rhythm. As described in chapter 3, mechanoreceptors in the skin contribute to proprioception (Edin & Abbs, 1991), and in the absence of this sensory input, perception of the initial and final positions of the fingers during typing can be impaired. Errors then increase with successive movements.

Movement Speed and Keyboard Skills

For skilled typists, the interval between successive keystrokes is typically 100–200 ms (Martin et al., 1996; Terzuolo & Viviani, 1980), but intervals as brief as 60 ms are not uncommon (van Galen &

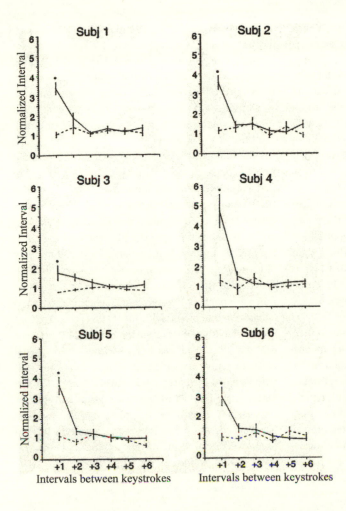

Figure 7.8. Intervals between keystrokes (+1 to +6 mean +SEM) following an erroneous keystroke during control trials (solid line) and following digital anesthesia (dashed line) of the index finger. The intervals are normalized with respect to the same intervals when the words were typed correctly. Asterisks denote a significant difference from the correctly typed keystrokes. Reprinted from Gordon & Soechting, 1995, with permission of Springer-Verlag and the authors.

Wing, 1984). The interval between successive keystrokes is usually 30–60 ms shorter if the preceding keystroke were made by the opposite hand (Salthouse, 1984). When keystrokes are made with fingers on the same hand, there is little opportunity to prepare for the next keystroke, whereas with the opposite hand the relevant finger can begin to move toward the next key at the same time as the preceding key is being struck, and it does so about 30 ms before the termination of the preceding keystroke (Larochelle, 1984).

The standard QWERTY keyboard is generally considered to have a suboptimal layout for touch typing because it overloads the left hand and the little fingers of both hands. The arrangement of the keys was dictated more by mechanical considerations, in particular avoiding the clashing of type bars, rather than facilitating human performance (Dvorak, 1943). There is no evidence, however, that faster keystroke performance is achieved with alternative keyboard designs, even though they may be preferred by users (Noyes, 1983). Analyses of the characteristics of typing by novice and skilled typists indicate that whereas the novice is limited primarily by cognitive constraints (e.g., processing text, identifying key position, planning finger movements), the main constraint affecting skilled typists' performance is motoric. Interstroke intervals for experts can vary by a factor of two or more and primarily depend on the keyboard location of the letters and the fingers used to type them. The cognitive effects are generally small for skilled typists, and factors such as word frequency have a relatively minor influence on experts' typing speed (Gentner, 1988). In addition, there is no relation between typing skill and the degree of comprehension of the material that is being typed (Salthouse, 1984). Also, the rate of typing is very similar for random words as compared to meaningful text (Terzuolo & Viviani, 1980; West & Sabban, 1982), which indicates that the semantic content of the material is not important in typing. However, typing rate does slow as the material to be typed approaches a random sequence of letters (West & Sabban, 1982). Finally, the rate at which text is read is generally about four times that with which it can be typed (240–260 wpm when reading as compared to 50–60 wpm when typing), and so it is not the input processes that are the limiting factor in maximum typing rates (Salthouse, 1984).

Sending Morse code and playing the piano also involve repetitive finger movements, and these can be performed at rates similar to those reported for typing (L. A. Jones, 1998). Experienced Morse code operators can send code at the rate of 20–30 wpm (ARRL, 2003); at this transmission rate, the duration of a dit, the short key closure, averages 52 ms (Tan, Durlach, Rabinowitz, Reed, & Santos, 1997). When playing the piano, accurate temporal control is essential for rhythmic performance, which means that the timing for a particular note is influenced by the timing of other adjacent and nonadjacent notes (Vorberg & Hambuch, 1978). The speed with which piano keys are played clearly depends on the timing constraints of the musical score, but at faster tempos internote intervals are often 80–100 ms (MacKenzie & Van Eerd, 1990). For brief periods of time, such as

when playing trills, pianists can produce 20–30 successive notes per second (one every 40 ms) with each hand (C. Palmer & van de Sande, 1995). To put these response times in the context of reflex responses, the most rapid spinal reflex in the intrinsic hand muscles is around 40 ms (Matthews, 1984), as defined in terms of the onset of stimulation to the mechanical response of the muscle.

The similarity in peak movement speeds recorded from expert typists and pianists suggests that they are performing near the mechanical and neural limits of the human hand and that the intervals between successive keystrokes are determined mainly by the layout of the keyboard and the physical constraints of the fingers and hands (Gentner, 1983). A summary of the average and peak interresponse intervals associated with these tasks is given in figure 7.9. These times are particularly impressive when compared to choice reaction times. Salthouse (1984) reported that the median interkey interval for a group of touch typists that he studied was 177 ms, whereas in a serial two-alternative choice reaction task, the same individuals had a mean reaction time of 560 ms. The latter is clearly a serial task because no stimulus appears until the occurrence of the preceding response. The parallel processing that is seen in expert typists, who prepare finger movements on each hand in advance, is not possible in the reaction-time task.

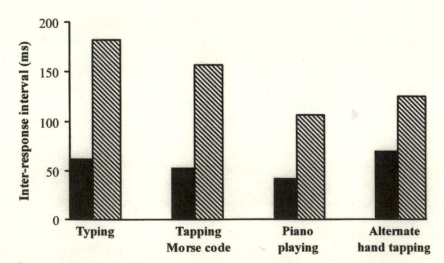

Figure 7.9. Shortest (black) and median (hatched) interresponse intervals for skilled practitioners performing a number of repetitive manual tasks. The typing and alternate hand tapping data are taken from Salthouse, 1984; the Morse code data are from Tan et al., 1997; and the piano-playing data are from MacKenzie & Van Eerd, 1990. Reprinted from L. A. Jones, 1998, with permission of MacKeith Press.

Force and Keyboard Skills

The forces produced during typing have been the focus of many studies of keyboard use, motivated by the increased frequency of cumulative trauma disorders involving the hand and arm in keyboard operators (Gerard, Armstrong, Franzblau, Martin, & Rempel, 1999; Martin et al., 1996). The activation forces for most electronic keyboards are low, typically less than 1 N (August & Weiss, 1992; Rempel, Dennerlein, Mote, & Armstrong, 1994), which is consistent with the recommended upper limit for actuation force of 1.5 N (Martin et al., 1996). The breakaway force, that is, the force at which tactile feedback is presented indicating that the key has registered, is much smaller than this and ranges from 0.04 to 0.25 N (Rempel et al., 1997). For skilled typists, the forces produced during typing do not vary significantly across the fingers, and the mean keystroke force is around 0.9 N (Rempel et al., 1997). The peak keystroke forces are usually much higher than this, and average 2 N (Armstrong, Foulke, Martin, Gerson, & Rempel, 1994; Martin et al., 1996). These forces are highly correlated with the velocity of the fingertip prior to key contact (Dennerlein, Mote, & Rempel, 1998), but are themselves quite variable, which suggests that during typing, force is a rather loosely controlled variable that is programmed simply to exceed the activation force of the key (Martin et al., 1996). When the forces produced by the fingers change as a consequence of varying the inertial mass of each finger, the average typing rate is unaffected, even with a threefold increase in mass. If only one finger is loaded, then the typing rate is affected in some participants (Terzuolo & Viviani, 1980). These findings indicate that the implementation of the movement sequence during typing can readily compensate for changes in the mechanical properties of the hand, provided that these are uniform.

The force used to depress a piano key has usually been measured in terms of the velocity of the keystroke, which in an acoustic piano determines the velocity with which the hammer hits the string. In contrast to typing, force is actively controlled by pianists as it determines sound intensity and contributes to the dynamics of the performance. Western tonal music represents pitch and duration explicitly in the musical score, but intensity and tone quality are only approximately specified, and so the performer has considerable freedom in using these ambiguities to interpret musical content (C. Palmer, 1997). This means that the interpretation of a particular musical piece will vary across performers and that there is no single ideal interpretation. In general, expert pianists apply more force than amateurs when playing the piano and produce more variability in the forces applied (Krampe & Ericsson, 1996). They also use variations in the force or intensity of the notes played by one hand to carry the melody of the music (Shaffer, 1980).

Finger Tapping

A non-prehensile motor skill that contains elements of both goal-directed aiming movements and keyboard skills is finger tapping. In this basic motor task, the index finger repeatedly taps a lever mounted on a microswitch, and the number of taps made in a specified interval of time is recorded (Peters, 1981). In some studies, an alternate tapping task has been used in which the left and right index fingers alternately tap keys as rapidly as possible (Salthouse, 1984). The most common measure of tapping ability is a global rate defined in terms of the average number of taps per unit time (e.g., 10-s trial with between 5 and 10 repetitions of each trial). In addition to the mean tapping rate, a number of task-defined temporal and kinetic variables have been studied. These include intertap interval, duration of key closure or dwell time, the interval between force peaks, and the variance and magnitude of the peak forces. The latter variables have primarily been the focus of studies investigating the basis of performance differences between the hands (e.g., Peters, 1980), which will be reviewed in chapter 8.

Finger tapping is often used as a neuropsychological test to assist in determining the laterality of damage to the cerebral cortex (Lansdell & Donnelly, 1977), as a test to evaluate performance asymmetries between the hands of normal healthy adults (Todor & Smiley-Oyen, 1987), and in developmental studies as a measure of motor skill (Wolff & Hurwitz, 1976). The results from normal healthy adults indicate that in right-handed individuals the right hand taps more rapidly than the left (Todor & Smiley-Oyen, 1987). However, in skilled right-handed typists, the speed of repetitive tapping is faster with the left hand as compared to their right hand, which suggests that with extensive training (4–5 years), the performance differences between the hands disappear (Provins & Glencross, 1968). It is of interest to note in this context that the more skilled the typist, the faster the repetitive tapping rate when measured with one finger or with alternate-hand tapping (Salthouse, 1984). The mean tapping rates of men are higher than those of women (Dodrill, 1979). The latter difference may be attributable to hand size, as there is a significant positive correlation between hand span and finger-tapping rate, and when the hands of men and women are matched with respect to size, the performance differences are no longer apparent (Dodrill, 1979).

Bimanual Music Skills

One class of non-prehensile skills that has received very little attention from the perspective of understanding human movement control is bimanual music skills; this encompasses playing stringed, wind, or percussion instruments. There have been some biomechanical studies of the patterns of joint movements involved in bowing different stringed instruments (Turner-Stokes & Reid, 1999) and of the postural changes that occur when playing wind instruments (Bejjani & Halpern, 1989). However, this research has primarily been conducted with the objective of understanding the factors that may contribute to some of the soft tissue and musculoskeletal problems that affect professional musicians (An & Bejjani, 1990).

Professional musicians who play stringed instruments have also provided an excellent subject pool for studying the mechanisms of plasticity in the motor and sensory cortices (Elbert et al., 1995; Kim et al., 2004). In this research, the representation of the individual digits has been mapped in the primary somatosensory areas, and the relation of these maps to musical experience and training has been studied. As described in chapter 3, Elbert et al. (1995) reported that the representation of the digits of the left hand, which are continuously engaged in fingering the strings when playing a violin or cello, is substantially enlarged in the cortices of string players, when compared to that of nonmusicians. Moreover, the enlargement of the cortical representation of the digits is related to the extent of their use in violin playing, with the representation of the little finger showing a greater relative increase in representation than the thumb, which grasps the neck of the violin. These studies have contributed to our understanding of cortical plasticity and should be complemented by further analysis of the manual skills acquired by expert string players.

Bimanual music skills, like most of the non-prehensile skills described in this chapter, require explicit training to develop expertise, take a long time to become an expert, but once acquired are fairly resilient to loss. There are features that are common to the performances of expert practitioners of these skills that differentiate their abilities from those of novices. In particular, the ability to plan ahead as the current movement is executed has been noted in expert pianists and typists. This enables them to execute finger-movement sequences more rapidly (typists) and consistently (pianists). Frequent and regular practice is, however, essential to the retention of these skills, as will be described in chapter 9.

8

End-effector Constraints

In medical practice, rubber fingers and gloves are occasionally used in the examination of internal organs as well as in surgery, when there is danger of infection. . . . touching via an intermediary ought to modify the tactual impression.

DAVID KATZ, 1925/1989

The manner in which the hand makes contact with the environment affects the type of information that can be extracted and the actions that can be performed. It is therefore important to expand our discussion of hand function now to consider the impact of the site and area of contact, the number of digits and hands involved, and the use of probes and other intermediate links. We will address each of these end-effector constraints as they relate to the hand-function continuum shown in figure 1.1, namely, tactile sensing, active haptic sensing, prehension, and non-prehensile skilled activities.

Tactile Sensing and Active Haptic Sensing

The nature of the end effectors that are used to contact and explore objects influences cutaneous sensitivity and perception in several ways. In this section, we consider a number of parameters that relate to the end effector(s) selected and to their consequences for human manual performance. Although the list below is not exhaustive, it addresses those parameters that research has shown to be most crit-

ical. These parameters are relevant to either tactile sensing and/or active haptic sensing, as we explain in the discussion that follows.

Site of Contact
Within Hand

Body locus strongly influences both cutaneous sensitivity and resolution. For most fingers, pressure sensitivity within the finger tends to be highest on the distal phalanx; however, it would be inappropriate to describe the sensitivity ordering as strictly proximodistal, inasmuch as the middle phalanx is less sensitive than the proximal phalanx (Weinstein, 1962). Vibrotactile sensitivity also varies with the region of the hand that is stimulated. For frequencies ranging from 0.8 to 400 Hz (Löfvenberg & Johansson, 1984) or from about 25 to 800 Hz (Verrillo, 1971), a proximodistal gradient is observed, with regions on the palm being least sensitive and the most distal portion of the fingertips being most sensitive. Within the hand itself, the dorsal surface is the least sensitive to vibrotactile stimulation (Wilska, 1954). In keeping with pressure and vibrotactile sensitivity, thermal sensitivity to warmth and cold is also markedly affected by body locus. Absolute thermal sensitivity to warming and cooling at 70 sites on the hand and forearm has been reported for two observers. Johnson et al. (1973) found that the back of the hand, the dorsal hairy skin of the fingers, and the thenar eminence (palm) are most thermally sensitive, and the fingerpads and the palmar pads at the base of the four fingers are the least sensitive. Thus, the proximodistal ordering of glabrous skin in terms of vibrotactile and, to a lesser extent, pressure sensitivity is reversed for thermal sensitivity (see also Stevens & Choo, 1998). It is possible that the greater thermal sensitivity on the back (as compared to the glabrous regions) of the hand is due to the fact that dorsal skin is thinner, as noted in chapter 2. Overall, the research indicates that the sensitivity with which individuals can detect static contact, vibration, warmth, and cold is markedly influenced by the region of the hand that is stimulated, although the specific pattern of results depends upon the form of stimulation.

With respect to the spatial-resolving capacity of the hand, the proximodistal ordering is both clear and consistent: The palm is the least spatially acute, followed by the more proximal part of the fingers, then by several sites on the distal fingerpads, each progressively closer to the tip of the finger (Craig,

1999; Craig & Lyle, 2001; Stevens & Choo, 1996; Vallbo & Johansson, 1978; Weinstein, 1968). As discussed in chapter 4, spatial acuity has been measured in terms of two-point touch, point localization, and grating-orientation thresholds. Whereas figure 4.7 highlights the proximodistal difference in two-point touch thresholds between the palm and the fingertips, figure 8.1 reveals a proximodistal trend within the index finger itself, based on the results of a two-alternative, forced-choice, tactile-orientation discrimination task (Craig, 1999). Participants were required to identify the orientation of gratings (i.e., along versus across) that were presented with a force of 1 N to the right index finger at four different sites: on the distal fingerpad at 5, 15, and 25 mm from the tip of the finger and in the center of the proximal phalanx. Spatial resolution was highest at the two sites closest to the fingertip, followed by the 25 mm site on the distal pad, and then the most proximal site near the base of the finger. In summary, the research indicates that people's ability to resolve fine spatial details systematically improves as stimulation is applied to regions of the hand along a proximodistal gradient.

To our knowledge, the temporal-resolving capacity of the hand, a topic that was discussed in chapter

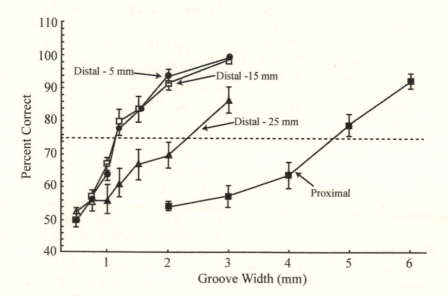

Figure 8.1. Percentage of correct discrimination of orientation as a function of the groove width of square-wave gratings. The groove width that corresponds to 75% performance accuracy is used to define threshold. The gratings were presented with 1 N of force to four sites on the right index finger: 5, 15, and 25 mm from the tip of the distal phalanx and a fourth site in the center of the most proximal phalanx. Reprinted from Craig, 1999, with permission of Taylor and Francis Ltd.

4, has not been compared across different hand sites. Such data would contribute to a more complete understanding of how the locus of stimulation within the hand influences sensory resolution.

Among Fingers

Using both the two-point discrimination and point-localization tasks described in chapter 4, Weinstein (1968) reported a nonsignificant trend in right-handed subjects for the index finger to be the most spatially acute, as compared with the middle and ring fingers of the right hand. More recently, Vega-Bermudez and Johnson (2001) employed the more sensitive grating-orientation task and discerning letter patterns to confirm that the spatial acuity of the index finger was significantly higher than that of the middle finger, which in turn was significantly higher than that of the ring finger. Less conclusive results have been obtained by Sathian and Zangaladze (1996) with the grating-orientation task, although the reason for the discrepancy with Vega-Bermudez and Johnson's results is not clear. We know of no experiments that have compared thermal sensitivity across the fingers.

People consistently perceive the roughness of surfaces to be greatest with the ring fingertip, followed in sequence by the middle and index fingertips. This effect can be simply demonstrated by using each of these fingers in turn to explore the surface texture created by the fingerprint whorls on the thumb (Lederman, 1976). While such differences in roughness perception may be related to differences in spatial acuity across the fingertips, they may also be the result of other fingertip differences, such as callus thickness.

Although there have been relatively few comparisons of sensitivity and resolution across the fingers, the small number of studies reviewed in this section indicates that the fingers do appear to vary in terms of their ability to resolve spatial details and with respect to the relative magnitude with which surface roughness is judged. In both tasks, the relative ordering of the fingers in terms of their performance tends to follow a systematic pattern, as described specifically for each task above. A more comprehensive appreciation of how this end-effector parameter may influence performance over the broader range of tasks outlined in chapter 4 awaits further research.

Between Hands

Summers and Lederman (1990) performed a meta-analysis of laterality differences in unimanual and dichhaptic tasks for all relevant studies between 1929 and 1986. (Dichhaptic tasks involve presenting two objects to the hands simultaneously.) Only a few of the studies considered included left- as well as right-handed subjects. During that period, hand superiorities in cutaneous sensitivity were rarely observed; that is, no significant differences were documented between the left and right hands in terms of two-point touch, point-localization, pressure, sharpness, vibrotactile, or tickle thresholds. More recently, Vega-Bermudez and Johnson (2001) have confirmed that there is no difference in spatial acuity between the hands. With respect to the few studies that have reported a hand difference, Weinstein and Sersen (1961) and Weinstein (1962) both showed that pressure sensitivity on the palm and fingers of the left hand of right-handed subjects is consistently higher than that of the right hand. With respect to vibrotactile sensitivity, Rhodes and Schwartz (1981) have shown that for frequencies ranging from 70 to 480 Hz, the left index finger is more sensitive than the right index finger.

The perception of various object properties has also been examined in terms of laterality differences with somewhat mixed results. For example, no differences between the hands have been observed when right-handed observers judge surface roughness (Lederman, Jones, & Segalowitz, 1984). Of the 117 spatial tasks that predicted a left-hand (i.e., right-hemisphere) advantage on a priori grounds, Summers and Lederman (1990) considered only 16 to be methodologically sound. These included form-board tasks, haptic illusions, haptic mazes, a conflict-drawing test, tactual face identification, spatial finger tapping, orientation, target tracking, and nonsense-shape matching tasks. Of those 16 tasks, 9 showed a left-hand superiority, 7 produced null results, and none showed a right-hand superiority. Of the 50 results that predicted a right-hand (i.e., left-hemisphere) advantage, only 11 were considered methodologically valid. These included sequential localization; sequential finger tapping; haptically perceived words, letters, and numbers; and verbally encoded shapes. Of those 11 studies, 5 showed a right-hand superiority, 6 produced null results, and none showed a left-hand superiority. Of the 27 acceptable studies, the large majority used right-handed participants. Summers and Lederman (1990) concluded:

> [A] very clear picture emerges that is in fact consistent with visual and auditory perceptual asymmetries. . . . there is a perceptual asymmetry in these studies, favoring the left hand on tasks that would seem to demand spatial mediation, and favoring the right hand in tasks that would seem to demand verbal mediation. These differential abilities are not robust, however, as there is good support for roughly as many null results as for predicted left- or right-hand superiorities. (p. 221)

Contact Area and Duration

On both the thenar eminence and the fingerpad of the right middle finger, vibrotactile sensitivity increases at the rate of approximately 3 dB per doubling of area for frequencies above 40 Hz (Verrillo, 1963). This is attributable to spatial summation. For contactor areas above 1 mm^2, spatial summation is greater on the thenar eminence than on the fingertip (Gescheider, Bolanowski, Pope, & Verrillo, 2002); however, for very small contactor areas and for lower frequencies, little spatial summation has been observed. Sherrick and Cholewiak (1986) subsequently noted that Verrillo (1963) did not partial out the effect of static force. On the basis of the evidence available at that time, they concluded that for vibrotactile stimulation, the spatial summation documented in tests of absolute sensitivity is small, at best. More recently, however, Brisben et al. (1999) have challenged this conclusion. In their study, observers sensed vibration either by grasping a vibrating cylinder (like the handle of a tool) or by means of a probe that was mounted on a linear motor with the 1-mm tip glued to the skin to ensure that it followed the sinusoidal motion of the probe when it was vibrated perpendicular and parallel to the skin surface. Vibrotactile sensitivity for the former was 23% lower at 40 Hz and 61% lower at 300 Hz, as compared to the latter condition. These results confirm that the area of skin contact and, hence, spatial summation is indeed an important factor in vibrotactile sensitivity, presumably mediated by FA II units.

Temporal summation of vibrotactile stimulation is affected by contact duration and, in some cases, contact area (e.g., Gescheider, Berryhill, Verrillo, & Bolanowski, 1999; Green, 1976; Verrillo, 1968).

More specifically, vibrotactile sensitivity grows with increasing stimulus duration or number of stimuli in a sequence, particularly for higher frequencies and larger-sized contactors. Presumably these effects are also mediated by the FA II system.

The thermal sensitivity of the skin to both warmth and cold is clearly enhanced by both spatial and temporal summation, although we have found no studies that directly involve the hand. The skin summates stimulus intensity over increasingly larger areas (Greenspan & Kenshalo, 1985; Kenshalo, Decker, & Hamilton, 1967). Doubling the contact area doubles thermal sensitivity up to a temperature that is close to the pain threshold (Greenspan & Kenshalo, 1985), indicating that spatial summation is almost complete; however, more recently, Green and Zaharchuk (2001) have challenged this interpretation (see chapter 4). In addition, the skin summates temporally over increasingly longer durations for both warmth (Stevens, Okulicz, & Marks, 1973) and cold (Bujas, 1938).

It is evident that the area and duration of skin contact on the hand critically influence vibrotactile and thermal sensitivity in terms of how both mechanical and thermal energy are summated across space and time.

Number of Digits Stimulated

Because the tips of the instruments used to test cutaneous spatial acuity and pressure sensitivity are smaller than a single fingertip, the number of digits is not highly relevant to these performance measures. However, when both the thumb and index fingerpads are simultaneously used to explore a surface actively, the magnitude of roughness perceived with the index finger is greater when two fingers are stimulated than when the index fingerpad is stimulated on its own (Verrillo, Bolanowski, & McGlone, 1999). These results suggest that a perceptual enhancement effect occurs due to spatial summation at these sites. Given the strong evidence for spatial summation of thermal stimulation as well, we would expect similar results for thermal sensitivity at threshold and suprathreshold intensity levels as the number of digits stimulated increases. In addition, the perceived heaviness that results from a weight being actively lifted by flexing the distal joint of a single digit increases when an adjacent finger lifts a weight simultaneously (Kilbreath & Gandevia, 1991, 1992).

Because this occurs for weights that constitute either a relatively small (3–5%) or much larger (20–25%) percentage of the total maximal voluntary force, the increase in perceived heaviness seems to operate across a wide range of muscle forces generated by the hand. With respect to the tactile processing of spatial characters, however, there is little empirical support for the notion that increasing the cutaneous field of view by displaying the information across two adjacent fingers improves either the perception of spatial patterns or tactile reading (e.g., Bliss, 1978; Lappin & Foulke, 1973; Loomis, Klatzky, & Lederman, 1991). This finding suggests that, unlike visual processing, passive tactile sensing of characters is limited to one character at a time. Although both spatial and temporal interaction effects occur, they are greater when patterns are presented within a single finger rather than across two fingers of the same hand (Craig, 1983b, 1985). Unlike tactile sensing of geometric features, whole-hand haptic exploration produces faster and more accurate identification of common objects classified by their geometric properties than does manual exploration that is limited to a single finger (Klatzky et al., 1993). Presumably, the decline in performance observed when a single end effector is employed is attributable to the marked reduction in cutaneous and kinesthetic cues during single-finger exploration.

In summary, the research literature reveals important ways in which the number of digits employed influences manual performance. The effects range from those related to simple sensitivity and resolution measures to the more complex interdigital interactions that affect pattern perception via tactile sensing and weight perception and object recognition via active haptic sensing.

Unimanual Versus Bimanual

Is there any advantage to using two as opposed to only one hand when haptically exploring objects and their surfaces? It is unlikely that both hands are necessary, or even helpful, for extracting information about surfaces and the material properties of objects, such as texture, compliance, and thermal characteristics. However, people generally switch from using one to two hands as the size of the object or display increases beyond that of a single hand. Clearly, the assessment of weight requires two hands when the object to be lifted is large or relatively heavy. In addi-

tion, both hands can be, and often are, used to explore the shape, symmetry, and size of an object when it is externally stabilized. Research has also documented a two-hand advantage in Braille reading and in tactile pattern-recognition tasks (Craig, 1985; Craig & Xu, 1990; Lappin & Foulke, 1973). Craig (personal communication, June 1991) attributes the two-hand superiority to greater processing flexibility resulting from extra resources for the second hand, which may serve as a separate limited-capacity processor. Note that these tasks may be performed using only the cutaneous inputs produced by the pattern being moved across passive fingers.

When the objects are fixed in place, it is possible for people to use either one or two hands actively to detect the presence of symmetry. The use of two index fingers to detect the bilateral symmetry of small (fingertip-sized), raised, two-dimensional shapes (figure 8.2A, B) and two hands to explore larger, unfamiliar, wooden, three-dimensional objects (figure 8.2C, D) improves performance relative to one-handed exploration (Ballesteros, Manga, & Reales, 1997). The observers were required to decide as quickly and accurately as possible whether a given shape was symmetric or asymmetric. Ballesteros et al. propose that parallel hand movements offer additional haptic spatial information relative to the body midline for use in judging symmetry within a body-centered frame of reference.

If the object is not fixed in place, one hand must stabilize the object while the other hand manually explores it (Lederman & Klatzky, 1987). Under such circumstances, each hand serves both executive and sensory functions simultaneously. Thus, in addition to its motoric functions, the stabilizing hand can provide the sensory information that is normally available when the hand statically grasps an object using an enclosure EP. An important avenue for future research on haptic perception involves understanding how the two hands are coordinated in space and time during manual exploration for purposes of successful identification of objects and their concrete properties.

Direct Versus Remote Sensing

When tactile sensing involves direct contact between the stationary hand and an object or surface, the full range of mechanical and thermal inputs is available

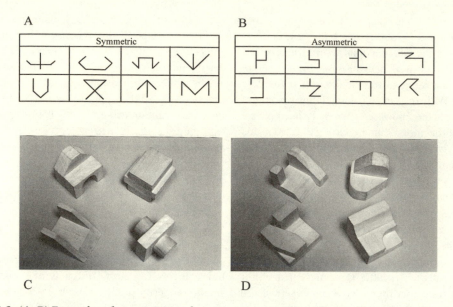

Figure 8.2. (A, B) Examples of symmetric and asymmetric raised-line shapes (fingertip-sized), respectively. (C, D) Examples of unfamiliar symmetric and asymmetric wooden objects, respectively. Reprinted from Ballesteros, Manga, & Reales, 1997, with permission of the Psychonomic Society.

for further processing (e.g., net forces, spatially distributed forces, vibrations, and temperature changes). Active haptic perception may involve direct contact with or without movement of the hand. When there is no movement, the mechanical and thermal inputs that are available during tactile sensing are available; however, when observers voluntarily move their hand(s) over an object or surface, additional kinesthetic feedback and information from corollary discharges become available.

Active haptic perception is also achieved through indirect contact by means of some intermediate link (e.g., 1-degree-of-freedom thimbles and probes and multidegree-of-freedom handles and grippers). Such tools, if not artificially augmented with force feedback (see chapter 10), offer only reduced tactile (i.e., static net forces and vibratory signals) and kinesthetic feedback.

In chapter 5, we emphasized the importance of manual exploration for haptic perception. We also briefly alluded to significant consequences for haptic object recognition when manual exploration is constrained. We consider this issue now in greater detail. Lederman and Klatzky (2004) compared the effects of imposing five different constraints on the haptic recognition of common objects that were defined primarily by their geometric features. Of the

five constraints, three direct-touch constraints were initially considered by Klatzky et al. (1993): limiting the number of end effectors during exploration, wearing a thick compliant finger cover, and splinting the end effector(s). Lederman and Klatzky (2004) subsequently imposed two additional constraints that both involved remote haptic exploration: wearing a rigid finger cover and using a rigid probe. In table 8.1, we consider the five constraints in terms of how each differentially restricts the normally available tactile, thermal, and kinesthetic inputs. Figure 8.3 shows the effects of these constraints, singly and in specific combinations, on the accuracy and speed with which common objects were recognized in these two studies.

Reducing the number of end effectors from five fingers to one finger (constraint 1) impaired accuracy only a little but increased response time substantially, presumably because of the need for sequential exploration. The effect of this constraint was to eliminate distributed tactile spatial inputs across multiple end effectors and to reduce kinesthetic spatial and temporal cues (see also Jansson & Monaci, 2004). When observers wore a thick compliant skin cover on their hands (constraint 2) as opposed to when they used their uncovered hand under each of three different conditions of manual exploration (i.e., whole hand

Table 8.1. Constraints on manual exploration and associated reductions of somatosensory cues

Constraint number	Nature of manipulation	Cutaneous			Kinesthetic, spatial, and temporal
		Spatial	Temporal	Thermal	
1	Reduced number of end effectors	X			X
2	Compliant covering	X	X	X	
3	Rigid finger splinting				X
4	Rigid finger sheath	X		X	X
5	Rigid probe	X		X	X

Reprinted from Lederman & Klatzky (2004), with permission of the Psychonomic Society.

splinted, whole hand unsplinted, and single finger splinted), recognition accuracy tended to decline and response time to increase, particularly in the single-finger-splinted condition. In each case, the compliant cover eliminated tactile (spatial and temporal) and thermal cutaneous cues, while leaving kinesthetic cues intact. Rigidly splinting (constraint 3) the bare finger affected performance to the same degree as using the bare unsplinted finger because the finger was outstretched in both cases. Such a constraint impaired accuracy relatively little, but considerably increased response times, presumably because observers were again forced to explore sequentially with limited access to the spatial and temporal kinesthetic cues normally provided when the finger molds to the object's contours. When a rigid finger sheath (constraint 4) on a single finger was used for exploration, both accuracy and response time were notably

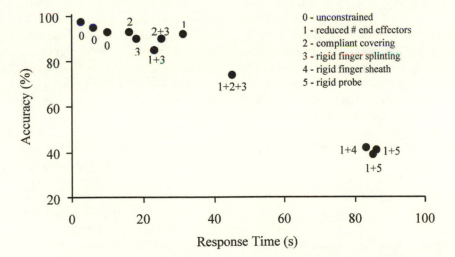

Figure 8.3. Recognition accuracy (%) as a function of response time (seconds) for a control and five different constraints on manual exploration, presented singly and in some combinations. The constraints are listed by both name and number in the legend. The data are drawn from the experimental conditions in Klatzky, Loomis, Lederman, Wake, & Fujita, 1993; and Lederman & Klatzky, 2004. The + sign indicates the existence of multiple simultaneous constraints. Reprinted from Lederman & Klatzky, 2004, with permission of the Psychonomic Society.

impaired, again because of the loss of spatial and temporal kinesthetic cues produced as the finger conformed to the object's contours. Finally, when subjects were required to use rigid probes with small- or large-diameter spherical tips (constraint 5), accuracy was markedly impaired, and response times were prohibitively slow relative to an unsplinted-single-finger control (Lederman & Klatzky, 2004). In this case, the impairments can be attributed to one or more of the following factors: elimination of spatially distributed force patterns on the skin (i.e., single-point force information), elimination of thermal cues, and the need to integrate and remember kinesthetic spatial information via sequential point contacts. Finally, we note that imposing multiple simultaneous constraints on manual exploration clearly impaired the accuracy and/or speed with which common objects were recognized compared to any single-constraint condition. The implications of constraining manual exploration are explored further in chapter 10, where we focus on more applied issues.

Last, we address a form of remote sensing that Turvey (e.g., 1996) has described as dynamic touch. As explained in chapter 5, this mode of perceptual exploration involves the use of a rod held statically in the hand or wielded for purposes of learning about the properties of objects (e.g., length, orientation, weight, and the position of the hand relative to object). Such a form of sensing is thus specifically dependent on sensory inputs arising from the kinesthetic system. While kinesthetic feedback is undoubtedly required for tool manipulation and remote haptic sensing, such an approach is necessarily limited by the fact that it does not take into consideration the valuable contribution of the cutaneous inputs (e.g., vibration) produced when the hand grasps and manipulates the tool in contact with the concrete environment (see Brisben et al., 1999; Johansson & Westling, 1984).

Prehension

Most prehensile activities involve grasping an object in the hand, and so the properties of the object (e.g., size, weight, texture, and shape) and the task objective usually determine how the object is held, the area of contact between the object and the hand, and the number of digits involved in the grasp. These will often change during the course of an activity in which the hand may switch from one posture to another in order to accomplish the task, such as picking up a pen and then writing. Detailed electromyographic studies of muscle activation during different tasks indicate that there are marked shifts in the patterns of muscle use as the hand changes from one grasp pattern to another (Long, Conrad, Hall, & Furler, 1970).

Site of Contact

There have been numerous attempts to develop classificatory systems that characterize the diverse range of postures that the hand adopts to perform different tasks (e.g., Cutkosky & Howe, 1990; Elliot & Connolly, 1984; MacKenzie & Iberall, 1994; Napier, 1956). In general, the posture of the hand determines the fingers that are used to make contact with the object and hence the site of contact. Most of these taxonomies distinguish between the two dominant prehensile postures, that is, the power grip and the precision grip, each of which may take several forms, as shown in figure 8.4. The power grip is used when force is the primary objective, such as when using a hammer. In this grip there is usually a large area of contact between the grasped object and the palmar surfaces of the fingers and palm, but little movement of the fingers; hence the grasp is stable and very resistant to slipping. Movement of the object held in a power grip is achieved by moving the whole hand, using the wrist and arm. In contrast, the precision grip involves grasping an object between the tips of the thumb and index finger, and sometimes also the middle finger, in such a way that there is precise control of the position of the object and the grasping forces.

Three types of precision grip, shown in figure 8.5, are commonly distinguished: the tip, palmar, and lateral pinches. The tip pinch is used when a small object is held between the pulp surfaces of the opposed thumb and index finger. In this grasp, the thumb and index finger may be flexed or extended, depending on the task objective. The palmar pinch is a prehensile pattern in which the pulp surface of the thumb opposes the fingerpads of the index and middle fingers. The lateral, or key, pinch involves contact between the thumb pulp and the lateral surface of the middle phalanx of the index finger. The most commonly used prehensile grips are the tip and lateral pinches. Although the maximum forces produced by

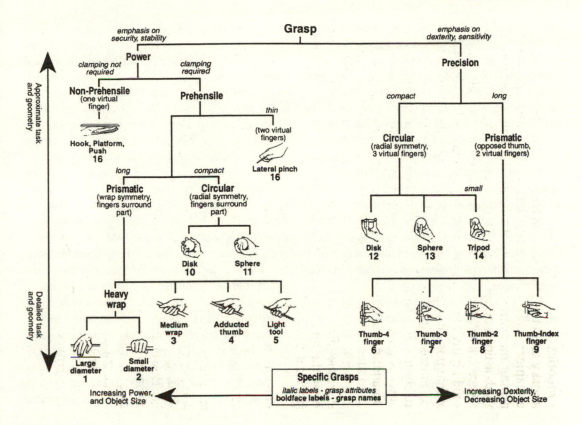

Figure 8.4. Taxonomy of grasps revised from Cutkosky & Wright, 1986. Reprinted from Cutkosky & Howe, 1990, with permission of Springer-Verlag and the authors.

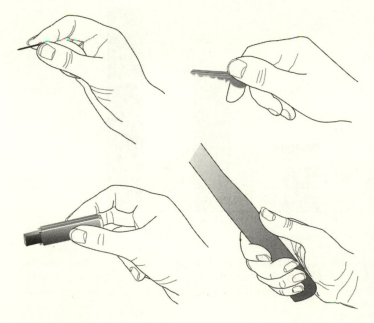

Figure 8.5. Most commonly used prehensile patterns. The tip pinch between the pulps of the opposed thumb and index finger is illustrated in the upper left; the lateral pinch in which the thumb pulp is in contact with the lateral surface of the middle phalanx of the index finger is shown in the upper right; the palmar pinch in which the pulp surface of the thumb opposes the fingerpads of the index and middle fingers is depicted on the lower left; and the power grip is shown on the lower right.

these prehensile grips can be large, ranging from 50 to 120 N depending on the grasp, age, and sex of the person (Mathiowetz et al., 1985), for most of the tasks in which these grips are used the forces generated at the fingertips are small. Swanson, De Groot Swanson, and Göran-Hagert (1990) estimated that a pinch force of 10 N is sufficient for performing most of the simple grasping tasks involving the fingers. At very low forces, the fingertip contact area increases sharply with increasing normal force, but by 1 N it is almost 70% of the contact area at 5 N (Serina et al., 1997), as described in chapter 2 and shown in figure 2.10. This means that if a normal force of 1–2 N at the fingertips is adequate to maintain stable contact between the fingers and an object, there is little additional tactile information available at higher forces as the increase in contact area is small.

If the hand does not have to support the object, then the finger forces used to explore an object depend on the property being perceived, as illustrated in figure 8.6. The contact forces used by people to estimate the roughness of raised-dot surfaces range from 0.8 to 1.6 N with an average of 1 N (Lederman, 1974; Meftah, Belingard, & Chapman, 2000), whereas the forces applied when vertically pressing the finger against a surface to perceive its temperature have been found to vary from 0.5 to 2 N, with an

average of 1.5 N (Ino et al., 1993; L. A. Jones & Berris, 2003). Much lower forces are used when people are asked to rate the surface friction of macroscopically smooth, flat surfaces (0.2–0.3 N; Smith & Scott, 1996) or to find a small feature in an otherwise flat surface (0.5 N; Smith, Gosselin, & Houde, 2002).

In addition to the power and precision grips, a number of other grips are used, which are generally described in terms of the posture of the hand (i.e., hook grip, interdigital grip, pliers grasp) or the shape of the object being held (e.g., cylindrical grasp, spherical grasp). Some of these grasps, such as the hook grip used to hold a suitcase and the platform grasp that a waiter may use to hold a tray, are nonprehensile as the hand is primarily providing support for an object. Because the hand usually adopts different grasps during the performance of a task, it has been difficult to develop taxonomies of hand function that can predict hand grasps from a specification of the task requirements and object geometry. It has been possible to predict the general types of grasps used to hold tools when performing specific actions, but these are highly dependent on the actions being performed and so are not applicable generally (Cutkosky, 1989).

A number of studies have addressed the issue of whether the selection of a specific hand configura-

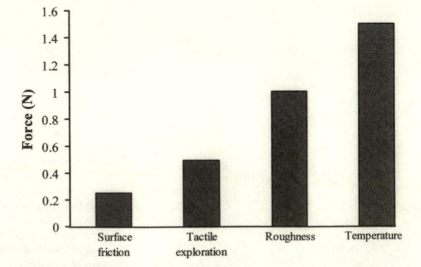

Figure 8.6. Average contact forces measured at the fingertip when people are required to make judgments about the following properties: surface friction (Smith & Scott, 1996), small feature in smooth surface (A. M. Smith et al., 2002), roughness (Lederman, 1974; Meftah et al., 2000), and temperature (Ino et al., 1993; L. A. Jones & Berris, 2003).

tion used to grasp an object reduces the degrees of freedom of the fingers. Arbib et al. (1985) proposed the concept of virtual fingers to describe the configuration of the fingers as they apply forces to a hand-held object. On one side, there is the thumb or palm of the hand and, on the other, a virtual finger that may be made up of one or more fingers that act in synchrony and are controlled as a unit. These two virtual fingers then interact with the object through an opposition space, which occurs between two hand sites that are defined in relation to a coordinate frame placed on the palm. The theory of virtual fingers provides a useful framework for classifying how objects are grasped and for considering how to control robotic hands (Iberall, 1997).

Although the concept was originally developed to describe the kinematic aspects of grasping (Iberall & MacKenzie, 1990), a number of experiments have supported the general concept of virtual fingers in terms of digit force control. For example, Baud-Bovy and Soechting (2001) found that the three-fingered tripod grasp can be viewed as a two-virtual-finger grasp involving one opposition axis. In their experiment, the sum of the index and middle finger forces always acted in the same direction independently of the orientation of the contact surfaces, and so the index and middle fingers could be replaced by a single virtual finger placed midway between the two real fingers. Using a different five-digit apparatus, Zatsiorsky, Gao, and Latash (2003) observed that the forces generated by four fingers combined (i.e., the virtual finger) were equal to that produced by the thumb. This scaling of finger forces was maintained independently of the torques being generated and the geometry of the object being grasped.

Number of Digits

The primary contributor to the versatility of the human hand is the opposable thumb, which is essential for the effective handling and exploration of small objects. Opposition of the thumb involves flexion, abduction, and medial rotation about the saddle joint between the trapezium and the first metacarpal, so that the surface on the thumb pad can make contact with the terminal phalanges of one or a number of other digits. The extensive area of contact between the fingerpads of the thumb and index finger is a uniquely human characteristic. It is important to remember, however, that it is the development and elaboration of the central nervous system and not the specialization of the hand that provides the substrate for human manual skill (Lemon, 1999).

It is estimated that there is a 40% loss in the functional capacity of the hand following amputation of the thumb (Reid, 1960; Swanson, 1964). This estimate is surprisingly close to King Canute's rate of compensation for the loss of a thumb in the 11th century, which was 30 solidi, representing 36% of the total amount allocated for loss of all of the fingers (Bertelsen & Capener, 1960; Napier, 1993). In King Canute's time, rates of compensation for the loss of digits were well established as they had been decreed in the 7th century by the first Anglo-Danish king of Kent, Aethelberht. At that time, the thumb was given the same value as all of the other fingers together. One thousand years later, there have been few changes in the ratings of the functional importance of the various fingers, as shown in table 8.2, with the exception of the ring finger (annularis), to which King Canute ascribed a high value (18 solidi),

Table 8.2. Compensation for loss of digits

Digit	King Canute (ca. 1000)[a] (solidi)	UK Ministry of Pensions (1947)[b] (%)	AMA (1993)[c] (%)
Thumb	30	30	40
Index	15	14	20
Middle	12	12	20
Ring	18	7	10
Little	9	7	10

[a]Bertelsen & Capener (1960).
[b]Ministry of Pensions (1947).
[c]American Medical Association (1993).

possibly because it had some mystical significance (Clarkson & Pelly, 1962).

The loss of a thumb in modern times is considered so disabling that reconstructive surgery involving transferring the great toe to the hand is occasionally performed so that the hand has some opposable capacity (Foucher & Chabaud, 1998; Robbins & Reece, 1985). This procedure was first attempted in 1897, with only limited success, by Nicoladoni, who transferred the second toe to the hand of a 5-year-old boy. Amputation of either the index or middle finger is considered to result in a 20% loss of hand function, whereas the functional capacity of the hand is reduced by only 10% when either the ring or little finger is amputated (Swanson, 1964). These ratings of functional loss are consistent with the observation that most manual tasks can be accomplished using three or four fingers, provided that one of these digits is the thumb. The three- or four-fingered hand is also the preferred design for most dexterous robotic hands that are capable of performing a variety of skilled tasks (L. A. Jones, 1997). From a robotic viewpoint, three fingers are the minimum required for a stable grasp by a manipulator (Brock, 1993; Howard & Kumar, 1996).

The grasp used to hold an object depends on the size and shape of the object and the intended use. When reaching for objects of varying size, people tend to use a three-fingered grasp, such as the tripod grasp, to pick them up and occasionally use a pinch grip for smaller objects. The size of the object does not affect the final spatial position of the thumb on the object but does influence the position of the other two digits involved in the grasp, which suggests that the control of these two digits changes with object size (Gentilucci et al., 2003). As the size and mass of the object increase, four rather than three digits are used to pick it up. These transitions from three to four fingers and then from one to two hands can be predicted from the mass and length of both the object and the hand reaching to grasp it (Cesari & Newell, 2000).

Studies of the coordination of finger forces during prehensile grasping with three, four, or five digits reveal that as object weight increases or surface friction decreases, the largest change in force comes from the index finger. Kinoshita, Kawai, and Ikuta (1995) reported that the percentage contributions of individual finger force to total grip force for the index, middle, ring, and little fingers were 42%, 27%, 18%, and 13%, respectively, and that these percentages remained relatively constant as the weight of an object increased from 200 to 800 g. In absolute terms, this means that the heavier the weight being lifted, the larger the index and middle fingers' contributions to grip force. As the grip mode changed from five to four to three fingers, the total grip force increased, but the force contributed by the index finger remained constant at around 43%, and so the forces produced by the ring and middle fingers increased. Finger forces also changed with the grasping surface, as has been described previously (Johansson & Westling, 1984), with the total grip force for a sandpaper surface being about 60% of that for plastic, which has a much lower surface friction. However, the relative contribution of each finger to the overall force remained essentially the same, independent of the surface texture (Kinoshita et al., 1995). A further study of finger forces during shaking tasks showed a similar pattern of results with the index finger again providing the largest grip force among the four fingers, regardless of the speed or direction of the shaking movements (Kinoshita, Kawai, Ikuta, & Teraoka, 1996). In all of these studies, the force produced by the thumb was equal to the sum of the opposing finger forces.

These findings suggest that the control of individual finger forces during grasping is achieved using a relatively simple scaling strategy that maintains constant the relative contributions of various fingers to total grip force. This strategy simplifies force control in a task having many degrees of freedom by creating fixed relations between the forces produced at the fingertips, as shown in figure 8.7, where the contribution of each finger to the total grip force and load force in the static phase of grasping is illustrated (Reilmann, Gordon, & Henningsen, 2001). As the contact area with an object decreases, due to a reduction in the number of digits involved in lifting, the grip force increases, which suggests that a larger contact area across the fingers may facilitate the fine adjustment of finger forces. The particular contribution of individual finger forces to the total grip force can, however, change under some conditions. For example, with heavier objects weighing 2 kg or more, it has been found that the middle finger produces the largest force followed in descending order by the index, ring, and little fingers (Radwin, Oh, Jensen, & Webster, 1992). Also, when the grasp configuration changes so that the thumb is no longer in

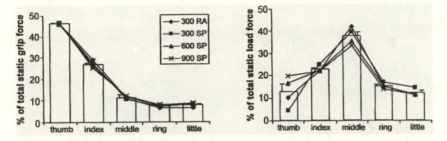

Figure 8.7. Contribution of each digit to the total grip force and total load force in the static phase of grasp (see figure 6.5) averaged across 16 subjects for four conditions: grasping surface textures of rayon (RA) or sandpaper (SP), and three object weights (300, 600, and 900 g). Reprinted from Reilmann et al., 2001, with permission of Springer-Verlag and the authors.

direct opposition to the fingers, which occurs when grasping a cylindrical object at the top with the whole hand, the thumb and ring and little fingers make the greatest contribution to the total grip force, and the index finger the smallest. As the number of fingers used to grasp the cylindrical object increases from two to five, the total grip force decreases by 30% (Kinoshita, Murase, & Bandou, 1996). This may result from a reduction in the safety margins employed at each finger due to the increased stability of the grip, or it may reflect the increased surface contact area with the object, which should facilitate force control.

When an object is grasped, lifted, and held with the whole hand, the time course of the forces produced by the fingers is similar to that described for grasping movements involving the thumb and index finger (see chapter 6). Santello and Soechting (2000) found that the normal forces produced by the fingers co-varied and that the forces exerted by each finger depended on the location of the object's center of mass. When the object was held in a stable position, the fluctuations in force, although small, were highly synchronized across the hand.

Bimanual Grasping

Many prehensile activities require that one hand stabilize the object while the other, usually the preferred hand, manipulates it. For most of these tasks, the coordination is primarily constrained by the physical properties of the object and the task objective. There is both temporal and spatial coordination of the hands to achieve these task goals, so that a disruption in the capacity of one hand to execute part of a movement

sequence is readily compensated for in the timing of the movements of the other hand (Wiesendanger, Kazennikov, Perrig, & Kaluzny, 1996). There is also anticipatory control during activities that require bimanual coordination. In a task involving dropping a ball with one hand into a receptacle held by the other, Johansson and Westling (1988) reported that there was an increase in prehensile forces 150 ms prior to the impact of the ball and that the receptacle was raised to meet the impact of the ball. These anticipatory actions are precisely timed, do not depend on visual feedback, and are scaled to meet the dynamics of the task (i.e., weight of the ball and distance it is dropped). Moreover, information obtained by one hand (the hand holding the ball) is available to the other hand so that appropriate adjustments in grip force can be made.

Some bimanual tasks, such as juggling, have very precise spatial, temporal, and physical constraints, with even the smallest variation in the performance of the two hands leading to errors and collisions between the objects juggled (Beek & Lewbel, 1995). At the same time, juggling patterns are intrinsically variable, and no two throws or catches are exactly the same (van Santvoord & Beek, 1996). To avoid errors, jugglers attempt to throw the objects (rings, balls, plates, hoops, clubs) as consistently as possible and to control variables, such as the angle and velocity of the release and the location and height of the throws (Beek & van Santvoord, 1996). Analyses of the trajectories of the balls and hands during juggling reveal that the focus of control is on reducing the spatiotemporal variability of the balls' trajectories rather than that of the hands. The balls are thrown to a relatively fixed height and caught in such

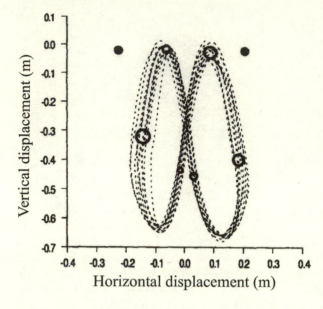

Figure 8.8. Example of the ball trajectories from an expert juggler in a spatially constrained condition in which the juggler attempted to throw the balls to a specific height (0.5 m) as indicated by two targets (solid circles) suspended from the ceiling. The top two open circles represent the mean vertical distance of the zeniths of the ball flights from the target location; the middle two open circles represent the mean spatial location for catches; and the bottom two open circles represent the mean spatial location for throws. Position data were filtered at 10 Hz. Reprinted from van Santvoord & Beek, 1996, with permission of Elsevier.

a way that variations in flight time are minimized, as illustrated for an expert juggler in figure 8.8. The performance of intermediate and expert jugglers can be distinguished on the basis of variability in spatial variables, such as the angle of release and the width of the elliptical hand loops, and not temporal variables, such as the height and velocity of the release of the ball (van Santvoord & Beek, 1996).

Haptic and visual information about the movements of the hands and balls is required for successful juggling, but expert performers appear to depend more on the sensations associated with contact between the hands and balls than do novice or intermediate jugglers, who rely predominantly on visual feedback (Beek & Lewbel, 1995). With extensive practice, it appears that the sight of the moving ball has calibrated the tactile and proprioceptive senses so that slight deviations in the angle or velocity of release can be detected from haptic feedback. This would account for the performance of expert jugglers who are capable of juggling for several minutes when blindfolded.

Remote Versus Direct Action

Prehensile activities involving object manipulation frequently involve the use of tools, and so the sensory information available to the human operator is limited by the physical dimensions of the tool and whether it is a passive tool, such as a screwdriver, or an active system, like a power drill. Many of these tasks involve stabilizing an object, such as a screw or piece of wood, with one hand, while the other hand holds the tool and applies force through the tool to the object. The neuromuscular control system readily adapts to the changes in the dynamics of the hand and the tool, so that the tool is treated as an extension of the limb in both a physical and perceptual sense, a point made by the philosopher Michael Polanyi:

> Our subsidiary awareness of tools and probes can be regarded now as the act of making them form a part of our own body. The way we use a hammer or a blind man uses his stick, shows in fact that in both cases we shift outwards the points at which we make contact with the things that we observe as objects outside ourselves. While we rely on a tool or probe, these are not handled as external objects. (Polanyi, 1958, p. 59)

This incorporation of an external object into the perceptual map of the human body, often referred to as the body schema, is described in countless contexts from playing a musical instrument to driving a backhoe (Wilson, 1998). The integration of the body with a tool has demonstrated neural correlates as shown by functional brain mapping (Obayashi et al., 2001) and neuronal recording studies (Iriki, Tanaka, & Iwamura, 1996) in monkeys. Both of these studies reported that when monkeys used a rake-shaped tool to retrieve food, the visual response properties of bimodal (tactile and visual) neurons in the intraparietal cortex were modified to include the hand-held tool, a response not seen when the monkeys simply reached for food with their hands. As described in chapter 3, the cortical representation of the body is dynamic, not static, and it appears that external objects such as tools can be incorporated into these cortical maps.

Non-prehensile Skilled Movements

How Independent Are the Fingers?

Most muscles controlling movements of the fingers and wrist act over many joints, and so the forces and movements produced by these muscles are not controlled independently, and coactivation is common (Schieber & Santello, 2004). The inability to flex a single distal joint in the hand without also moving adjacent digits is often observed when playing the piano or typing (Fish & Soechting, 1992). It reflects both neural factors related to the recruitment of multidigit motor units in extrinsic hand muscles and to mechanical linkages between anatomically distinct muscles and portions of muscles with several tendons (Kilbreath, Gorman, Raymond, & Gandevia, 2002). As described in chapter 2, three of the extrinsic muscles in the forearm, the flexor digitorum profundus, flexor digitorum superficialis, and extensor digitorum communis, connect to each of the four fingers, and within these muscles there are multidigit motor units (Kilbreath & Gandevia, 1994; Reilly & Schieber, 2003). The degree to which each of these muscles can produce force and movement in an isolated finger depends on the extent of compartmentalization within the muscle. The limitations imposed by these mechanical linkages is perhaps best illustrated by the 19th-century surgical practice of dividing the accessory tendons that bind the ring finger to the neighboring fingers in pianists (Parrot & Harrison, 1980), so that the ring finger had an increased range of independent movement. Häger-Ross and Schieber (2000) quantified the degree of independence of the fingers in a motor task requiring isolated finger movements and reported that the thumb and index finger are the most independent digits and the ring and middle fingers the least independent.

In terms of force production, a number of factors, both neural and mechanical, limits the maximum forces that can be produced when the fingers act synergistically. This means that the maximum force produced by a finger decreases as a function of the number of other fingers concurrently generating peak forces. The effect of concurrent activation is considerable, as shown by the 37% increase (range: 22–56%) in the maximum forces produced individually by the fingers as compared to the maximum pinch strength measured when all four fingers flex simultaneously (Ohtsuki, 1981a). The decrease in the maximum force that an individual finger can produce with concurrent activation of one or a number of adjacent fingers is illustrated in figure 8.9. This decline in each finger's maximum force as more fingers are involved in the task is assumed to reflect a limit in the central neural drive delivered to a set of effectors, so that the more effectors involved, the less the share received by each effector and the larger the deficit in force production (Li, Latash, & Zatsiorsky, 1998). Finger forces are also influenced by the simultaneous activity of the other hand. When the left hand generates a maximum force while the individual fingers of the right hand are producing forces, the average decrease in finger force is 13% (Ohtsuki, 1981b). If maximum forces are produced with a varying number of fingers in both hands, the decline in force observed during one-handed tasks is further exacerbated, with decreases in force averaging around 12%. This decline in force depends on the distribution of fingers across the two hands and is larger for asymmetrical than symmetrical tasks (Li, Danion, Latash, Li, & Zatsiorsky, 2001).

The mechanical and neural coupling between the fingers that influences maximum force production also affects the ability to produce isolated forces independently with the fingers. Reilly and Hammond (2000) measured the degree of independence in force production by measuring the extent of the coactivation of other fingers in the hand when participants were instructed to produce a flexion force (ranging from 40 to 80% of the maximum finger force) with a target finger. They found that the extent of coactivation was least when the thumb produced the target force and increased progressively when the index, middle, little, and ring fingers generated the target force. These findings show that the ability to

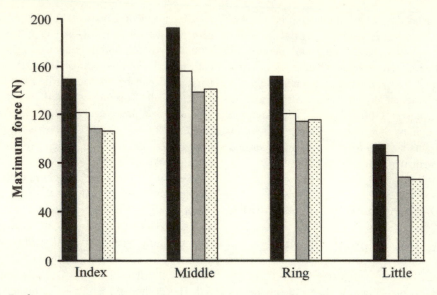

Figure 8.9. Decline in maximum force with concurrent activation of adjacent fingers. Solid bars represent the maximum force produced by each finger when activated alone; white bars are the forces produced by the same finger when an adjacent finger is simultaneously producing a maximal force; gray bars represent the force produced by the finger when two additional fingers are also producing a maximal force; and dotted bars represent the force produced by the finger when all four fingers are maximally activated. Data are the means from 3–10 participants and are from table 1 in Ohtsuki, 1981a.

confine force production to a single digit is limited at higher finger forces. Even at the considerably smaller forces used in many functional activities, coactivation of adjacent fingers can occur (Kilbreath et al., 2002).

It is not only the motor apparatus of the hand that functions as a unit, but its sensory feedback systems also do so. Afferent discharges originating in cutaneous and muscle mechanoreceptors are distributed to various groups of motor neurons in the spinal cord. This means that the sensory signals arising from cutaneous receptors in one digit, such as the index finger, influence not only the motor commands sent to that finger but also those sent to adjacent digits, such as the thumb. When these sensory inputs are eliminated or enhanced, there are significant changes in the motor commands generated, as revealed by changes in the perceived heaviness of weights. Gandevia and McCloskey (1977) demonstrated these effects using a weight-matching task in which the weights supported by the index fingers on the right and left hands were matched in subjective magnitude while the sensory inputs arising from the thumb of one hand were altered. They found that a weight lifted by flexing the index finger was perceived to be heavier when the

thumb was anesthetized and lighter when it was electrically stimulated, as compared to control conditions. The loss of cutaneous inputs from the thumb and index finger following anesthesia also influences the maximum forces that can be produced by the fingers. Augurelle et al. (2003) found that when the thumb and index finger were anesthetized, the maximum pinch force declined by over 25%. These findings indicate that there is functional convergence of the sensory and motor systems involved in cooperative activities of the hand, even when the sensory inputs do not arise from the fingers stimulated by the motor activity.

Coupling of Movements During Typing and Piano Playing

Kinematic analyses of typing indicate that when letters are typed in isolation, there is little within-person variability in the finger and wrist kinematics from one keystroke to the next when the same key is repetitively struck. Individual keystrokes are performed in a stereotypical manner kinematically, regardless of the word or phrase being typed, but the time required to produce a keystroke does depend on the context in which the character appears (Ange-laki & Soechting, 1993). For a given letter, this may range from 130 to 370 ms, depending on the word being typed (Salthouse, 1984, 1986). Movements of the fingers of one hand are often highly correlated, and fingers adjacent to the digit typing a character frequently flex and extend or abduct and adduct together (Fish & Soechting, 1992). This linkage between fingers is greatest for the middle and ring fingers (Soechting & Flanders, 1997), as would be predicted from other studies of the degree of individuation of finger movements (e.g., Häger-Ross & Schieber, 2000). It is remarkable that even though almost all of the degrees of freedom of the fingers, thumb, and wrist are in motion when a letter is typed, stereotypical patterns of finger movement occur that are idiosyncratic, in that they vary from person to person, but are consistent within people and are relatively invariant.

In piano playing, as was noted in typing, there is little trial-to-trial variability in the kinematics, and so movement sequences are assembled stereotypically. Unlike typing, however, there are anticipatory changes to the trajectories of the fingers, but these appear to occur only one note in advance (Soechting, Gordon, & Engel, 1996). The anticipatory modification of the fingers' movements occurs at times ranging from 165 to 500 ms prior to striking the key and varies depending on the individual pianist and the musical score being played (Engel, Flanders, & Soechting, 1997). This difference between the two tasks may reflect the temporal constraints on piano playing, which are not present when typing.

Bimanual Movements

Bimanual rhythmic movements, such as tapping with two fingers synchronously or at some specified rate (multifrequency tapping) or moving the hands to separate targets, have been the focus of numerous studies of the organizational principles governing voluntary movements (e.g., Kelso, 1995; Mechsner, Kerzel, Knoblich, & Prinz, 2001). This research has examined the spatial, temporal, and intensive characteristics of coordination between the hands. When people make rapid movements with their hands to different targets, the initiation and termination of the movements are both usually simultaneous, even when the targets have marked differences in terms of their spatial properties. Although the hands move at different speeds to different spatial locations, the times to peak acceleration and deceleration are very similar, as was shown in figure 7.5, and so movement duration is kept constant (Kelso et al., 1979). The temporal and spatial coupling of movements that is usually a feature of these tasks has been interpreted as a constraint underlying the coordination of the hands (Semjen & Summers, 2002).

In contrast to these simple, rhythmic, bimanual movements in which the motions of the hands tend to become synchronized (Kelso, 1984), during typing and other skilled bimanual activities, the hands move simultaneously but often independently of one another. Flanders and Soechting (1992) reported that during typing, bilateral finger movements are uncorrelated and that motion of one hand does not affect the time course of the movement of the other hand or the variability of movements. Similarly, when playing the piano, the pianist is usually required to play different patterns of notes with each hand at the same time, and these typically involve different movement trajectories, different levels of intensity, and different timing patterns. If the music is polyphonic (i.e., it has different voices), then each hand may be required to play two voices that maintain their separate identities in the performance. In addition, it has been noted that when errors are made by expert pianists with one hand, they can be corrected by adjusting the timing of that hand alone, without affecting the timing in the other hand (Shaffer, 1981). These findings suggest that the constraints that underlie coordination of the hands are task dependent and that, with training, the timing resources available to the expert performer allow considerable independence in controlling the movements and forces produced by each hand.

Hand Preference

The hand that is habitually used to hold a tool when performing a skilled task, such as picking up a small object with tweezers or cutting with scissors, is usually referred to as the preferred hand (Oldfield, 1971). For about 90% of the population, this is the right hand, and the other 10% are either left-handed or ambidextrous (Corballis, 1991; Lansky, Feinstein, & Peterson, 1988; McManus, 2002). Handedness is usually defined in terms of the hand that is used most frequently to perform a number of activities, such as writing, drawing, throwing a ball, holding scissors, and cutting with a knife (McManus, 2002). It is not, however, a dichotomous variable but is distributed along a continuum ranging from strongly right-handed to strongly left-handed individuals. Left-handedness is slightly more prevalent in men than women, with about five left-handed men for every four left-handed women, and is more common in younger than older samples of people as social pressures to change handedness in children (i.e., to be right-handed) have diminished over the past 100 years (McManus, 2002).

Studies of the differences in the performance characteristics of the preferred and nonpreferred hands indicate that handedness is not related simply to a greater independence of finger movements or force control in the preferred hand. There is no evidence of systematic differences between the hands with respect to the range of motion (Mallon et al., 1991), degree of independence of finger movements (Häger-Ross & Schieber, 2000; Kimura & Vanderwolf, 1970), or control of individual finger forces (Reilly & Hammond, 2000). Although the maximum force produced by the preferred hand is usually greater by 3–10% than that produced by the nonpreferred hand in right-handed people, for left-handed people there is much greater variability between the hands in strength, with many left-handers having a stronger right hand (Armstrong & Oldham, 1999; Petersen, Petrick, Connor, & Conklin, 1989). This difference between right- and left-handed individuals reflects a general finding on handedness, namely that, as a group, left-handed people are more heterogeneous.

A variety of activities has been used to evaluate the differences in the performance of the two hands, ranging from pegboard tasks in which people pick up small objects one at a time and place them into target holes as quickly and accurately as possible, to repetitive finger-tapping tasks (Elliot & Chua, 1996). On many of these measures of manual skill, the preferred hand performs the task more quickly and consistently and with fewer errors than the nonpreferred hand. Even with prolonged practice, this performance asymmetry between the hands remains on some tasks (Peters, 1981). For example, on tasks in which people have to pick up and place pegs in holes, the difference in the performance of the left and right hands of right-handed people is related to the accuracy requirements for placing the peg, with the left hand being slower because it makes more small cor-

rective movements than the right hand as the peg is placed in its final position (Annett, Annett, Hudson, & Turner, 1979).

For finger tapping, the superiority of the right hand of right-handed people results from more consistent intertap intervals and less variability in the forces produced by the right hand (Todor & Kyprie, 1980; Todor & Smiley-Oyen, 1987). Analyses of the finger movements made during tapping have also shown that the right hand exhibits shorter intertap intervals and dwell times (key closure) than does the left hand, which indicates that the movement reversal phase from finger flexion to extension is executed more efficiently by the preferred hand (Todor & Smiley, 1985). This suggests that the ability to time the onset and offset of muscular contractions precisely is superior for the right hand in right-handers. It is not known whether these findings also apply to left-handed people, as left-handers generally use their nonpreferred hand more frequently than do right-handers and tend to be less strongly lateralized on many measures of manual skill (Steenhuis, 1996).

Kinematic and kinetic analyses of limb dynamics during pointing movements have revealed differences in the facility with which limb segment dynamics are controlled in the right and left arms of right-handed people. Bagesteiro and Sainburg (2002) have shown that muscle torques are better coordinated in the shoulder and elbow joints of the right arm than the left of right-handed individuals, with the result that the right arm executes a pointing movement with much lower torques than the left arm at similar speeds and with similar final position accuracies. When the intersegmental dynamics of the arm are changed by adding an inertial load, the right arm of right-handed individuals adapts more effectively to the novel inertial load than does the left arm (Sainburg, 2002). These results have been interpreted as indicating that there is more efficient control of the intersegmental dynamics of the preferred arm in right-handed individuals, perhaps reflecting the more extensive experience in controlling the preferred arm (Bagesteiro & Sainburg, 2002; Sainburg & Kalakanis, 2000).

On well-practiced tasks that involve both hands, such as typing, there is little or no difference between the hands of right-handed typists in typing speed and error rates, and in fact there is a slight tendency to perform better with the left hand (Provins & Glencross, 1968). These findings presumably reflect the extensive experience of typists, whose work always involves both hands, and the slightly superior left-hand performance has been attributed to the more extensive practice that the left hand receives in normal typing. As noted in chapter 7, in typing standard English on the QWERTY keyboard, the left hand is favored in the allocation of letters between the hands with estimates suggesting that it types more than 57% of the total number of keystrokes (Dvorak, 1943).

Hand-skill asymmetry has also been studied in professional musicians, who usually use both hands to play a keyboard or stringed instrument. In tapping tasks, right-handed musicians demonstrate significantly less right-hand superiority than right-handed nonmusicians (Jäncke, Schlaug, & Steinmetz, 1997). The difference reflects the superior tapping performance of the left hand in musicians as compared to nonmusicians and is specific to this motor skill.

Jäncke et al. (1997) found that there was no difference between right-handed musicians and nonmusicians in the performance of writing skills, such as tracing lines and dotting circles and squares. They further noted that there was a negative correlation between the age at which musical training commenced, but not the duration of training, and hand-performance asymmetry as measured on the tapping task. These findings suggest that intensive skill training with both hands reduces the asymmetry in manual performance and that early hand-skill training improves the performance of the nonpreferred hand.

In the next chapter, we examine how the sensory and motor capacities of the hand change from infancy to old age and show the importance of practice in maintaining adequate hand function in old age.

9

Hand Function Across the Lifespan

The only habit the child should be allowed to contract is that of having no habits; let him be carried on either arm, let him be accustomed to offer either hand, to use one or other indifferently.

ROUSSEAU, 1762/1961

Hand function is not immune to the declines in performance with increasing age that have been systematically documented across the senses. In the current chapter, we consider what is known about changes in hand function from infancy through later life. We start by examining how the hand physically develops with respect to its anatomy and the effects of maturation of the central nervous system on hand use. We then review the developmental literature, again as this pertains to the four categories used to depict the sensorimotor continuum of human hand function (figure 1.1).

Physical Aspects Pertaining to Hand Development

The anatomical features of the hand are well demarcated at 10 weeks' gestation, and a primitive grasp reflex in response to traction across the palm can be seen at 12 weeks' gestation (Brown, Omar, & O'Regan, 1997). As the child develops, the size of the limbs gradually increases. The length and width of the infant's hand has increased on average by a factor of 200–300% by the age of 18 years (Flatt &

Burmeister, 1979; Hajnis, 1969). The changes in hand size in early childhood influence the manipulative abilities of the child and affect whether one or two hands are required to grasp an object (Newell, McDonald, & Baillargeon, 1993).

Functional maturation of the central nervous system is reflected in the progression of white matter myelination. It occurs in a fixed sequence that depends in part on the anatomical level but also on the function of the neural pathways. At the cortical level, myelination occurs first in the sensory areas of the visual, auditory, and somatosensory cortex and then in the motor cortex. The association cortex does not become fully myelinated until the second decade of life. Although the intracranical corticospinal fibers are probably myelinated before birth, there is no significant myelination of these fibers within the spinal cord at birth. Full myelination of corticospinal fibers in the spinal cord does not occur until 8–10 years (Müller, Hömberg, & Lenard, 1991). The conduction velocity of these corticospinal axons in the human neonatal spinal cord has been estimated to be about 10 m/s and increases markedly from birth until 3–4 years, when it is approximately 40 m/s. It does not reach the adult values of 50–70 m/s until the child is around 8 years of age (Khater-Boidin & Duron, 1991). In contrast, the conduction velocity of sensory fibers in the median and ulnar nerves that innervate the hand has been estimated to reach adult levels by 12–18 months of age (Desmedt, Noel, Debecker, & Nameche, 1973). The speed with which the peripheral sensory and motor pathways function has important functional consequences and explains why some of the adaptive reflex responses seen in adults, such as increases in grip forces when an object begins to slip between the fingers (described in chapter 6), cannot be used effectively by young children.

There are relatively few recent studies of the histological changes that occur in the skin of the hand with advancing age, which has resulted in an excessive reliance on biopsy studies conducted in the 1960s and 1970s. A number of studies have shown that there are changes in the number, morphology, and distribution of mechanoreceptors in the hands of older adults (e.g., Bolton, Winkelmann, & Dyck, 1966; Bruce, 1980; Cauna, 1965; Quilliam & Ridley, 1971). Using skin samples taken from cadavers aged 62–83 years, Bruce (1980) found that there was an age-related reduction in the number of Meissner's corpuscles, with an estimated decrease of between 23 and 34% across different sites in the hand as compared to the number of corpuscles identified in amputated fingers from young adults. The morphology and distribution of these mechanoreceptors also changed with age, becoming more irregular and located away from the epidermis. These results are similar to those of Bolton et al. (1966), who used skin biopsy specimens from the little finger to determine the concentration, distribution, and morphology of Meissner's corpuscles. In the latter study, the concentration of corpuscles decreased from an average of 24 per mm^2 at 22 years of age to 8.4 per mm^2 at 78 years. Pacinian corpuscles also change in morphology and distribution with age. Histological studies of palmar digital skin have shown that Pacinian corpuscles gradually decrease in number with age, and those that survive undergo transformations, either increasing in size or remodeling as new corpuscles are formed on the axons of older ones (Cauna, 1965). In contrast to these changes in Meissner's and Pacinian corpuscles, Cauna (1965) observed that Merkel's corpuscles and the free nerve endings in the dermal papillae do not change radically with respect to structure or frequency in postnatal life. There do not appear to be any reports of the effects of age on the morphology or

density of Ruffini corpuscles in skin, which may reflect the relative infrequency with which these are found in the hand, as described in chapter 3 (Paré et al., 2003). In addition to the histological evidence of age-related changes in cutaneous mechanoreceptors, several studies have shown that there is a loss of large, myelinated axons in peripheral nerves and a slowing of sensory nerve action potentials with age (e.g., Bouche et al., 1993). All of these changes are important to understanding the mechanisms involved in the decline of sensory and motor functions in the elderly, which is discussed in this chapter.

In older adults, the skin on the dorsum of the hand undergoes morphological changes and becomes thinner and, consequently, more fragile and drier (Roberts, Andrews, & Caird, 1975). The latter presumably results from the reduced eccrine sweat gland output (Silver, Montagna, & Karacan, 1965) and the associated decrease in the moisture content of the stratum corneum of the skin (Potts, Buras, & Chrisman, 1984). There is a reduced and slower rate of sweating in older individuals, with a significantly diminished total output of sweat (Montagna, 1965). A reduction in sweat with increasing age may reduce the friction between hand and object, thus increasing the chance that an object may slip within the person's grasp.

One of the most common changes in the musculoskeletal system in the elderly is the reduction in muscle mass (by 25–45%), which is associated with a decline in maximum strength (Mathiowetz et al., 1985). Beyond the age of 60 years, there is a progressive loss in the number of motor units (the contractile element) in muscles due to the death of alpha motoneurons in the spinal cord. Motoneurons innervating fast-twitch muscle fibers appear to be most affected in this process (Larsson, Sjödin, & Karlsson, 1978). This is accompanied by a process of reinnervation of some of the denervated muscle fibers by surviving motor units, which leads to the emergence of larger and slower motor units (Larsson & Ansved, 1995). Such a process means that the muscles of older adults have fewer motor units than those of younger adults and that these motor units tend to be larger and slower (Carmeli, Patish, & Coleman, 2003). There are also morphological and pathological changes in the joints, with osteoarthritis and rheumatoid arthritis common in aging fingers and joints, which collectively influence the functional capacity of the hand in the elderly.

We turn now to the developmental changes that occur in manual performance with respect to the sensorimotor continuum of human hand function (figure 1.1).

Tactile Sensing

Sensitivity and Acuity

Pressure Sensitivity

To our knowledge, developmental changes in the sensitivity of infants and children to pressure have not been systematically investigated as yet. What we do know is that tactile pressure sensitivity is consistently lower in older individuals. This decline has been confirmed when stimulating the index fingerpads of males and females ranging in age from 19 to 88 years. In this study, Thornbury and Mistretta (1981) used a two-interval, forced-choice task with a double-staircase procedure in which participants were required to decide in which of two intervals the pressure stimulus was applied. Pressure sensitivity declined by a factor of 1.4 across the age range examined. In addition, older individuals (mean of 72 years) are less sensitive by a factor of 2.5 at three sites on the little finger relative to younger controls (18–20 years) (Bruce, 1980). Pressure sensitivity to single indentations (ramp and hold) on the thenar eminence is also lower in older (55–84 years) than in younger (19–31 years) individuals (Kenshalo, 1986).

Vibrotactile Sensitivity

Does vibrotactile sensitivity also decline with increasing age? Gescheider, Bolanowski, Hall, Hoffman, and Verrillo (1994) varied both stimulus frequency and contactor size in order to examine the effects of aging on the vibrotactile sensitivity of the four psychophysical channels that mediate vibratory sensations on the glabrous skin of the hand (Bolanowski et al., 1988), as discussed in chapters 3 and 4. Participants were required to decide which of two intervals contained the stimulus, which appeared in each interval equally often. The 75% correct-detection level was used to determine sensitivity. In one experiment, frequencies spanning a range from 0.4 to 500 Hz were presented to the thenar eminence of younger (mean of 23.6 years) and older (mean of 73.6 years) men and women.

Collectively, the results indicate that, with age, there is a general decline in sensitivity within all four channels. However, the decline is greater in the P channel (mediated by Pacinian/FA II units) than in the NP I, NP II, and NP III channels (presumably mediated by Meissner/FA I, Merkel/SA I, and Ruffini/FA II units, respectively). Recall that the model assumes that when more than one psychophysical channel is activated, the most sensitive one mediates vibrotactile sensation (chapter 4). These patterns are evident in figure 9.1, which additionally reflects the fact that women are more sensitive to vibrotactile stimulation than are men. Gescheider, Bolanowski, et al. (1994) have suggested that the loss in sensitivity with increasing age may be the result of an associated reduction in receptor density (Cauna, 1965). The P channel should be most affected inasmuch as it is capable of spatial summation, that is, of integrating neural activity across many receptors. Note that the magnitude of temporal summation, also mediated by the P channel, declines with increasing age as well (Gescheider, Beiles, Checkosky, Bolanowski, & Verrillo, 1994). It is possible that this decline in performance may also be the result of the reduced number of Pacinian corpuscles in the hand.

The absolute difference threshold, or difference in amplitude between two stimuli for correct discrimination 75% of the time, for a 250-Hz vibrotactile stimulus increases with age at a number of sensation levels (Gescheider, Edwards, Lackner, Bolanowski, & Verrillo, 1996), as shown in figure 9.2A. Presumably, this pattern is the result of the number of Pacinian corpuscles decreasing with age (Cauna, 1965). Since a general sensory decline with aging is commonly documented across all of the senses, it is not surprising that an older group discriminates vibrotactile amplitudes more poorly than does a younger group. What might seem initially puzzling, however, is that when these values are converted to relative difference thresholds (i.e., Weber fractions, or the proportional change in amplitude detected between a pair of stimuli), discrimination is unaffected by age with the exception of stimulus pairs close to absolute threshold (figure 9.2B). Thus, despite clear age differences in the absolute difference thresholds, the relative difference thresholds are equivalent. Presumably, this is because the difference thresholds for each age group are calculated relative to that group's absolute threshold. It would

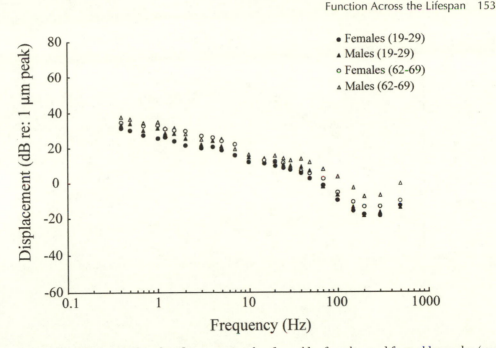

Figure 9.1. Means of eight young females, five young males, four older females, and four older males (age ranges in parentheses) for the detection of vibrotactile signals varying in frequency delivered through a 2.9 cm² contactor to the thenar eminence. Reprinted from Gescheider, Bolanowski, Hall, Hoffman, & Verrillo, 1994, with permission of Taylor and Francis Ltd. and the authors.

seem to be a valuable exercise to assess whether Weber fractions for other forms of discrimination also fail to show impairment with increasing age.

Finally, we note an age-related decline in the perceived magnitude of a 250-Hz suprathreshold stimulus presented at varying intensity levels (Verrillo, Bolanowski, & Gescheider, 2002). The subjective magnitudes are consistently and substantially lower in a group of older (mean of 68.6 years) than in a group of younger (mean of 23.5 years) participants. The difference in magnitude is 16.5 dB. This age-related impairment may be attributed to a reduction in the number of Pacinian corpuscles with aging and, possibly, to morphological changes in the receptors and in skin compliance, both of which occur with age (Cauna, 1965).

Thermal Sensitivity

Regardless of age, absolute sensitivity to cold tends to exceed that to warmth (Stevens & Choo, 1998). With increasing age, decreases in thermal sensitivity are more notable in the body's extremities (i.e., the hand and, especially, the foot) as compared to more central

areas, such as the face. When warm or cold stimuli are applied to the tip of the index finger and to the thenar eminence of participants ranging in age from 18 to 88 years, sensitivity to warmth decreases at the rate of 0.02°C/annum on the index finger and, more slowly, at 0.002°C/annum on the thenar eminence. Corresponding values for cold are 0.014°C and 0.002°C/annum, as shown in figure 9.3 (Stevens & Choo, 1998).

Spatial and Temporal Acuity

Gellis and Pool (1977) evaluated spatial acuity by measuring two-point touch thresholds at three major locations on the hand: the fingertips, thenar and hypothenar (palmar) areas, and dorsum (back). The age of participants ranged from 7 to 86 years. Thresholds were consistently lowest in the third decade of life and highest in the ninth decade. Stevens and Patterson (1995; also Stevens & Choo, 1996) further confirmed and clarified the decline in tactile spatial acuity on the (index) fingertip with increasing age. Acuity was assessed in four different ways on the finger: two-point gap discrimination (in transverse and longitudinal directions), point local-

A

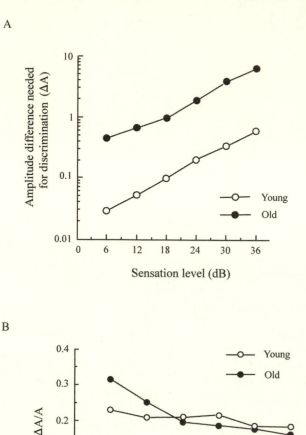

B

Figure 9.2. (A) Mean difference threshold, expressed as the amplitude change (micrometers) in peak displacement of the skin (ΔA) required for discrimination and plotted as a function of the sensation level (dB) of a stimulus of fixed amplitude. (B) Mean relative difference threshold, expressed as the amplitude change needed for discrimination divided by the amplitude of the stimulus ($\Delta A/A$) and plotted as a function of sensation level (dB). Reprinted from Gescheider, Edwards, Lackner, Bolanowski, & Verrillo, 1996, with permission of Taylor and Francis Ltd. and the authors.

ization, line-length discrimination, and line-orientation discrimination. The results are shown in figure 9.4. Panels showing two-point gap discrimination (longitudinal; panel B), length (5 mm base; panel E), and the overall mean finger acuity (panel G) only involve adults ranging in age from 20 to 86 years. The remaining panels report data that also include children between 8 and 14 years of age. Overall, the study showed differences among the four acuity dimensions tested with respect to size: from smallest to highest acuity, the dimensions are ordered as length, locus, orientation, and gap. However, the sensitivity of the fingertips of participants between 20 and 80 years of age declined similarly at the rate of approximately 1% per annum. A similar rate of

decline across all four acuity dimensions suggests a common mechanism, which the authors propose could involve thinning of the shared mediating mechanoreceptor network.

Van Doren, Gescheider, and Verrillo (1990) have used a temporal gap-detection task to investigate the effect of increasing age on tactile temporal acuity. Vibratory stimuli were presented to the right thenar eminence of participants varying from 8 to 75 years of age. A silent temporal gap (varying from 8 to 256 ms in duration) was bounded by a pair of 350-ms vibrotactile stimuli consisting of either band-limited noise (250–500 Hz) or 256-Hz sinusoids. A second stimulus had no gap and lasted the same total duration. A two-alternative, forced-choice, tracking par-

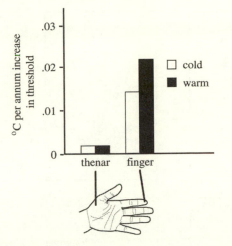

Figure 9.3. Hand maps showing average estimated rise per annum in warm and cold thresholds for the thenar eminence and fingertips, computed using the regression provided in table 2 of Stevens & Choo, 1998.

adigm was used in which participants were required to detect in which of two intervals the gap stimulus occurred. For any run, the gap remained constant, while the amplitudes of flanking and no-gap stimuli were varied using the tracking procedure. Note that the investigators did not just vary the size of a temporal gap. Rather, they measured how intense the stimuli flanking a gap must be for observers to detect the presence of the gap; the smaller the gap, the more intense the surrounding stimuli must be to detect the gap. As expected, tactile temporal acuity (measured as the gap-detection threshold at 75% correct in dB SL) increased with increasing gap duration, although the acuity for short-duration gaps was lower with noise than with sinusoids. Temporal acuity with flanking noise significantly decreased with age for gap durations of 8, 16, 32, and 256 ms; moreover, the rate of improvement with lengthening gap durations was greater with increasing age. Overall, age had less effect on temporal acuity with sinusoids (i.e., only for a duration of 64 ms), and the rate of improvement with lengthening gap durations slightly decreased. Different results with noise and sinusoids suggest the engagement of different underlying processes, although it is also possible that the result is due to a difference in stimulus complexity. It is notable that the results of the study by van Doren et al. eliminate one possible

mechanism of detection: an integrator of fixed time constant that operates independently of intensity level. However, other interpretations remain viable, including a version of the multiple look model (e.g., Viemeister & Wakefield, 1991), which has been used to explain temporal summation and discrimination in hearing.

Active Haptic Sensing

The effects of age on tactile sensing occur as well when active haptic sensing is used. In this section, we address what is known about the effects of aging on the performance of a variety of perceptual tasks that involve active haptic sensing.

Object Recognition and Classification

Piaget and Inhelder (1948) reported that children aged 2.5–3.5 years can easily recognize common objects by touch. Unfortunately, their experimental methodology was not precisely controlled. Since that time, several additional studies have collectively shown that haptic object recognition of both familiar and unfamiliar objects begins around 2–3 years of age and improves rapidly up to about 5 years. For example, Bigelow (1981) showed that, as early as 2.5 years of age, children are able to identify haptically an average of 5.6 out of a total of 7 miniature common objects (e.g., spoon), and that by 5 years their performance is almost perfect. Morrongiello, Humphrey, Timney, Choi, and Rocca (1994) evaluated haptic object recognition of common objects by blindfolded children aged 3–8 years. Both speed and accuracy improved with increasing age. In addition, this study revealed that object parts that are diagnostic of object categories contributed to haptic object identification, especially in the older children. Finally, Bushnell and Baxt (1999) have shown that, like adults, 5-year-old children are excellent at recognizing unfamiliar as well as common objects. They concluded: "The finding that children's haptic object recognition is excellent even with unfamiliar objects contrasts with the conclusions of prior researchers that children do not encode haptically perceived information very precisely or do not retain it very well" (p. 1871), which had been suggested by some researchers (e.g., Abravanel, 1972; Bryant & Raz, 1975).

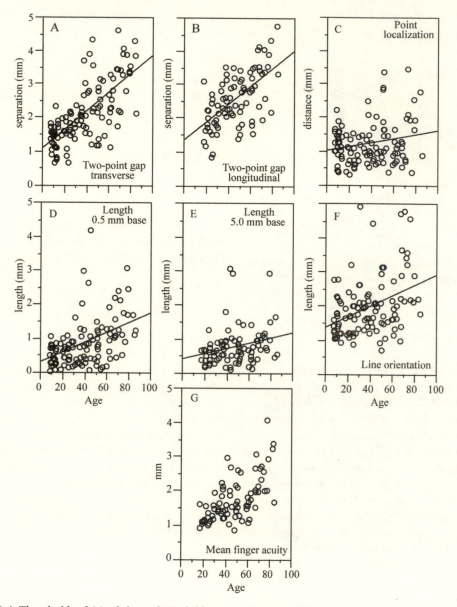

Figure 9.4. Thresholds of 89 adults and 26 children as a function of their ages for 6 spatial acuity tasks and for the overall task means. Note that B, E, and G show the results from the 89 adult observers; the other graphs show the results of all 115 observers, including both children and adults. The figure omits eight individual thresholds greater than 5 mm: three from B, one from C, and four from F. Reprinted from Stevens & Patterson, 1995, with permission of Taylor and Francis Ltd.

Turning now to the other end of the lifespan, we ask how well older people can haptically recognize common objects. Despite marked declines in tactile sensitivity, tactile acuity, and fine motor control with increasing age, Jessel (1987) found that an older group of observers (65–89 years) haptically recog-

nized common objects with relatively little decrement in performance relative to a group of young adults (18–23 years). Recognition accuracy was 88% and 93%, respectively. In addition, the older group required only an additional 0.5 s to recognize each object. Presumably common objects can be successfully recognized on the basis of relatively coarse differences along a number of object dimensions (e.g., roughness, hardness, shape, size, weight) that remain accessible despite declining sensory and motor capacity.

To date, very little research has focused on developmental trends regarding how multiattribute objects are haptically processed. Berger and Hatwell (1993, 1995, 1996) examined this issue in a series of experiments that considered whether observers haptically process objects more globally (i.e., on the basis of the overall similarity of all featural dimensions) or more analytically (i.e., on the basis of similarity with respect to only one featural dimension). L. B. Smith and Kemler (1977) had previously shown that while young children visually process multidimensional objects globally, adults tend to adopt a feature-based rule, which requires analytic (as opposed to global) processing. Berger and Hatwell (1993, 1995) asked how haptics compared developmentally to vision. They used free-classification tasks, one of which is shown and described in figure 9.5.

Unlike the developmental patterns observed in earlier vision studies, all three age groups (5, 9, and 22 years) classified multiattribute objects more by their separate featural dimensions (indicating analytic processing) than by their overall similarity (indicating global processing). Berger and Hatwell argued that the highly specialized nature of EPs (Lederman & Klatzky, 1987) is critical to the more frequent use of analytic strategies with touch than with vision. Unlike vision, haptic information about object properties is usually acquired sequentially and is often incomplete, partly because many exploratory procedures are highly specialized, regardless of age; moreover, with children, manual exploration is often incomplete. The specialized nature of EPs may also account for the developmental shift from preferring roughness to preferring size as a classificatory variable. That is, lateral motion, which is highly specialized for texture, was used very frequently by the 5-year-olds; in contrast, enclosure and contour following, which are highly specialized for size, were both rarely executed. The authors attributed instances of global processing in young children to perceptual factors resulting from the execution of broadly sufficient EPs (e.g., enclosure), which provide information about multiple object properties (Lederman & Klatzky, 1987). In contrast, the adults chose to perform a sequentially executed

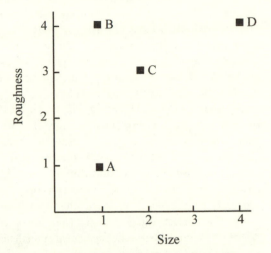

Figure 9.5. Example of one type of stimulus tetrad used to test for a preference for using dimensional-versus identity-similarity rules in a free object-classification task. Participants are initially presented with a standard object and then asked to select one of the remaining three comparison objects with which "it goes better." If participants selectively attend to either component dimension (i.e., both dimensional rules), then they should also judge that the standard object B goes best with either object A (if size is weighted more strongly) or object D (if roughness is weighted more strongly). If, however, participants choose to match by overall similarity (identity rule), then they should judge that object C goes better with the standard object B. Reprinted from Berger & Hatwell, 1993, with permission of Elsevier.

series of highly specialized EPs. Such EPs provide the most precise information about a single property (Lederman & Klatzky, 1987), as opposed to information about multiple properties that is provided when a broadly sufficient EP (e.g., enclosure) is executed. Berger and Hatwell further argued, therefore, that the nature of the manual exploration performed by adults cannot account at the perceptual level for the fact that adults (compared to children) clearly attended to both dimensions, as opposed to only one. They concluded that instances of overall similarity (global processing) judgments by adults must have resulted from integrating the perceptual information about separate properties extracted via highly specialized EPs into a global object representation during a subsequent, higher-level decisional processing stage. This second stage does not occur visually

(Berger & Hatwell, 1996). Collectively, the Berger and Hatwell studies are important because they highlight how the haptic and visual systems process multiattribute objects with regard to global versus analytic strategies, how such strategies change with age, and the relevance of manual exploration.

Schwarzer, Kufer, and Wilkening (1999) extended the developmental investigation of how multiattribute objects are processed by using a haptic category-learning task, which unlike the free-classification tasks used by Berger and Hatwell, required observers to learn the correct category-grouping principle, with feedback provided after each trial about the correct categorization. The task was designed to permit the use of either global or analytic processing strategies by allowing the observer to define a category in terms of the overall similarity of multiple object features or in terms of only one of those features. Children 3–9 years and adults performed the task using very high levels of analytic responding, as previously demonstrated using the same task with vision (e.g., Ward, 1989; Ward & Scott, 1987) and with haptic free-classification tasks involving texture/size and, to a lesser extent, texture/compliance (Berger & Hatwell, 1993; 1995; 1996). To the limited extent that global heuristics were adopted in this task, they were most frequently used by the adults (20%). A developmental trend was observed in which material-based (especially texture) learning changed to geometric-based (especially shape) learning with increasing age, as previously shown by Berger and Hatwell (1993), with a preference for texture changing to one for size.

Morrongiello et al. (1994) have addressed the issue of processing multiattribute objects from a different perspective by examining haptic identification of common objects in children ranging in age from 3 to 8 years. The common objects belonged to one of four categories: normal-sized, miniaturized small objects, miniaturized large objects, and oversized objects. The blindfolded children were required to name each common object as quickly and as accurately as possible following haptic exploration. Recognition accuracy increased with age and was related to developmental changes in the choice of manual exploration strategy. More specifically, the degree to which search was exhaustive tended to increase with age. In addition, the types of errors changed; whereas the younger children tended to confuse global form, the older children tended to confuse both global shape and local parts. The authors proposed that the age-related patterns of confusion errors indicate that children's haptic object representations change from being based primarily on global shape (younger children) to including both global shape and local parts (older children). Children of all ages had greater difficulty identifying miniaturized objects.

A more comprehensive understanding of how multiattribute objects are processed across the ages is very important for developing models of haptic object processing and for their application to different problem domains, as discussed in chapter 10. Some of the discrepant results among the few existing studies may be caused by the varied demands of the experimental tasks and because cross-study comparisons are based on critical concepts, such as "global," "holistic," "analytic," "dimensional," "part," and "feature," which have not always been similarly defined.

Perceiving Properties of Surfaces and Objects

Recently, there has been a noticeable increase in attention paid to the haptic perception of material (e.g., texture, compliance, thermal properties, and weight) and geometric (e.g., shape) object properties by infants and young children. Conclusions drawn from much of the infant work have relied on results based on cross-modal or bimodal perception, as opposed to intramodal haptic perception. In keeping with the focus of this book, wherever possible we have chosen to consider research that pertains explicitly to the haptic perception of object properties.

Texture

Initially limited to the results from a few diverse and somewhat indirect assessments, Bushnell and Boudreau (1991) tentatively concluded that by 6 months of age infants are sensitive to coarse differences in surface texture. They based their conclusion on several different studies. For example, Bushnell, Weinberger, and Sasseville (1989; see also Bushnell & Weinberger, 1987) showed that 12-month-old infants (and, to a lesser extent, 9-month-old infants) leaned forward more to examine haptically the nonvisible backs of dowels covered in fur, sandpaper, or bumps than backs that were smooth. Bushnell et al.

(1989) proposed that the differences in behavior could have been produced by haptic sensitivity to differences in surface texture. Ruff (1984) reported that infants 6, 9, and 12 months of age manually explored rough surfaces more than smooth objects. Additional evidence for haptic texture discrimination as early as 6 months of age is provided by Steele and Pederson (1977; see also Pederson, Steele, & Klein, 1980), C. F. Palmer (1989), and Lockman and McHale (1989), who showed changes in behavior (e.g., manual or looking) associated with coarse differences in texture.

Although previous studies had suggested that at younger ages infants may not be capable of perceiving texture haptically, more recent studies have pushed back the estimated age by assessing texture differentiation in infants 4–6 months of age (Morange-Majoux et al., 1997) and in neonates (Molina & Jouen, 2003). In the Morange-Majoux et al. study, the infants were allowed unconstrained tactual exploration of a large horizontal cylinder, whose surface alternated in "smooth" and "rough" sections, although details concerning the two textures were not provided. The texture variation was discernible haptically but not visually. Analysis of the videotaped hand movements indicated that infants increased the frequency and length of active exploration (compared to static contact), particularly from 4 to 5 months. Whichever hand was used, purposive exploration was described as being most often in the form of an adultlike lateral motion EP, which has been shown to be optimal for obtaining precise texture information (Lederman & Klatzky, 1987). It should be noted that manual exploration was not scored explicitly in terms of the occurrence of any other adultlike EPs described by Lederman and Klatzky. Nevertheless, all infants actively explored the boundaries between the rough and smooth sections. The right hand performed the lateral motion EP more often than did the left hand, particularly on the rough texture; in contrast, the left hand performed this EP predominately at the texture boundaries as early as 4 months of age. Collectively, these findings suggest that texture discrimination may occur in infants as early as 4 months.

Most recently, Molina and Jouen (2003) assessed texture perception in neonates (3–5 days old) by recording the manual activity (hold time, hand-pressure frequency) of the right hand in the presence (test interval) or absence (pre- and posttest intervals) of a second object held simultaneously in the left hand. On half of the trials, the texture density of the two objects matched (rough-rough, smooth-smooth); on the other half, it did not (smooth-rough, rough-smooth). Manual activity was measured when the object was placed in the right hand prior to, during, and after a second object was placed in the left hand. Although holding times decreased from pre- to posttest, they did not significantly change when the texture density was altered. However, there was a striking increase in the frequency of squeezes during the test interval for the nonmatching (compared to the matching) condition, as shown in figure 9.6. The frequency of squeezes was determined by recognizing peaks in a time series of measured positive hand pressures (described in further detail in the paper). These results directly suggest that even neonates may be capable of differentiating coarse variations in surface texture, although a visual contribution to performance cannot be excluded.

More generally, conclusions must be tempered regarding the age of onset of haptic perception of texture by the fact that most of these studies have permitted the infant to look at, as well as touch, the objects. We conclude that, on the basis of the haptic-only study by Morange-Majoux et al. (1997), by 4 months of age infants are capable of differentiating coarse variation in texture; it is even possible, although inconclusive at this time, that such an ability may occur in the early postnatal period.

Compliance, Temperature, and Weight

We know of no infant studies that have focused specifically on intramodal haptic perception of compliant features (e.g., object hardness, flexibility). As with the investigation of haptic texture perception, studies of perceived hardness have used naturalistic tasks that include vision as well as touch. For example, Gibson and Walker (1984) showed that 12-month-old infants prefer to look at either a rigid or flexible cylinder, depending on which of the two they had previously handled in the dark. Moreover, during the earlier period without vision, the infants tended to squeeze the flexible object and bang the rigid object against a table, suggesting that the infants did indeed differentiate the objects by compliance. Similar results were obtained by Lockman and Wright (1988) and by C. F. Palmer (1989), who studied infants ranging in age from 6 to 12 months.

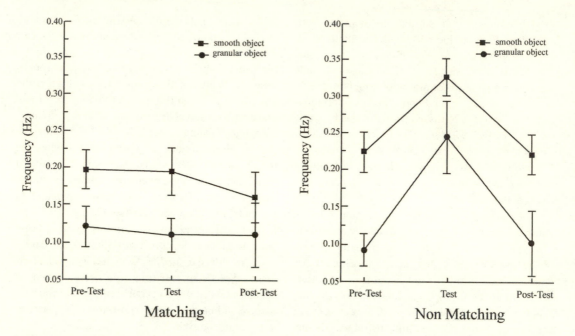

Figure 9.6. Frequency of right-hand squeezes prior to, during, and after the test period. During the test period, the left hand was presented with a second object that was either the same (matching) or different (nonmatching) from the right-hand object in texture density (smooth versus granular). Reprinted from Molina & Jouen, 2003, with permission of John Wiley & Sons.

A study by Rochat (1987) concluded that newborns, two-, and three-month-olds could also differentiate a rigid object from an elastic object. However, their dependent measure was frequency of squeezing, which may have simply reflected the occurrence of a grasp reflex. Bushnell and Boudreau (1991) concluded, "[I]t seems from 6 or 7 months on, infants may haptically discriminate hard objects from flexible or compressible ones" (p. 147). However, the capacity of younger infants to discriminate hardness remains uncertain.

Two studies confirm that, at 6 months of age, infants are capable of differentiating objects that are either warm or cool. Bushnell, Shaw, and Strauss (1985) and Bushnell, Weinberger, and Sasseville (1989) familiarized 6-month-old infants with cylinders that felt either warm or cool to the touch. In both cases, the infants were initially familiarized with one of the two objects (warm or cool) and then presented with either the familiar object or the unfamiliar object (cool or warm). The infants touched the unfamiliar objects longer than the familiar ones, indicating that infants as young as 6 months can differentiate warm and cool objects by touch. We know

of no studies that have investigated thermal perception in infants younger than 6 months.

The infant literature has only indirectly investigated the haptic perception of weight. Using a methodology similar to one used by Bushnell et al. (1985), Ruff (1982, 1984) showed that neither 9- nor 12-month-old infants touched objects whose weight was familiar longer than those whose weight was unfamiliar to them. However, banging, as opposed to other hand movements, such as fingering, rotation, transfer, mouthing, which could involve moving the object into or out of the mouth, reorienting it within the mouth, or rubbing it against the gums, and dropping, tended to occur most frequently with a change in weight; banging also occurred more frequently for objects varying in weight than in texture, and less so (although the difference was not statistically significant) for objects that varied in shape. Thus, we can tentatively exclude the possibility that banging was used exclusively to generate sounds. C. F. Palmer (1989) also documented differences in the ways in which both 9- and 12-month-old infants handled (i.e., waved) light versus heavy objects; 6-month-olds did not show such behavioral differences. Mounod and Bower (1974) provided additional, less-direct evidence in a study in which 9.5-month-old infants appropriately adjusted their arm tension to anticipate the object's weight from its visual appearance, whereas infants 6, 7, and 8 months of age failed to do so. Based on studies that have generally only indirectly assessed haptic weight perception by infants, Bushnell and Boudreau (1991) concluded that this perceptual capacity appears to emerge somewhat later than the perception of texture, compliance, and thermal properties, that is, around 9 months of age at the earliest.

To our knowledge, only a handful of studies have examined haptic weight perception by children beyond infancy. For example, Robinson (1964) examined the susceptibility of children aged 2–10 years to the well-known haptic size-weight illusion described in chapter 5. The younger children were initially trained to discriminate between weights so that they understood what "heavier" meant. Robinson reported a size-weight illusion across the entire age range, although the magnitude decreased with age. Robinson proposed that the decrease related to an improvement in the ability to discriminate weight with increasing age. In contrast, Pick and Pick (1967) found that the magnitude of the illusion actually increased with age from 6 to 16 years. This result may be interpreted as indicating that, with increasing age, observers are more strongly influenced by changes in object volume (see Ellis & Lederman, 1993) or, as Amazeen and Turvey (1996) have suggested, in moment of inertia. Pick and Pick suggested that the apparent discrepancy with Robinson's data may lie in the fact that the developmental decrease observed by Robinson occurred mainly for subjects younger than 4 years, whereas their subjects were 6 years and older. By this age, it is unlikely that preliminary training would on its own have significantly improved the children's weight discrimination. Clearly, additional research is needed.

Orientation, Size, and Shape

We have failed to find any studies that have investigated the haptic perception of orientation by infants. Gentaz and Hatwell (1995) found that children aged 7 and 9 years made larger errors than adults in discriminating line orientation, with no difference between the two younger ages. Although all three age groups demonstrated an "oblique" effect (i.e., the orientation of vertical and horizontal lines is identified more accurately than that of lines presented at 45° or 135°), adults showed a larger oblique effect than did the two younger groups.

To date, evidence of haptic size perception has been obtained in all infant ages tested. For example, Bushnell (1978) showed that 15-month-olds could easily detect a visual-haptic discrepancy between a small and a large cube. In addition, C. F. Palmer (1989) reported that 6-, 9-, and 12-month-old infants all used more unimanual exploration when exploring small as opposed to large objects. We have found no studies that directly assess haptic perception of size in younger infants. However, Bushnell and Boudreau (1991) have suggested that differences that are characterized in the following section on shape perception as "topological" may also be considered as differences in size. By this definition, infants only 2–4 months of age can haptically differentiate size. Empirical confirmation is provided by two studies (Streri, 1987; Streri & Spelke, 1988), which will be discussed in the next section on haptic shape perception. Bushnell and Boudreau (1991) tentatively conclude that infants may be able to differentiate variations in object size before they can differentiate changes in other object dimensions, such as texture, compliance, and shape, although the recent study by Molina and Jouen (2003) tentatively suggests that texture perception may occur even earlier.

We turn now to haptic shape perception. Bushnell and Boudreau (1991) have categorized infant responses to the shape of stimulus objects in terms of those that vary topologically (e.g., stimuli with or without a hole), featurally (e.g., stimuli with edge discontinuities versus smooth contours), or configurally (e.g., stimuli with the same type of feature but that vary in number or spatial arrangement). According to this classification scheme, infants as young as 2–4 months are capable of haptically differentiating topological shape differences. The results of studies by Streri (1987) and Streri and Spelke (1988), initially referred to in the section on size, are both relevant. They showed that 2-month-old infants can differentiate between a solid disk and a disk with a hole in it by touch (Streri, 1987) and that 4-month-old infants can haptically differentiate between two rings that are either connected or unconnected (Streri & Spelke, 1988; for further details, see below on manual exploration).

The earliest infant research suggests that from about 6 months of age, infants can differentiate shapes based on featural differences, such as sharp angles, edges, and protrusions versus smooth curves (e.g., Brown & Gottfried, 1986; Ruff & Kohler, 1978; but see Bryant, Jones, Claxton, & Perkins, 1972). However, research by Streri and Molina (1993) and Lhote and Streri (1998) indicates that a coarse shape difference can be discriminated by 2-month-old infants. Most recently, Streri, Lhote, and Dutilleul (2000) presented evidence suggesting that this capacity may even be present in newborns. They showed that newborns decrease the length of time that they hold a cylinder (or prism) presented across successive trials; they subsequently increase holding time when presented with an object that has a different shape, that is, a prism (or cylinder). The authors concluded that even newborns are capable of haptically differentiating a coarse difference in shape. The newborns primarily used a reflexive grasp, which would make available coarse shape information. There was no evidence that newborns performed a contour-following EP, which is necessary (as well as optimal) for adults to obtain precise shape details (Lederman & Klatzky, 1987). Rather than detecting configural shape, it is more likely that newborns discriminated shape featurally by detecting the presence (prism) or absence of an edge (cylinder). Adults can manually detect the presence (or absence) of an edge based on a very brief initial contact, typically lasting about 20 ms, and in one case lasting only 4 ms (Lederman & Klatzky, 1997). To resolve this ambiguity, one could use object pairs with both members possessing edges or neither possessing edges. For now, we will classify such discrimination as being based on featural differences.

It is not until infants are about 15 months, however, that they can haptically perceive shape differences in terms of their overall spatial configuration. (Bushnell, 1978; Bushnell & Weinberger, 1987; Landau, 1990). In the Bushnell studies, the stimulus pairs contained the same kind of feature (e.g., angles, edges) but varied in terms of the number and position of the features. We regard the ability to detect differences haptically in the overall spatial configuration as the best assessment of shape perception. The existing evidence suggests that configural shape perception emerges relatively later in infant development.

Bushnell and Boudreau (1993) have graphically summarized the infant literature in terms of a timeline for the development of haptic perception of various object properties in newborns and infants. An updated version is presented in figure 9.7. In keeping with Bushnell and Boudreau (1993), we have only included the time course for shape perception based on spatial configuration. Details pertaining to the associated studies are offered in table 9.1. We offer a caveat that these timelines, particularly as they relate to newborns, should be viewed with some caution, inasmuch as it has been difficult to ensure that performance differences are determined by perceptual differentiation as opposed to motoric differences attributable to variations in object affordances.

The results of the more-recent infant literature would appear to question earlier work by Piaget and Inhelder (1948), who claimed that recognition of topological forms only begins at about 3.5–4 years of age. This was observed to occur later than for common objects which, as we have previously noted, vary in more ways than just shape. The recognition of geometric (i.e., in Bushnell and Boudreau's terms, configural) forms developed still later, at about 4–4.5 years, with explicit differentiation in terms of angles and dimensions not beginning until about 4.5–5.6 years. Thus, the order in which topological and configural shape perception develops is similar to that documented in the more-recent infant literature, but the earlier studies suggested that it begins much later. In a more carefully controlled replication of the earlier study by Piaget and Inhelder (1948), Hoop (1971) included nursery school children aged 3.5–5.5 years, in 6-month increments. The most consistent age trends indicated that the youngest children recognized significantly fewer common objects and topological objects than the oldest children. Common objects were recognized more accurately than either topological or geometric objects in all but the youngest age groups. But only the 5.5-year-old children recognized the topological objects better than the geometric objects. The marked discrepancy above with respect to when haptic shape perception begins is likely due to the greater difficulty of the task and stimuli used with young children (more matching objects from which to choose, smaller featural differences) than those used with neonates and infants. However, the latter studies with neonates and young infants indicate that although haptic shape perception clearly improves with age, it does begin considerably earlier than initially claimed.

Two studies have collectively addressed the development of haptic shape perception from early to late

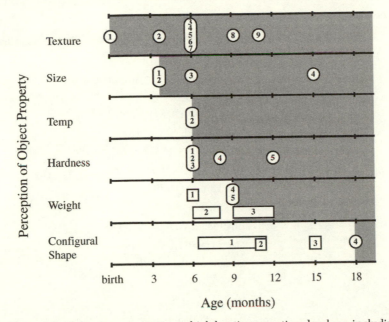

Figure 9.7. An illustration of the time course over which haptic perception develops, including studies originally discussed in Bushnell & Boudreau, 1991, and updated to include more recent work. For each object property, citations are placed approximately according to the age of the infants studied. The left-hand (white) portions indicate that either these ages have not yet been studied (no relevant citations) or that the citations listed (numbers enclosed in rectilinear boxes) have yielded null results. Right-hand portions (shaded) indicate ages for which there is positive evidence for property discrimination (numbers enclosed in rounded boxes) or null results (numbers enclosed in rectilinear boxes). The numbers are listed with the associated studies in table 9.1. Adapted from Bushnell & Boudreau, 1993, with permission of the authors.

in life. As part of their test for neurological deficits, Benton, Hamsher, Varney, and Spreen (1983) reported the normative performance of control groups ranging in age from 15 to 71–80 years on a nonverbal test of tactile form perception. All but one subject were right-handed. A match-to-sample, two-dimensional, form-perception task was used: Subjects explored each standard form haptically with one (left or right) hand and then selected the best match for the form from a set of comparison shapes presented visually. The geometric shapes were made of fine sandpaper. For purposes of age comparisons, they also reported the corresponding normative results obtained earlier by Spreen and Gaddes (1969) for children aged 8–14 years in which preferred and nonpreferred hands were each used on the same form-perception test. The one-hand mean performance averaged across boys and girls improved with age from 8 through 14 years by about 19%, at which point the accuracy approached the corresponding scores for observers aged 15–50 years in the later

Benton et al. (1983) study. There was only a slight decline in accuracy (~6.5%) from 51 to 70 years.

Manual Exploration

To date, relatively few studies have specifically focused on the development of manual exploration. Streri and Spelke (1988) showed that 4-month-old infants use the results of manual exploration to perceive the unity and boundaries of objects. These infants grasped two rings, one in each hand, out of their field of view. In one condition, the rings were rigidly connected; in a second condition, they could be manipulated independently of each other. Infants discriminated between the two pairs of hand movements (i.e., they made different movements, which suggests they discriminated between the two objects, not the movements). In the first condition, they responded as if they were contacting a single connected object with the hands. In the second condition, they responded as if they were manipulating

Table 9.1. Studies listed in figure 9.7 by number

Property	Number	Study	Earliest age (months)
Texture	1	Molina & Jouen (2003)	Birth
	2	Morange-Majoux, Cougnot, & Bloch (1997)	4
	3	Rose, Gottfried, & Bridger (1981)	6
	4	Steele & Pederson (1977)	6
	5	Ruff (1984)	6
	6	Lockman & McHale (1989)	6
	7	C. F. Palmer (1989)	6
	8	Bushnell (1982)	9.5
	9	Bushnell & Weinberger (1987)	11
Size	1	Streri (1987)	4
	2	Streri & Spelke (1988)	4
	3	C. F. Palmer (1989)	6
	4	Bushnell (1978)	15
Temperature	1	Bushnell, Weinberger, & Sasseville (1989)	6
	2	Bushnell, Shaw, & Strauss (1985)	6
Hardness	1	Pederson, Steele, & Klein (1980)	6
	2	Lockman & Wright (1988)	6
	3	C. F. Palmer (1989)	6
	4	McCall (1974)	8
	5	Gibson & Walker (1984)	12
Weight	1	C. F. Palmer (1989)	Not up to 6
	2	Mounod & Bower (1974)	Not up to 8
	3	Ruff (1984)	Not up to 12
	4	C. F. Palmer (1989)	9
	5	Mounod & Bower (1974)	9.5
Configural shape	1	Bryant, Jones, Claxton, & Perkins (1972)	Not up to 11
	2	Bushnell & Weinberger (1987)	Not up to 11
	3	Bushnell (1978)	Not up to 15
	4	Landau (1990)	18

two separate objects. Streri, Spelke, and Rameix (1993) subsequently showed the differential perceptual effects of actively controlled versus passive movement of object parts during haptic object perception by infants aged 4.5 months. When the infants were allowed to move actively two rigidly connected handles that formed part of an unseen object assembly, they perceived a single connected object. In contrast, when the two handles were similarly moved relative to each other, but passively by the experimenter, the infants perceived two disconnected objects. The authors concluded, "[F]or infants as for adults, haptic perception is enhanced by the active production of surface motion" (p. 251).

Taking an alternate perspective on manual exploration, Bushnell and Boudreau (1991) have noted that the order in which the haptic perception of various object dimensions initially develops in infants is determined in large part by the temporal sequence in which the associated exploratory procedures (Lederman & Klatzky, 1987) become motorically possible. Morange-Majoux et al. (1997) confirmed that, as early as 4 months, infants are capable of discriminating coarse variations in texture. In keeping with Bushnell and Boudreau's hypothesis linking manual and haptic perceptual development, infants' hand movements proved similar to the adult EP known as lateral motion, which is optimal for extracting texture. Lederman and Klatzky (1987) noted that the choice of end effector did not influence the association between EP and object property, although certain end effectors do provide more precise information than others. In keeping with this observation, when infants actively explored the textured cylinder, although they used both their whole hand and fingertips at 4 and 5 months, by 6 months of age the infants primarily used their fingertips. Presumably, this is in keeping with the fact that myelination of the pyramidal tract occurs around 4 months (cited in Morange-Majoux et al., 1997), which is important for organizing distal movements that underlie the development of manual exploration. Furthermore, a lateral-motion EP executed with the fingertip (compared to the palm) provides the most precise information about texture because of the relatively high density of mechanoreceptors. As mentioned earlier, Molina and Jouen (2003) measured the frequency of squeezes in matching and nonmatching texture conditions using neonates. It is well known that neonates execute grasps reflexively. However, they executed significantly more squeezes in the nonmatching than in the matching texture condition, suggesting that neonates could differentiate a coarse change of surface texture. Lederman and Klatzky (1987) have shown that while not optimal for providing precise texture information, the enclosure EP (grasp) does provide coarse information about surface texture.

Streri, Lhote, and Dutilleul (2000) used a habituation paradigm with neonates to show that they grasp a novel object longer than a familiar object. These authors also suggested that the reflexive grasp may afford the additional perceptual functions of an enclosure EP. Accordingly, an enclosure EP could provide sufficient, but not optimal, information about shape to neonates. We must, however, keep in mind that edge detection would also permit the neonate to differentiate the two stimulus shapes (prism versus cylinder), but such success would not be based on true haptic perception of the object's configuration. The same may be said with respect to the findings of Streri and Gentaz (2003).

In all of the neonatal and infant studies reported above, manual exploration was data-driven, by which we mean that hand movements were guided by the stimulus and its properties. The developmental study of knowledge-driven EPs would constitute a significant addition to this literature. In one study, Klatzky, Lederman, and Mankinen (2005) focused on whether preschool children aged 4 years and 7 months could make appropriate judgments about tool function using touch and/or vision. The children were asked to decide and report verbally whether a tool (spoon or stick) could be used to perform a particular task (move a candy or stir sugar or gravel, respectively). The spoons varied in size and the sticks varied in compliance, such that the tool would or would not function appropriately. The children proved to be sensitive to the characteristics of the tool and to the constraints on its function. Moreover, their judgments were attained through appropriate perceptual analysis: In keeping with the geometric/material distinction highlighted in chapter 5, the children used vision, not touch, to explore the size (a geometric property) of the tool. In contrast, they chose to explore the mixing tool haptically. More specifically, they performed a pressure EP, which has been shown in adults to be optimal for learning about compliance, a material property (Lederman & Klatzky, 1987). The children pushed, squeezed, bent, and applied force in other ways, but they did not execute other specialized EPs or touching behaviors that were less than optimal for learning about the relevant property of the tool.

Finally, we briefly recall a third way in which manual exploration has been investigated. The Morrongiello et al. (1994) study on haptic object recognition in children aged 3–8 years showed that not only are older children faster and more accurate, but they also manually explore common objects with greater thoroughness.

Overall, we conclude that manual exploration appears to be intimately linked to the development of the haptic perception of objects and their properties. However, to date, there has been relatively little, if any, consideration of the role of hand movements in the haptic perception either of children beyond the age of infants or of older adults.

Prehension

Within the first year of life, the infant's ability to use the hand to explore the environment undergoes remarkable changes, as both the motor and sensory systems mature. There have been many studies about reaching and grasping in infants and young children and about how different aspects of these movements mature with age. At the other end of the spectrum, complaints of impaired dexterous manipulation and difficulties in performing everyday tasks, such as tying shoelaces and fastening buttons, are common in the elderly (Cole, 1991). These functional impairments are associated with reduced tactile acuity (Gellis & Pool, 1977) and with prolonged response times in object manipulation tasks (Agnew & Maas, 1982). In this section, we will review the changes in reaching and grasping across the lifespan.

Reaching to Grasp

Infants and Children

The newborn infant has a limited repertoire of hand and finger movements, which are either endogenously generated or elicited in response to external stimulation. If the palm of the hand is stroked, the fingers flex in unison in what is known as the grasp reflex. If the dorsal surface of the hand is stimulated, this reflex can be inhibited or the hand can open reflexively (Twitchell, 1970). Claims that newborns are capable of opening their hands to fit the size of an object presented to them (e.g., Bower, 1972) have not been supported in more-recent quantitative studies of infant reaching and grasping (Hofsten, 1982; Hofsten & Rönnqvist, 1988). It appears that, in the earlier studies, the movements were usually imprecise and not necessarily related to the presence of visual objects. However, it has been shown that newborn infants (10–24 days old) will deliberately move their hands to counteract external forces applied to their wrists in order to keep their hands in the field of view and that they move their arms more when they can see them (van der Meer, van der Weel, & Lee, 1995). These movements are presumably important precursors to the development of reaching as they assist the infant in developing a frame of reference for action.

Studies of reaching in infants and young children usually involve presenting an object in front of the children, which they then reach for and grasp. As with many developmental studies, task constraints can have a significant effect on the behavior manifested. In studies of reaching, the position and size of the object and its base of support, that is, whether it is stabilized by an experimenter, attached to a rod, or resting on a table, influences the grasping behavior observed. In their study of infant reaching and grasping, Newell, Scully, McDonald, and Baillargeon (1989) presented an object in the outstretched hand of the experimenter and reported that 71% of infants reached and grasped at 4 months. In contrast, Fetters and Todd (1987) presented a stationary object on a table in front of the infants and observed that grasping is often unsuccessful at 7 months. Under these latter conditions, the rate of success in grasping the object increased significantly (90%) by 8–9 months of age (Fagard, 1998). An additional task constraint that may influence the type of prehensile response seen in young children and that has varied in different experimental studies is the postural position of the child (i.e., seated, reclined, prone, or supine) and whether the child is able to maintain an upright seated posture without support (Newell & McDonald, 1997). Infants who have achieved postural control and are able to sit on their own tend to reach toward objects with one hand, whereas children who have yet to attain postural control (nonsitters) use their arms symmetrically so that the hands meet in the midline (Rochat, 1992).

The capacity to reach for objects develops gradually from birth, but even neonates (1–2 weeks) are capable of moving their arms toward an object of interest presented in front of them (Hofsten, 1982). The movements are made while the neonate is looking at the object, and although the movements never result in grasping and the hand rarely makes contact with the object, they are all directed toward the object and bring the hand into the visual field of view. In some of these "prereaching" movements, the hand actually slows down as it approaches the object. Hofsten (1984) conducted a developmental study of the form and frequency of these movements in a cohort of 23 infants from birth through 19 weeks of age. He found that there was a gradual decrease in the frequency of prereaching movements directed toward the object and that the frequency of the movements reached a minimum at about 7 weeks of age. The number of these movements then increased dramatically; at the same time, there was a change in the posture of the hand as the arm moved forward. In the early weeks, the hand was open, that is, the fingers tended to be extended as the arm extended as part of an extension synergy. By about the 7th week, this pattern changed, and infants made a fist as the arm extended, which indicates the beginning of independent control of the arm and hand. At around 16–20 weeks of age, the infants reached toward and grasped the object, showing that the two systems are somewhat differentiated at this age.

Reaching skills undergo considerable changes in infants from 15 through 36 weeks of age, particularly with respect to the fluidity of the movement. In infants aged 12–16 weeks, the reaching response consists of a series of submovements (as judged by zero crossings in the acceleration record), which results in a long, fragmented movement trajectory that covers less than half of the total distance in the first part of the trajectory (Hofsten, 1979). By 36 weeks of age, the total movement path has decreased considerably, and the initial movement now covers more than 70% of the total distance, so that the hand is brought near the target with a single movement (Hofsten, 1991). This latter pattern resembles that seen in adults.

At 3–4 months of age, voluntary prehension begins to appear as infants learn to grasp objects by flexing all of their fingers against the palm in a palmar grasp (K. J. Connolly & Elliot, 1972). At this age, changes in hand orientation as a function of the orientation of the object start to develop when the child reaches for an object (Hofsten & Fazel-Zandy, 1984). By 4–5 months, infants can reach successfully for an object and grasp it when it comes in contact with the hand (Hofsten, 1984), and by 6–7 months they begin to close the hand prior to contact with the object. In a detailed study of grasping in infants aged 5–13 months, Hofsten and Rönnqvist (1988) used LEDs to measure the positions of the thumb and index finger as infants reached for and grasped spherical wooden objects (1.5–3.5 cm diameter) mounted on a movable rod placed in front of them. They observed that in 5- to 6-month-old infants, 75% of the grasps occurred within 100 ms of encountering the object so that the closing of the hand was well timed in relation to contact with the object. Even at this age, the approach of the hand

and grasping of the object were coordinated actions, and visual cues were used effectively to control the grasp (i.e., to estimate the distance and direction of movement). As has been observed with adults, these infants planned their grasps during the reaching phase of the movement with the hand starting to close during the ongoing movement. Infants aged 13 months integrate reaching and grasping more consistently than do younger children, who initiate the grasp earlier in the reaching phase and at a greater distance from the object. In addition, 9- and 13-month-old children begin to adjust the aperture of the hand for object size, but 5- to 6-month-old infants do not change their hand aperture consistently, as shown in figure 9.8. Even in the older children, hand aperture is not scaled in proportion to object size. In the younger children, this may be due to visual limitations in perceiving differences in the sizes of the objects presented or may reflect an inability to use visual information to preprogram grip size. Although older infants open their hands as they approach the object, their responses are less well tuned to object size than is seen in adults, with the hand open more fully for all target sizes presented (Hofsten & Rönnqvist, 1988).

As the size of the object being grasped increases, young children use two hands to reach and grasp it. The point at which two hands are used to grasp an object can be predicted from hand size. Newell, Scully, Tenenbaum, and Hardiman (1989) observed that, compared to adults, children (3–5 years) use two-handed grasping more frequently than one-handed grasping (60% versus 40% of trials). For cubes larger than 82 mm, children generally use two hands, whereas for adults the point of shift to two-handed grasping is a cube of 142 mm. However, both adults and young children use a single finger with the thumb to pick up very small objects. In a further study of grasping in 5- to 8-month-old infants and in adults, it was found that the probability of using one or two hands to pick up an object was very similar in both groups when the results were scaled to hand size (Newell, McDonald, & Baillargeon, 1993). Figure 9.9 shows the relation between the ratios of object diameter to hand length and the number of digits used to grasp an object, for infants and adults. The 50% level for switching from one to two hands occurs at approximately the same object-hand value of 0.75–0.8 for both infants and adults. These findings demonstrate the important

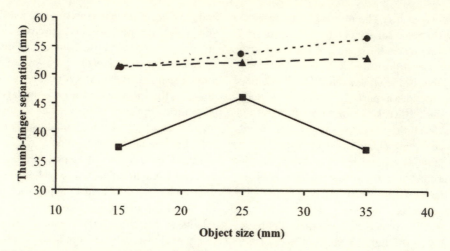

Figure 9.8. Mean maximum opening of the hand during approach when reaching for an object as a function of object size and age in the following age groups: 5- to 6-month-old infants (squares), 9-month-old infants (triangles), and 13-month-old infants (circles). Redrawn from Hofsten & Rönnqvist, 1988, with permission of the American Psychological Association and the authors.

role of body scale in the selection of grasp patterns over the lifespan.

Prehensile skills continue to be refined during childhood, even though the essential elements of coordinating reaching and grasping are present in 3- to 4-year-olds. A cross-sectional study of 4- to 12-year-old children revealed that even though movement time and normalized hand velocity were constant across this age range, three developmental changes were evident (Kuhtz-Buschbeck, Stolze, Jöhnk, Boczek-Funcke, & Illert, 1998). First, there was a tighter temporal coupling between grip formation and hand transport in older children; second, movements became more regular and uniform with age; and finally, when reaching in the absence of visual feedback to a remembered location of the tar-

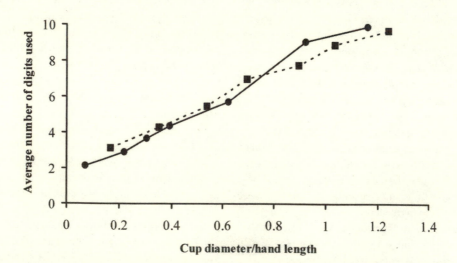

Figure 9.9. Average number of digits used to grasp an object when object size is scaled to hand length for infants (dashed line with squares) and adults (solid line with circles). Redrawn from Newell et al., 1993. Reprinted with permission of the authors and John Wiley & Sons.

get, the adjustment of finger aperture to the object's diameter was evident only in the older, 12-year-old children. Younger children often missed the target on the first attempt at reaching without visual guidance and were clearly much more dependent on visual cues when reaching.

The Elderly

In the elderly, the primary focus of research on motor performance has been on measurements of reaction time and movement duration, both of which increase in people over 60 years of age (Welford, 1988). These age-related changes in performance during middle age and late adulthood have been documented for many tasks that involve a speeded component and affect all domains, namely, motor, perceptual, and cognitive (Salthouse, 1985). It has therefore been assumed that they result from a single common cause and are mediated by a reduction in processing speed, although there are different patterns of age-related decline in different cognitive tasks.

Kinematic analyses of reach-to-grasp movements in the elderly show only subtle differences when compared to those of younger individuals. In older individuals, as in younger adults, there is parallel activation of the transport and manipulation components when reaching for an object, with the kinematic patterning of both components being influenced by the characteristics of the object to be grasped. There are no age differences in the time-to-peak wrist velocity or the time-to-peak aperture of the hand as the hand reaches to grasp an object (Carnahan, Vandervoort, & Swanson, 1998). However, movement duration is longer for older participants, and the peak acceleration and deceleration of the arm are lower as the hand reaches for the object (Bennett & Castiello, 1994). It has been proposed that the longer deceleration time of elderly people reflects a conservative movement strategy, in which more time is allocated for placing the fingers precisely on the object being grasped.

Grasping

Infants and Children

The patterns of hand use in early childhood have been the focus of many developmental studies, ranging from the naturalistic observations of the late 19th and early 20th centuries (e.g., Dearborn, 1910; Shinn, 1893) to the seminal work of Gesell (1928) and Halverson (1931), who conducted experimental studies with groups of young children to map the development of prehension. They viewed maturation as the regulatory mechanism in the process of development and so assumed that there was an orderly, and relatively invariant, sequence of prehensile development as the nervous system matured. More recently, the relation between motor and cognitive development in infancy has been of interest to those studying prehension. The focus of this research has been on how changes in cognitive capacity during infant maturation influence the ability to perceive and interact with the environment (Connolly & Elliott, 1972). This approach has also assumed a relatively orderly sequence of development of prehensile skills. Another perspective from which infant motor development has been conceptualized is the dynamic systems approach (Thelen & Smith, 1994), which proposes that new motor skills result from the interactions of the developing nervous system with perceptual processes and the environment.

The sequence of prehensile development progresses from reflexive neonatal grasping (described earlier), through grasping initiated by contact with an object (i.e., tactile stimulation), to visually guided grasping. A schematic representation of this progression of hand use in infants from no contact to a superior forefinger grasp is shown in figure 9.10. This sequence is not invariant and should not be rigidly tied to the chronological ages described below. There are variations in the onset of different patterns of finger coordination, and examples of omissions and reversals in the sequence of acquiring fine manipulative skills are found in the developmental literature (e.g., Newell & McDonald, 1997). The palmar grasp in which all fingers close together around an object is evident in infants around 16 weeks of age, and by about 28 weeks the thumb is used in opposition to the other fingers when grasping (hand grasp in figure 9.10). In some studies (e.g., M. C. Jones, 1926), thumb opposition has been reported to occur in younger children (15 weeks). However, in that study, it was not present in all children at this age; moreover, the properties of the object used to elicit the grasping response were not described. By about 36–40 weeks, thumb and finger movements begin to be independent, and infants can now grasp objects between the tips of the thumb and index finger in a prehensile grip (superior forefinger grasp in figure 9.10). Most of these early grips do not allow for movement of the fingers, and so they are called rigid grips (Connolly & Elliot, 1972). Between the ages of 1 and 3 years, children develop a wide range of prehensile patterns that permit movements of the fingers while grasping in what are described as flexible grips (Manoel & Connolly, 1998).

There have been several studies of the development of prehensile force control in young children using the instrumented grip objects described in chapter 6 (e.g., Forssberg, Kinoshita, Eliasson, Johansson, Westling, & Gordon, 1992). In these studies, children grasp a version of the object that has been modified so that it has a wider base of support, which enables it to be placed on a table in front of the child or, for the younger children, in the palm

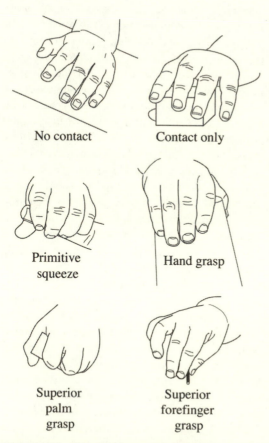

Figure 9.10. Schematic representation of the progression of hand use in infants from no contact to a superior forefinger grasp, as originally described by Halverson, 1931.

No contact

Contact only

Primitive squeeze

Hand grasp

Superior palm grasp

Superior forefinger grasp

of the experimenter's hand. The child grips the object between the thumb and index finger and lifts it from the table while the vertical movement and the grip and load forces are recorded. Some younger children often use the middle finger in addition to the index finger for additional support when lifting (Forssberg, Eliasson, Kinoshita, Johansson, & Westling, 1991). In adults, there are parallel increases in the grip and load forces as the object is grasped prior to lifting, and the entire loading phase is programmed in one force-rate pulse. A specific pattern of muscular activity and associated force development characterizes each phase of the grip and lift movement, and these are programmed to achieve specific goals. Children below the age of 2 years do not demonstrate this parallel force programming, instead generating grip and load forces sequentially. They often produce a negative load force initially by pressing the object against its support surface, increasing the force in steps with multiple force-rate pulses until the object is lifted from the support surface, as illustrated in figure 9.11 (Forssberg et al., 1991). By the onset of a positive load force, the grip force is already large, and a high grip force is maintained throughout all phases of the lifting task. For children younger than 2 years, the peak grip force can be twice that used by adults (Forssberg et al., 1991). These high grip forces result in large safety margins (as defined in chapter 6), which are being used to prevent slips. Up to 3–4 years of age, children generate a larger safety margin for rough materials than is required to prevent slips (Forssberg, Eliasson, Kinoshita, Westling, & Johansson, 1995). This suggests that they are unable to use tactile information to adapt the grip force to the friction conditions. However, the high safety margins may also be related to the large fluctuations in grip force (see figure 9.11) seen in young children.

In contrast to adults, children younger than 6 years show considerable variability in the grip and load forces produced as they repeatedly lift an object, and they generate excessive oscillating grip forces during the static phase of lifting, as shown in figure 9.11 (Forssberg et al., 1991). The higher safety margins adopted by children under 5 may compensate for these fluctuations in force and so reduce the probability that the object will be dropped. It is also known that the automatic adjustments in force that are triggered by tactile afferent responses in adults when an object begins to slip (Johansson & Westling,

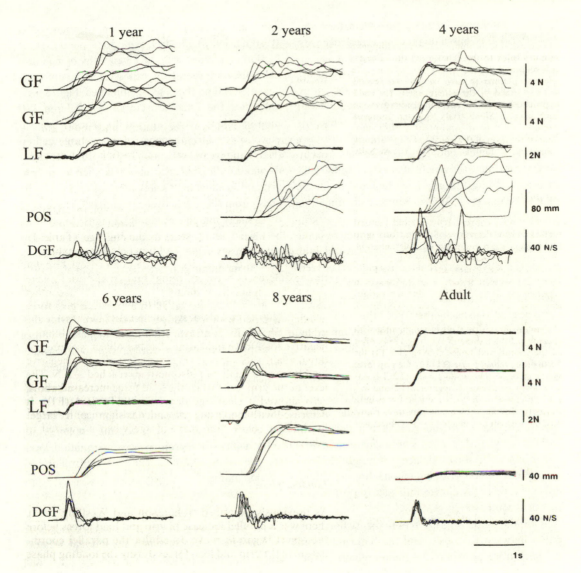

Figure 9.11. Superimposed records of lifts performed by a representative child in each age group and by one adult. Grip force (GF) for the thumb and index finger, load force (LF), vertical position (POS), and grip force rate (DGF) are shown as a function of time. The weight of the object was 200 g, and the grip surfaces were suede. The position trace is absent for the 1-year-old children as they sat in their parents' laps while the test object was presented to them in the palm of the experimenter's hand. This modified procedure precluded monitoring position data. Reprinted from Forssberg et al., 1991, with permission of Springer-Verlag and the authors.

1984) are much weaker or even absent in young children (Evans, Harrison, & Stephens, 1990), and so this mechanism is not available in children to compensate for slips.

It is not only the magnitude of the forces that differentiates the grasping patterns of young children from those of adults, but also the duration of the grasp. Children take much longer to grasp and lift an object than do adults. The time from the first contact with the object to the increase in load force is twice as long in children younger than 3 years than in adults. Younger children often make several contacts with the object before they are able to establish a stable grasp and initiate the loading phase, which may reflect the ineffective shaping and closure of the hand during reaching, which was described above

(Hofsten & Rönnqvist, 1988). The anticipatory control of grip force based on previous experience with the object emerges between 1 and 2 years of age but is still immature until 4–6 years (Forssberg et al., 1992). This means that it takes longer for young children to build up an internal representation of an object's weight based on experience and to use this information to assist in programming their grasping forces. Visual cues about an object's size are used by adults to estimate the weight of an object and so help in programming the grip force required to lift an object (Gordon et al., 1991a). Children are unable to use size information for anticipatory control of grip forces before the age of 3, even though they respond to differences in weight by 1.5–2 years (Gordon et al., 1992). This suggests that the association between size and weight involves further cognitive development in which the properties of different objects must be related to each other. When this association does emerge, children are much more influenced by it than are adults, even if the size cues are not meaningful (Gordon & Forssberg, 1997). These findings suggest that there is a separation between the visual perception of size for the purposes of perception and action, as related to grasp formation. Studies of the size-weight illusion in young children indicate that children as young as 2 years are influenced by the size of objects in their judgments of weight (Robinson, 1964). When grasping objects, however, they do not use these size cues to anticipate the grip forces that are required.

The adaptation of grip forces to the friction condition between the object and the hand relies on feedback from cutaneous mechanoreceptors in the fingerpad, as described in chapter 6 (Johansson & Westling, 1988). Children begin to respond to these friction conditions around 18 months of age provided they are repeatedly exposed to the same surface material on the object being lifted. Figure 9.12 illustrates the force trajectories of children (20 months and 8 years) and adults when lifting an object covered in sandpaper or silk. Young children (less than 5 years) use much higher grip forces than do adults for less slippery materials, such as sandpaper, and so adopt high safety margins to prevent the object from slipping between the fingers. The sizes of these safety margins gradually decrease in children between 1 and 5 years as the child matures and is able to use tactile friction information more effectively to modulate grip force (Forssberg et al., 1995).

In summary, the control of prehensile force in children undergoes considerable refinement as the sensory and motor systems develop, but it is not until the age of 11–15 years that the anticipatory control of grip force reaches maturity. Prior to this, young children are able to compensate for their limited prehensile abilities by using higher grip forces than are required to prevent slipping and by prolonging the movement phases associated with grasping and lifting objects. These strategies become less important as the ability to integrate sensory information develops and as the neural mechanisms controlling prehension mature.

The Elderly

Impairments in grip force and fine manual dexterity are frequently reported in the elderly. Maximum

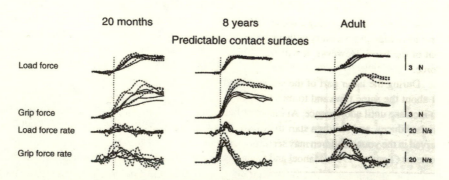

Figure 9.12. Force trajectories from an instrumented test object during lifts in which the grip surface changed for children (20 months and 8 years) and adults. The force trajectories are from lifts with sandpaper (N = 4; solid line) and silk (N = 4; dashed line). Reprinted from Forssberg, 1998, with permission of MacKeith Press.

hand-grip and palmar-pinch forces decrease as a function of age after approximately 40 years as shown in figure 9.13. This reflects the significant reduction in muscle mass with age, known as sarcopenia, which can range from 25% to 45%, as well as changes in the number and contractile properties of motor units (the contractile elements) in muscles (Galganski, Fuglevand, & Enoka, 1993). It appears that distal muscles are more affected by age than are proximal muscles, in that there is a greater decline in the strength of hand-grip muscles than of the elbow flexors (Viitasalo, Era, Leskinen, & Heikkinen, 1985). Older individuals are less able than younger people to maintain submaximal finger forces at a constant amplitude (Galganski et al., 1993; Ranganathan, Siemionow, Sahgal, & Yue, 2001) and have higher error rates than young people when tracking a visual target by generating pinch forces (Lazarus & Haynes, 1997).

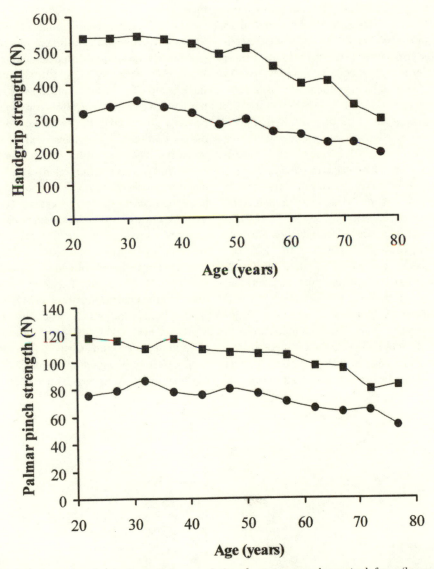

Figure 9.13. Maximum hand-grip strength (upper plot) and maximum palmar pinch force (lower plot) of the right hand of men (squares) and women (circles) as a function of age. Data are the means of between 21 and 31 participants in each of 12 age groups between 20 and 75+ years. Data are from tables 2 and 5 in Mathiowetz et al., 1985.

Tasks that involve manipulating small objects take longer to perform, particularly after the age of 60. This slowing of motor performance cannot be attributed to a general slowing of function, as it appears to affect fine manual skills much more than gross manipulation (C. D. Smith et al., 1999). Changes in the grasp forces used to hold objects have been reported in the elderly and are presumed to be related to this decline in dexterous manipulative abilities. Cole and his colleagues (Cole, 1991; Cole & Beck, 1994; Cole, Rotella, & Harper, 1999) and Kinoshita and Francis (1996) have used an instrumented test object similar to that developed by Johansson and his colleagues (described in chapter 6) to study the control of finger grip forces in the elderly. In all of these studies, the grip forces used by older people (69–93 years) were found to be more than twice as large as those used by younger adults (18–32 years), and the safety margins of older participants were significantly higher than those of younger individuals (see figure 9.14). The large safety margins used by the elderly may compensate for reduced tactile acuity on the fingertips (Stevens, 1992), which would result in an impaired ability to encode friction information at the skin-object interface (Cole et al., 1999). The friction at this interface is encoded by fast adapting afferents with small receptive fields (FA I), with Meissner's corpuscles being the putative ending (Vallbo & Johansson, 1984). Histological studies have revealed that these receptors decrease substantially in number in the elderly and that they undergo morphological changes (Bolton et al., 1966; Cauna, 1965), as described in the introduction to this chapter.

Age-related changes in skin properties, such as friction, must also be taken into account in explaining the higher safety margins adopted by older people when grasping small objects. Kinoshita and Francis (1996) found that the coefficient of friction of their elderly participants was 67% to 83% that of the younger individuals, indicating that the fingers of the older people were more slippery than those of the younger adults. This presumably results from the reduced eccrine sweat gland output (Silver, Montagna, & Karacan, 1965) and the associated decrease in the moisture content of the skin that has been described in the elderly (Potts, Buras, & Chrisman, 1984).

The relation between manual dexterity and tactile spatial acuity on the fingertips has been investigated in several studies with older adults. Tremblay, Wong, Sanderson, and Cote (2003) found that in their group of 33 older adults (aged 60–95 years), there was a significant positive correlation ($r = .66$) between the spatial tactile threshold, measured using a grating-orientation test, and manual dexterity scores, measured on the grooved-pegboard test. These findings were interpreted as indicating that impaired tactile sensation contributes to the decline in manual dexterity in the elderly. However, the strength of this relation does seem to depend on the motor task used to evaluate dexterity and the measure of tactile acuity selected. Cole, Rotella, and Harper (1998) found that impaired tactile sensations, as defined in terms of pressure thresholds, did not affect their participants' ability to grasp and lift a small sphere as long as the properties of the object being grasped and lifted were predictable. The grooved-pegboard test used by Tremblay et al. (2003) is a more demanding prehensile test than the grasp-and-lift task, as people have to grasp and ori-

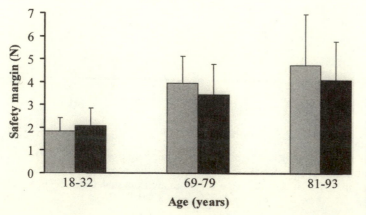

Figure 9.14. Safety margins used by young and old adults when grasping surfaces covered with sandpaper (gray bars) and silk (black bars). Data are the means and standard deviations from 10 participants in each age group and are from table 2 in Kinoshita & Francis, 1996.

ent small pegs for insertion into holes. Each peg has a key along one side and must be rotated to match the groove in the hole into which it is being inserted. These more-exacting task requirements presumably place a higher demand on tactile feedback from the fingertips, and if this is deficient, people perform the task less efficiently.

Psychometric Studies of Prehension

Infants and Children

A number of standardized tests exist to evaluate development in young children, such as the Bayley Scales of Infant Development (Bayley, 1969), the Peabody Developmental Motor Scales (Folio & Fewell, 1983), and the Miller Assessment for Preschoolers (Miller, 1982). These tests include items that evaluate hand function in terms of grasping, hand use, eye-hand coordination, and dexterity. They are primarily used to identify developmental problems in high-risk infants and young children, and so they focus on detecting abnormalities in hand use. The tests usually consist of items that are assessed as being performed or not performed.

In school-aged children, hand function has been evaluated using tests of fine manual dexterity, such as the Purdue pegboard test (Tiffin, 1968). In this test, the individual picks up a peg from a well at the top of a board, places it in a hole on the board, and continues to do so for 30 s as quickly and accurately as possible. The performance of each hand individually and both hands together are evaluated in terms of how many pegs are placed in the holes within the time limit. There are also assembly tasks that involve placing washers and collars on the peg using both hands. Gardner and Broman (1979) measured the performance on this test of more than 1300 children aged 5–16 years and found that scores increased monotonically with age and had not reached adult levels at 16 years. In this study and one involving individuals aged 14–19 years (Mathiowetz, Rogers, Dowe-Keval, Donahoe, & Rennells, 1986), girls consistently obtained higher scores (i.e., placed more pegs) than boys. This test has been used with younger children (2–6 years), who also demonstrated improved performance with increasing age and higher scores with the preferred (right) hand, although this was more evident in the older age groups (Wilson, Iacoviello, Wilson, &

Risucci, 1982). The ability of children to perform functional activities, such as writing and manipulating small objects, has been evaluated using the Jebsen test of hand function, a timed test comprising seven subtests: printing a sentence, turning over index cards, picking up small common objects and placing them in a container, stacking checkers, simulated feeding, moving large empty cans, and moving large weighted cans (Jebsen, Taylor, Trieschmann, Trotter, & Howard, 1969). The time taken to perform this test decreases from 6–7 years until 10–11 years, at which age the level of performance is comparable to that of adults (Taylor, Sand, & Jebsen, 1973).

The Elderly

Most studies of hand function in the elderly have used standardized dexterity tests or functional activities to evaluate performance. On tests involving reaching and picking up small objects and then placing them in holes, such as the Purdue pegboard test, the elderly are slower than younger adults. In a three-year longitudinal study of 264 60- to 90-year-olds (mean age: 72 years), Desrosiers, Hebert, Bravo, and Rochette (1999) reported that over the 3-year period, there was an 8% decline in performance on the Purdue pegboard test and that gross manual dexterity declined 13%, as measured using the box-and-block test (a test consisting of moving the maximum number of blocks within 60 s from one side of a box to the other, across a partition). It is of interest that the decline in manual performance over the 3-year period was related to the initial score obtained by an individual, but not to age, that is, the initial score on a test was a better predictor of the score obtained 3 years later by the same person than the person's age. On other tests, such as the Jebsen test of hand function, performance changes with age, with older people taking longer to complete the test than younger individuals. The largest increase in completion time for this test occurs in people aged 80–89 years (Hackel, Wolfe, Bang, & Canfield, 1992).

Non-prehensile Skilled Movements

Tapping and Pointing

The main class of non-prehensile movement that has been studied across the lifespan is finger tapping, a standard measure of motor speed. A variety of finger-tapping tasks have been used to study motor development and laterality in children and manual speed in the elderly (Nagasaki, Itoh, Maruyama, & Hashizume, 1989; Salthouse, 1984; Wolff & Hurwitz, 1976). These include tapping as fast as possible with a stylus on a contact plate, tapping with one or two digits (usually the index finger) on a key as rapidly as possible, or following the steady rhythm of a metronome by tapping with the fingers. The tapping speed of younger children (5 years) is significantly slower than that of older children (6–9 years), which reflects both a longer dwell time (the delay before initiating the next response) and a prolonged deceleration phase as the finger approaches the key in the younger children (Schellekens, Kalverboer, & Scholten, 1984). When children are asked to maintain a steady tapping rhythm with their left and right index fingers alternately, using the beat of a metronome, girls (6–12 years of age) are more accurate than boys in following the rate of the metronome and are better at maintaining a steady beat. In both girls and boys, the right hand is steadier than the left (Wolff & Hurwitz, 1976). These manual asymmetries and sex differences become less evident after the age of 12 years.

Pointing movements have been studied in school-aged children (5–12 years) with the objective of understanding how proprioceptive and visual cues are integrated during movements and what factors affect the precision with which the movements are executed. Hofsten and Rösblad (1988) studied manual pointing movements using a task in which the children were required to look at a target dot on a table and then place a pin underneath it as precisely as possible. The availability of visual and proprioceptive cues was controlled in different experimental conditions. In some trials, only visual cues were available to locate the target. In others, proprioceptive cues were used to locate the target by placing the finger on it and then participants closed their eyes while they reached underneath the table to place the pin on the target. Finally, both visual and proprioceptive cues were available when one finger remained in contact with the target, at which the child looked, while the other hand placed the pin underneath it. In all conditions, pointing errors decreased with age, with most of the decline occurring between the ages of 5 and 9 years. Performance was significantly better when visual cues were available, and the addition of proprioceptive cues did not

result in an increase in pointing accuracy. In addition, when only proprioceptive cues were available, errors were considerably higher (by about 5–8 mm) than when the target location was defined visually. The superiority of visual control in pointing accuracy found in this study is consistent with findings in adults and has been reported in other studies with children using different tasks (e.g., Pick, Pick, & Klein, 1967).

Studies of the accuracy of aiming movements using Fitts' task (described in chapter 7) have been undertaken in school-aged children (Connolly, Brown, & Bassett, 1968; Hay, 1981) but not in very young children. In adults, the time taken to reach a target is determined by the amplitude of the movement and the accuracy requirements or endpoint tolerance for the movement (see chapter 7). Fitts' law does apply to children over 5 years of age, and between the ages of 5 and 12 years, movement time on this task decreases. This presumably reflects an improved ability to control the muscle forces required to transport the arm between the two targets. There is also a decrease in the slope of the function relating the index of difficulty (ID) to movement time (MT) with age (see figure 9.15), which has been interpreted as indicating that not only are the movements more rapid in older children, but they become more accurate for any given distance (Hay, 1981).

Pointing movements have been studied in older individuals (55–85 years), who were required to touch a target location with a pen held in their hands after perceiving the target either visually or kinesthetically. In the visual condition, participants first saw the target and then, with their eyes closed, attempted to touch the target; in the kinesthetic condition, the target location was first perceived only kinesthetically by passively moving the participants' hands to the target location and then back to their laps. The participants then actively returned their hands back to the perceived target location. Using this task, Lovelace and Aikens (1990) found that there was no effect of age on the accuracy with which people reached for the target, independently of how its location was perceived. These findings suggest that proprioception in the arm and hand is relatively well preserved in healthy older adults.

Kinematic analyses of rapid aiming (tapping) movements in elderly participants indicate that they spend more time in the target approach or deceleration phase than do younger adults (Cooke, Brown,

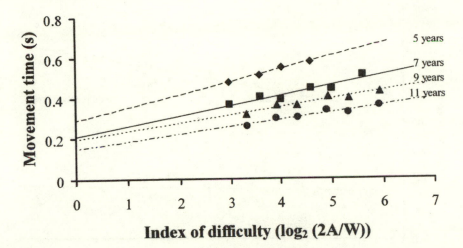

Figure 9.15. Mean movement time as a function of the index of difficulty (ID) of the movement as defined in terms of movement amplitude (*A*) and target width (*W*) in children 5, 7, 9, and 11 years of age. There were 12 children in each age group. Linear regression lines relating ID to movement time are also shown for each age group. Adapted and reprinted from Hay, 1981, with permission of the Helen Dwight Reid Educational Foundation.

& Cunningham, 1989), as was noted for reaching movements. Older people are also more affected than younger individuals by increasing the accuracy demands of the task (e.g., making the targets smaller) and make discrete adjustments in the trajectory of the hand. This suggests that they are more dependent on sensory feedback during movement execution and do not rely on feed-forward or on-line processes to achieve fast and accurate performance, as do younger people (Pohl, Winstein, & Fisher, 1996). As a consequence, the aiming movements of older people become slower, less continuous, and more variable (Morgan et al., 1994; Yan, Thomas, & Stelmach, 1998).

Typing and Piano Playing

A few studies of older skilled pianists and typists have focused on analyzing whether highly practiced motor skills are affected by advancing age, like other motor abilities. Salthouse (1984) studied a group of skilled typists aged 19–72 years and found that although there was a significant increase in choice reaction time and a decrease in tapping rate with age, as had previously been reported (e.g., Welford, 1988), there was no decline in typing speed with age. The relation between age and typing speed (i.e., interkey interval) and choice reaction time is shown

in figure 9.16. The median interkey interval in typing was nearly identical in typists 20 and 70 years old, whereas the choice reaction time increased by 80–150 ms. There was, however, a significant effect of age on eye–hand span, that is, the span between the character being visually fixated and the character being typed; typists in their 60s had between 254 and 264 ms additional preparation time relative to typists in their 20s, as defined by the product of the eye–hand span and the median interkey interval. This suggests that older typists may compensate for their declining perceptual-motor speed by more extensive anticipation and overlapping of upcoming keystrokes (Salthouse, 1984). It is also possible that the performance of the older typists at younger ages was already superior to that of the younger typists to whom their present performance was being compared and that the older typists' skills were better preserved because of a slower age-related decline. A longitudinal study of typing skills would be required to address this hypothesis.

A similar pattern of results was observed in a study of young and old expert and amateur pianists aged 20 to 68 years (12 pianists in each of the four groups). Krampe and Ericsson (1996) reported that performance on music-related tasks showed age-related decline for amateur pianists, but not for experts. Measures of musical performance included

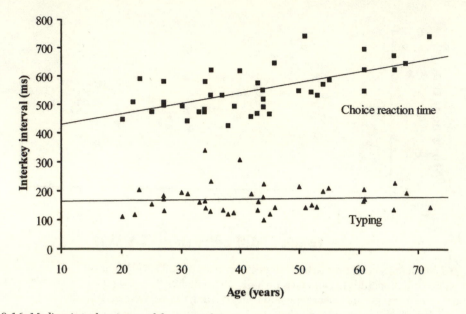

Figure 9.16. Median interkey interval for normal typing on choice reaction time tasks as a function of the age of the typist. Each data point represents a single typist, and the solid lines illustrate the regression equations relating age to interkey interval. Redrawn from Salthouse, 1984, with permission of the author.

movement coordination within and between hands, error rates, consistency of musical phrasing, and variation in loudness (force). On these measures, older expert pianists did not demonstrate any significant decrements in skilled musical performance, as compared to younger expert pianists. This elite performance had to be actively maintained, however, as task performance was related to the amount of practice in which the elderly expert pianists engaged. This measure, and in particular the amount of practice during the last 10 years, correlated most strongly with individual differences in performance, and the relation was strongest in the older expert pianists.

Krampe and Ericsson (1996) also found that the choice reaction times of older pianists were significantly longer than those of younger pianists, consistent with the findings from older skilled typists (Salthouse, 1984). However, the tapping performance of both younger and older expert pianists was superior to that of amateurs on all three measures of tapping speed (left index finger, right index finger, alternate index fingers), and no effect of age was found for either amateur or expert pianists. The lat-

ter result is surprising in that tapping speed has consistently been reported to be affected by age, even in skilled typists (Salthouse, 1984). It suggests that the skills required for rapid finger tapping are related to those acquired and maintained by expert pianists and that a sufficiently high level of deliberate practice in later adulthood preserves fine manual skills.

These studies on pianists and expert typists suggest that there is not an inevitable decline in all manual skills with age and that overlearned motor skills, such as piano playing and typing, can be preserved, if they are practiced regularly and frequently (Krampe & Ericsson, 1996). The specificity of this preservation is surprising, in that related motor tasks, such as pressing a key in response to a cue (choice reaction time) or finger tapping for skilled typists, are affected by advancing age. It has been proposed that expert pianists and typists acquire specific mechanisms (e.g., the enhanced eye-hand span noted in expert typists) that mediate their performance and permit them to overcome the general processing limitations that increase with age and affect their performance on other, unskilled tasks (Krampe & Ericsson, 1996).

10

Applied Aspects of Hand Function

In Japan today, there are so many new data entry devices that young people are called oyayubi sedai, the Thumb Generation.

TENNER, 2004

The results of fundamental research on human hand function offer valuable information about the capacities and limitations of the human user. Such information may be directed toward the solution of a number of real-world problems that depend on the sense of touch and manual dexterity. In chapter 10, we address five application domains that have been selected because of the critical mass of scientific results that are relevant to these topics. These include the evaluation of hand function and rehabilitation, language communication with the hand, sensory communication systems for the blind, haptic interfaces, and exploring art by touch. In each case, we begin with a description of the application domain, followed by consideration of design issues and possible implications of selected research findings.

Evaluation of Hand Function and Rehabilitation

The clinical evaluation of the hand is of critical importance in determining the extent of functional loss in people with traumatic injuries or diseases and in assessing the outcome of various surgical and rehabil-

itative procedures. Several aspects of hand function are typically included in assessment protocols, and these can be grouped into three broad domains. First, the integrity of the musculoskeletal system is measured in terms of the maximum force generated by different muscle groups and the range of motion of individual joints. Second, tactile, thermal, nociceptive, and proprioceptive sensations may be assessed; and third, the functional capabilities of the hand are evaluated (L. A. Jones, 1989). The particular tests chosen to evaluate hand function vary with the clinical condition and with which aspects of hand use it affects (Callahan, 2002; Dannenbaum & Jones, 1993; Varney, 1986). For example, in a person with rheumatoid arthritis, the focus of an evaluation will be on the musculoskeletal system and the functional aspects of the hand, whereas in an individual with a peripheral nerve injury, a comprehensive evaluation of sensory function would be undertaken together with an assessment of the functional capabilities of the hand.

Muscle and Joint Function

The maximum forces developed by muscles controlling the hand depend on the particular grasp adopted, which is determined by the task being performed, as described in chapter 8. A loss in grip strength is associated with a number of different neurological and musculoskeletal conditions, and so an assessment of hand-grip strength is generally included in hand evaluations as a test of gross motor power (Binkofski, Seitz, Arnold, Classen, Benecke, & Freund, 1996; Boatright, Kiebzak, O'Neil, & Peindl, 1997). Several large-scale studies have provided comprehensive normative data on the grip strength of healthy children (Ager, Olivett, & Johnson, 1984; Mathiowetz, Wiemer, & Federman, 1986; Newman, Pearn, Barnes, Young, Kehoe, & Newman, 1984) and adults (Hanten et al., 1999; Mathiowetz et al., 1985; Schmidt & Toews, 1970; Teraoka, 1979). These studies have indicated the importance of considering the sex, age, and hand preference of the individual when interpreting grip-strength data in clinical populations. They have also shown that although height and weight are positively correlated with grip strength (Hanten et al., 1999; Newman et al., 1984; Schmidt & Toews, 1970), the influence of these variables is considerably smaller than that of either sex or age. The average grip strength of women is approximately 60% that of men, and for both sexes grip strength reaches a maximum during the third decade of life and declines thereafter with increasing age (Hanten et al., 1999; Mathiowetz et al., 1985; Teraoka, 1979). The average difference between the hands in strength is approximately 10% in both left- and right-handed people (Schmidt & Toews, 1970; Thorngren & Werner, 1979; Woo & Pearson, 1927). It has also been shown, however, that the grip strength of the nonpreferred hand is equal to that of the preferred hand in approximately 25% of normal individuals (Bornstein, 1986) and that almost 50% of left-handed people have equivalent or greater strength in their nonpreferred hand (Armstrong & Oldham, 1999; Petersen et al., 1989). Differences between the hands in strength must therefore be interpreted with caution if disability or loss of function is defined in terms of such a discrepancy.

The maximum forces produced in the three precision grips shown in figure 8.5 (i.e., tip, palmar, and lateral pinch) are also frequently evaluated in clinical hand assessments. The peak forces generated with the palmar- and lateral-pinch grips are about 40% greater than that produced with the tip pinch, as illustrated in figure 10.1 (Mathiowetz et al., 1985). As would be expected, the factors that affect hand-grip strength, namely, age and sex, also influence the maximum forces generated with the three precision grips. The maximum forces produced with the power and precision grips are used relatively infrequently, and so can-

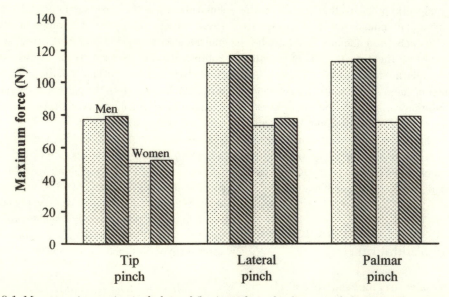

Figure 10.1. Mean maximum tip pinch, lateral (key) pinch, and palmar pinch forces for the left (dotted bars) and right (striped bars) hands of men ($n = 108$) and women ($n = 104$) between 20 and 39 years old. Data are from tables 3, 4, and 5 in Mathiowetz et al., 1985.

not be used to predict in which manual activities an individual can engage. It has been estimated that a grip force of approximately 40 N is sufficient for performing up to 90% of daily activities and that a pinch force of 10 N is adequate for accomplishing most of the simple grasping activities that involve the digits (Swanson, De Groot Swanson, & Göran-Hagert, 1990).

As part of the evaluation of the musculoskeletal system, the passive and active ranges of motion of the joints of the hand are measured using a goniometer. As with muscle-strength measurements, guidelines have been published that specify both the starting position of the hand during these assessments and the placement of the goniometer on the joint being evaluated (American Academy of Orthopedic Surgeons, 1965; Norkin & White, 1985). The data recorded can then be interpreted with reference to published normative data (American Academy of Orthopedic Surgeons, 1965), and the degree of severity of any impairment can be classified according to accepted definitions of disabilities (American Medical Association, 1993; Swanson, 1964).

Sensory Testing

Tactile sensibility of the hand is evaluated using a number of sensory tests, including many of those described in chapter 4. The most commonly used are two-point discrimination, pressure sensitivity, and haptic object recognition (L. A. Jones, 1989). Other procedures, such as vibratory threshold testing and tests of thermal sensitivity, are generally used with particular clinical conditions. For example, vibratory testing is often conducted in people with peripheral nerve injuries (Szabo, Gelberman, Williamson, Dellon, Yaru, & Dimick, 1984) but is less informative in individuals with cortical lesions. In this latter group, the perception of vibration can be preserved even when there are deficits in two-point discrimination, haptic object recognition, and position sensitivity (Derouesné, Mas, Bolgert, & Castaigne, 1984; R. T. Ross, 1991). The limited usefulness of vibratory testing in patients with cortical lesions was commented on in 1927 by the English neurologist Gordon Holmes, who noted: "The tuning fork is one of the least useful tests in cortical disease, for, as Head has shown, the appreciation of vibration is possible through the thalamus alone" (1927, p. 418).

Spatial acuity on the hand is frequently evaluated using the two-point discrimination test. Although normative data are available for two-point thresholds on different regions of the hand, such as the fingertips and the transverse axis of the palm (e.g., Corkin, Milner, & Rasmussen, 1970; Weinstein, 1968), most clinical sensory testing has focused on the palmar surfaces of the distal phalanges because of the high concentration of tactile mechanoreceptors and the importance of this region to functional use of the hand (Gellis & Pool, 1977; Wynn Parry & Salter, 1976). Two-point discrimination thresholds can vary as a function of the force applied when one point is in contact with the skin as compared to two points and with differences between examiners in the magnitude of forces applied. Brand (1985) has shown that even in skilled clinicians these forces can vary from 0.04 to 0.4 N. It should also be kept in mind, as noted in chapter 4, that many researchers dating back to Tawney in 1895 have argued that two-point discrimination is not a measure of tactile spatial acuity (Craig, 1999; Johnson & Phillips, 1981; Johnson, Van Boven, & Hsiao, 1994). This statement is based on the changes in two-point discrimination thresholds that occur with repeated testing and on the differences in thresholds that result from changing the criteria for judging the presence of two stimuli versus one stimulus. It appears that observers use intensive cues rather than spatial information to discriminate one from two points, that is, a single point can feel sharper than two points (Craig & Johnson, 2000). Several other measures show promise in assessing the ability to resolve fine spatial detail, including gap detection and grating orientation, both of which were described in chapter 4. These measures do not appear to be contaminated by intensive cues and seem to reflect the underlying neurophysiology of the skin. In clinical populations the grating-orientation test has been shown to provide a reliable index of recovery (e.g., Van Boven & Johnson, 1994).

A variation of the two-point test that involves moving the caliper or two-point aesthesiometer along the surface of the finger in a proximal-to-distal direction has been used to assess sensory function in people with peripheral nerve injuries (e.g., Patel & Bassini, 1999; Wei, Carver, Lee, Chuang, & Cheng, 2000). Dellon (1978, 1981) advocated using these dynamic stimuli to evaluate the integrity of fast adapting mechanoreceptors in the skin. With both static and dynamic stimulation (i.e., stationary and moving two-point tests), the FA I receptors are stimulated, although moving stimuli clearly provide a longer period of activation for all receptors. For this reason, it is not surprising that they are easier to detect (Louis, Greene, Jacobson, Rasmussen, Kolowich, & Goldstein, 1984). In people with median or ulnar nerve injuries, Dellon (1978) found that moving two-point thresholds decreased more rapidly than static two-point thresholds during the period of recovery following nerve surgery and showed a better overall recovery, which suggests that moving two-point thresholds were a more sensitive measure of improvement. Consistent with these results on the sensitivity of the test, Robertson and Jones (1994) found that in people with cortical lesions resulting in sensory loss, moving two-point thresholds could be elevated when static two-point thresholds were within normal limits. These findings are in agreement with the observation that different cortical mechanisms are involved in processing moving and static tactile stimuli. Together, these results suggest that evaluating the capacity to discriminate moving tactile stimuli is a useful adjunct to the assessment of static thresholds.

The ability to localize a point of stimulation on the hand is an important component of sensory function and is often impaired in people with peripheral nerve injuries (Hallin, Wiesenfeld, & Lindblom, 1981) and in those with cortical lesions involving the parietal cortex (Rapp, Hendel, & Medina, 2002; Schwartz, Marchok, Kreinick, & Flynn, 1979). In the case of peripheral nerve injuries, the mislocalization of tactile inputs results from misdirected axon regrowth in the peripheral nerve, which causes a disruption of the spatial ordering of projections from different regions of the hand to the cortex (Lundborg, 1993; Wall, Kaas, Sur, Nelsen, Felleman, & Merzenich, 1986). As can be seen in figure 10.2, when tactile stimulation is applied to reinnervated skin after median nerve injury and repair, the stimulus is felt not only in the area of actual stimulation but in other skin areas as well (Hallin et al., 1981). For people with cortical lesions, the mislocalization of tactile stimuli is presumed to reflect the reorganization of the somatosensory representation, which occurs subsequent to the cortical damage (Rapp et al., 2002). Tactile localization is evaluated by asking people either to point to the locus on the skin that was just touched with a blunt probe (Nakada, 1993) or to indicate whether two successive perceptible stimulations were at the same or two different locations on the hand (Corkin et al., 1970; Weinstein, 1968). Performance is scored in terms of the error of localization or the minimal distance that can be discriminated. Localization is most accurate on the distal pads of the fingers with errors averaging 1.5–1.8 mm and is least accurate in the middle of the palm (Nakada, 1993).

Pressure sensitivity is evaluated using the Semmes-Weinstein monofilaments, which are an adaptation of the von Frey hairs developed in the late 19th century. The psychophysical method of limits (see chapter 4) is used to determine pressure sensitivity on the fingertips, and the thresholds measured are then compared to normative data (Bell-Krotoski, 2002; Corkin et al., 1970; Weinstein, 1968). In a long-term study of sensory recovery in people with

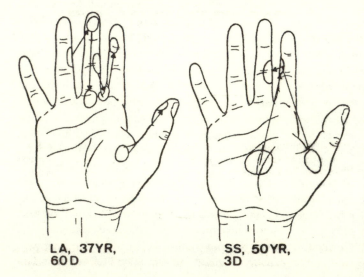

**LA, 37YR,
60D**

**SS, 50YR,
3D**

Figure 10.2. Patterns of spatial mislocalization of tactile stimulation in two individuals with traumatic median nerve injury and repair. The initials under each hand identify the person; their ages are given in years; and D refers to the delay from injury to nerve suture. Circular areas mark skin regions where the tactile stimulus was applied, and arrows point to the secondary areas where the stimulus was also localized. Reprinted from Hallin et al., 1981, with permission of Elsevier.

digital nerve injuries that had been surgically repaired, Poppen, McCarroll, Doyle, and Niebauer (1979) found that pressure thresholds were mildly correlated with two-point discrimination thresholds ($r = .66$). This is not surprising in that the two tests measure different aspects of tactile sensation.

Haptic object recognition, often called tactile gnosis or stereognosis, is a standard component of many neurological examinations and is included in many hand assessment protocols (e.g., Brink & Mackel, 1987; Moran & Callahan, 1984). As described in chapter 5, object recognition by active haptic sensing requires the integration of multiple sources of information, particularly the material and geometric properties of the object. Two methods have been used to assess haptic object recognition in clinical testing. First and most commonly, an object is placed in the unseen hand, and the person is required to name the object. The time taken to recognize the object is recorded in addition to the number of objects correctly identified. These scores are then compared to the normal response times, which are typically around 1–2 s for common objects (Klatzky et al., 1985). The second method uses a haptic matching procedure to evaluate object recognition. With this method, the person palpates an object in one hand and then attempts to find a replica of it in an array of objects. Haptic matching is more difficult than simply naming the object and has mainly been used in studies of hand function in people with cerebral lesions, who may have language or memory impairments that limit their ability to name an object (Semmes, 1965; Weinstein, 1962).

Deficits in haptic object recognition are frequently associated with elevated thresholds on the two-point discrimination and pressure sensitivity tests in people with peripheral nerve damage (Dellon & Kallman, 1983; King, 1997) and with cortical lesions that affect hand function (Robertson & Jones, 1994). A number of terms are used to describe these deficits in object recognition consequent to cortical damage, including astereognosis, tactile apraxia, and tactile paralysis. In contrast to the situation in people with peripheral nerve damage, in individuals with cerebral lesions, impairments in haptic object recognition can occur that cannot be attributed to the severity of their primary sensory loss (Caselli, 1993; Reed, Caselli, & Farah, 1996). In these cases, which are usually associated with posterior parietal lobe lesions, sensory thresholds on the hand are nor-

mal or only mildly impaired, but there is a profound deficit in recognizing objects haptically, which appears to result from an impairment in tactile shape perception and manual exploration strategies (Pause, Kunesch, Binkofski, & Freund, 1989). A kinematic analysis of the hand movements used by individuals with parietal lobe lesions as they explored objects haptically revealed that there was a decrease in the frequency and regularity of finger movements as compared to normal volunteers, even though repetitive finger movements could be performed at normal rates (Binkofski, Kunesch, Classen, Seitz, & Freund, 2001). Moreover, there was a significant negative correlation ($r = -.92$) between the size of the exploration space used to identify the object and the score on the test, where a higher score represented better performance. This means that the individual finger movements were less effectively controlled in people with parietal lesions, and the resulting tactile information was not accurately synthesized into a representation of the object. In contrast, the finger movements of people with anterofrontal lesions were comparable to those of normal individuals on all measures except regularity, which was defined in terms of the dominant frequency of the finger movements.

It would be interesting to examine the manual exploratory procedures (Lederman & Klatzky, 1987) of people with sensory loss due to other etiologies, such as peripheral nerve lesions, in order to see what strategies they use to identify objects and their properties. Improper selection of EPs has been shown to limit the type of peripheral sensory information available (Klatzky et al., 1987; Lederman, Klatzky, & Reed, 1993; Reed, Lederman, & Klatzky, 1990). These studies may prove useful as an adjunct to the sensory reeducation techniques that are used with people who have had peripheral nerve transection and repair (Carter-Wilson, 1991; Mackinnon & Dellon, 1988). Sensory rehabilitation attempts to retrain people to interpret correctly the "abnormal" sensory input arising from the hand due to the degeneration of receptors, erroneous reinnervation, and the establishment of discontinuous receptive fields—all of which can result from the nerve injury (Carter-Wilson, 1991).

The proprioceptive system is often evaluated in people with cortical or subcortical lesions (e.g., Jeannerod, Michel, & Prablanc, 1984) but is much less frequently tested in individuals with damage to the peripheral nervous system. The most commonly used tests are passive movement detection and weight discrimination, although other procedures, such as compliance discrimination (Roland, 1987), have also been used. Passive movement testing involves having the examiner passively flex or extend the proximal or distal joint of each digit a number of times (usually 10), and on each trial the person indicates in which direction the digit was moved. Normative data are available for this test, and from the scores it is possible to determine whether there is a slight, moderate, or severe impairment in proprioception (Corkin, Milner, & Taylor, 1973). Weight discrimination (i.e., barognosis) is less frequently assessed, although it was a common element of neurological exams early in the 20th century (Head, 1920). The task typically involves asking the person to indicate which of two weights that are lifted successively by the same hand is heavier (Victor Raj, Ingty, & Devanandan, 1985) or to sort objects varying in weight according to their perceived heaviness (Caselli, 1993). In contrast to the passive movement test, weight discrimination is an active process and so may be a better indicator of the status of the proprioceptive system.

Functional Tests

A number of tests have been developed to evaluate the functional capacity of the hand (e.g., Jebsen, Taylor, Trieschmann, Trotter, & Howard, 1969; H. B. Smith, 1973) as defined in terms of how the hand is used to perform a set of activities (e.g., picking up objects, writing, turning cards, stacking checkers, simulated feeding). The motivation for developing these tests arose from clinical studies of hand function in which it was found that the results from tests of sensory abilities, such as two-point discrimination and pressure sensitivity, could not be used to predict the functional capacity of the hand. This discrepancy between the results from sensory tests and the functional capacity of the hand was commented on by Moberg (1958, 1962), a Swedish hand surgeon, who then devised tests that related more to hand function than to sensation. The picking-up test in which blindfolded subjects pick up common objects and place them in a container is one such test. Moberg (1962) reported that in patients with peripheral nerve injuries, performance on the picking-up test was positively correlated with two-point discrimination thresholds but not with the results on any other test that he used to measure sensory function.

The picking-up test became the model from which many other hand function tests were devised (e.g., Dellon, 1981; Ng, Ho, & Chow, 1999). Usually each hand is tested separately, and the test is performed with and without vision. Performance is evaluated in terms of the time taken to complete the test or the number of responses made within a given time period, as described in chapter 9. Information regarding how the hand executes these tasks in terms of the kinematics of the movements and the forces produced is usually not considered (L. A. Jones, 1989). Unfortunately, the laboratory-based testing systems that have been used for studying reaching and grasping (described in chapter 6) have not been developed into more generally available clinical tests, although they have been employed in studies of specific patient populations (e.g., Forssberg, Eliasson, Redon-Zouitenn, Mercuri, & Dubowitz, 1999; Hermsdörfer et al., 2004 [unilateral brain lesions]; Müller & Dichgans, 1994; Nowak et al., 2002 [cerebellar lesions]; Robertson & Jones, 1994 [left-hemisphere cortical lesions]).

Language Communication with the Hand

As described in chapter 7, the hand can be used as a communication device when other sources of sensory information are limited due to disability or environmental limitations (Kendon, 1993; Klima & Bellugi, 1979). The contexts in which the hand is used to communicate range from the movements used to converse with others in sign language to scanning the raised-dot patterns of Braille letters with the fingertips so that text can be read. The hand can also be used to understand speech by placing it on the talker's face and monitoring the articulatory motions associated with the speech production process. This method, known as Tadoma, is believed to have been developed by a Norwegian teacher in the 1890s and was first used in the United States in the 1920s by Sophia Alcorn to teach two deaf and blind children, Tad Chapman and Oma Simpson, after whom the technique is named (Alcorn, 1932; C. M. Reed et al., 1985). The physical characteristics of speech production that are sensed by the Tadoma reader include airflow, which is felt by the thumb on the lips; the laryngeal vibration on the speaker's neck; and the in-out and up-down movements of the lips and jaw. The latter two features are detected by the fingers (Reed, Durlach, Braida, & Schultz, 1989). The information that the hand provides to the central nervous system of the Tadoma reader (see figure 10.3) is multidimensional but of sufficient resolution that extremely subtle changes in airflow or lip movement can be detected and interpreted in terms of a linguistic element. The speech reception capabilities of people who are both deaf and blind and who are extensively trained in the use of Tadoma are comparable to those of normal subjects listening to speech in a background of noise, at a speech-to-noise ratio of approximately 0–6 dB (C. M. Reed, 1995; Reed et al., 1985). Tadoma has also been shown to be superior to any artificial tactile display of speech that has been developed to date (C. M. Reed et al., 1992) and has demonstrated the possibilities of speech communication through the tactile domain.

Another tactile communication method, which is more commonly used in individuals who are both deaf and blind, is the haptic reception of sign language. This method is generally used by deaf people who acquired sign language before becoming blind. The deaf and blind individual perceives the signs produced by the signing hand(s) by placing a hand or hands over the signer's hand(s) and passively tracing the motion of the signing hand(s). The communication rates obtained with haptic reception of sign language are lower than those achieved with visual reception of signs (1.5 signs/s, compared to 2.5 signs/s), and errors are more common with tactile reception (Reed, Delhorne, Durlach, & Fischer, 1995). Nevertheless, the haptic method of communication is effective in deaf and blind individuals who are skilled in its use, and the levels of accuracy achieved make it an acceptable means of communication.

The performance of the hand as it communicates using the various systems described above and in chapter 7 can be compared in terms of words/s or in terms of information transfer (bits/s; Shannon & Weaver, 1963). The latter is defined as the increase in information that follows the transmission of a signal, which is simply a measure of how much more the receiving system knows about the state of the transmitting system after a signal is received than before (Sheridan & Ferrell, 1974). A comparison of the rates of communication and of information transfer for a number of different communication methods is shown in figure 10.4 (Reed & Durlach, 1998). As can be seen in the figure, the highest rates of information transfer (20–30 bits/s) are achieved when reading, listening to spoken English, or watching the gestures of American Sign Language (ASL). Methods of communication that involve interpreting sequential presentations of letters, such as vibrotactile presentation of Morse code or finger spelling, are very inefficient (1.4–8.1 bits/s) when compared to speech and sign language. Despite this limitation, they may be very accurate (Reed & Durlach, 1998; Reed, Durlach, & Delhorne, 1992).

The information transfer rates associated with different methods of producing language, ranging from speaking to signing to stenography, are shown in figure 10.5. The typical information transfer rates for the various methods of production vary from approximately 2 bits/s for Morse code to 23 bits/s for speech (Reed & Durlach, 1998). The output rates for communicating with the hand are generally lower than those achieved when speaking, although both signing and typing using a shorthand machine keyboard (i.e., stenography) achieve output rates that are close to those for speech.

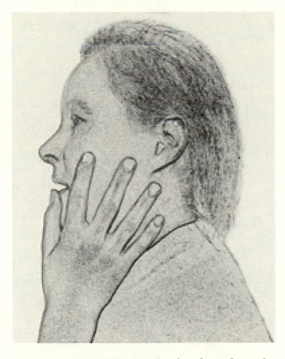

Figure 10.3. Position of the hand used in the Tadoma method to understand speech with the fingers sensing movements of the lips and jaw, vibration of the larynx, and airflow from the mouth.

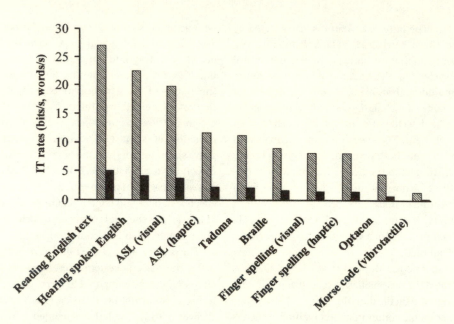

Figure 10.4. Typical information transfer (IT) rates (bits/s) for various methods of receiving language (striped bars). Normalized transmission rates (words/s) are also shown (black bars). ASL = American Sign Language; the Optacon is a reading device for the blind that scans text and displays it on the index fingertip using an array of vibrators. Data are from table 1 in Reed & Durlach, 1998.

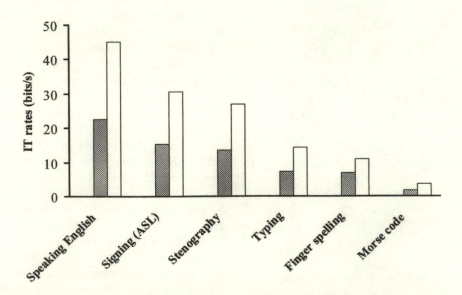

Figure 10.5. Typical (striped bars) and maximal (white bars) information transfer (IT) rates in bits/s for various methods used to produce language. Data are from table 3 in Reed & Durlach, 1998.

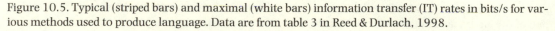

Sensory Communication Systems for the Blind

Tangible Graphics

Graphic displays use a variety of point, line, and areal (regional) symbols to provide information spatially in the form of pictures, maps, pie charts, or graphs. Although most commonly explored by eye, tangible graphic displays are meant to be read by hand as a method of providing spatial information to users with limited or no vision. Tangible graphic displays have been employed as supplements to Braille for classroom learning of a variety of disciplines, such as geography, the social sciences, physics, biology, and mathematics. They have also been used as tactile aids, allowing blind individuals to plan and navigate their environment, thereby reducing the need for a sighted companion. Evidence suggests that relief maps were created for the visually disabled more than 300 years ago, toward the end of the 17th century (Eriksson, Jansson, & Strucel, 2003). Eriksson et al. note that the oldest extant map, which is housed in Vienna, was probably produced for Maria Theresia von Paradis, a singer and composer who was blind. It is a standard print map, with borders, cities, and rivers produced in relief. National borders are embroidered; rivers are shown using finely stitched pewter thread; and cities, towns, and villages are symbolized with buttons of different sizes. Despite the maps' potential value, many visually impaired individuals have never explored a tactile map or other tangible graphic format (see Eriksson et al., 2003; Schiff & Foulke, 1982). Limited production methods (described below) and infrequent access to such aids have no doubt contributed significantly to this situation.

How might information be displayed to make it easily interpretable by hand rather than by eye? Until recently, tangible graphic displays have typically consisted of static two-dimensional arrangements of raised symbols with little, if any, variation in the third dimension. Material cues have rarely been used. A few examples are shown in figure 10.6: the human digestive tract (panel A), a line graph (panel B), and a relief map of the world (panel C).

The effective design of tangible graphic displays for the blind raises a number of important issues that relate to the inherent capabilities and limitations of manual sensing, as discussed in chapters 4 and 5. General principles based on scientific results can be used to guide the creation of tangible graphic displays and to anticipate the relative success with which different graphic formats (e.g., overlays, underlays) may be used. For further information on tangible graphics, the reader should consult Schiff and Foulke (1982). A more recent booklet devoted specifically to tactile maps by Eriksson et al. (2003) provides both a practical discussion of production issues and an overview of the research and development of tactile maps.

How does haptic processing constrain the effective use of current static displays? As just noted, tangible graphic displays have usually consisted of statically presented, raised contours, such as those shown in figure 10.6, with little if any variation in either material cues (e.g., vibration, friction, texture, compliance, thermal properties) or three-dimensional structure. The difficulty with which such current displays are typically processed reveals several inherent limitations in tactile and haptic processing, which were highlighted in chapters 4 and 5. The haptic system is relatively inefficient with respect to the speed and accuracy with which a spatial layout of raised two-dimensional contours can be processed (Lederman et al., 1990); in contrast, the haptic system is particularly effective at processing variations in material characteristics and, although still somewhat limited, is better with three-dimensional (compared to two-dimensional) geometric information (Klatzky et al., 1985).

In chapter 5, we addressed the reasons for the considerable discrepancy in the skill with which people haptically recognize common objects (Klatzky et al., 1985) as compared to their raised-line depictions (Lederman et al., 1990). Early on, those who designed and produced tangible graphic displays had little, if any, access to scientific information regarding human haptic processing. Of course, there was relatively little to offer. It is only recently that scientists have recognized that the haptic system can process highly complex information and clearly deserves to be comprehensively studied. With limited understanding of the sense of touch per se, it is not surprising that designers typically chose to produce tangible graphic displays as raised replicas of visual representations, inappropriately assuming that the visual and haptic systems always function similarly.

Although the visual system can extract valuable information from edges for the purposes of recognizing

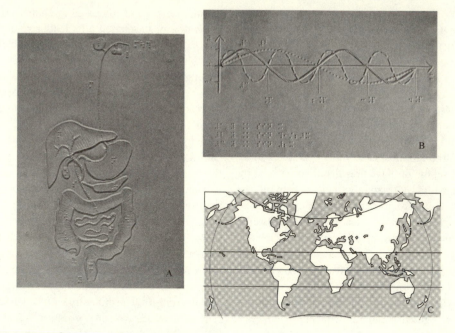

Figure 10.6. Examples of raised-line graphics. (A) Human digestive tract; (B) a line graph. Adapted from Schiff & Foulke, 1982. (C) Relief map of the world. Reprinted from Eriksson, Jansson, & Strucel, 2003, with permission of the Swedish Braille Authority.

objects and perceiving spatial layout, the haptic system is far less skilled at doing so, for reasons discussed in chapters 4 and 5. First, the spatial acuity of the fingertips is inferior to that of the fovea, which means that the haptic system is relatively limited in its ability to process very fine spatial details. Second, static raised-line tangible graphic displays force the haptic system to process two-dimensional geometric features based on the information extracted by the exploratory procedure known as contour following (Lederman & Klatzky, 1987). Although contour following is necessary for obtaining and integrating spatial details into an overall form, it is still inaccurate and unacceptably slow. The heavy memory load imposed on users restricts perceivers' ability to integrate the sequential inputs into a global spatial representation (B. Jones, 1981; Klatzky & Lederman, 2003a). Third, it is difficult for users to group or parse the spatially confusing two-dimensional information contained in raised-line displays. Fourth, although whole-hand exploration significantly improves haptic object identification and shape perception by making available multiple inputs distributed across the hand and by providing three-dimensional geometric information (Lederman et al., 1993; Lederman &

Klatzky, 2004), such haptic information is spatially coarse compared to what is typically available to the visual system.

Inasmuch as the haptic system is seriously limited in its ability to process spatially distributed edge information, graphic representations that demand the integration of precise spatial details, such as raised-line drawings of common objects (e.g., figure 10.6A), are likely to be apprehended poorly. Recall that Lederman et al. (1990; also see chapter 5) showed that both blindfolded and sighted subjects recognized a set of such common object depictions slowly and inaccurately. Although good performance has been achieved (e.g., Heller, Calcaterra, Burson, & Tyler, 1996), the conditions under which such levels are observed tend to be highly restrictive (e.g., limited number of stimulus objects; invariant stimulus size, orientation, and shape; categorical precuing). Practice with a specific exemplar can no doubt improve performance, but it would be more desirable ultimately if users could access an internationally accessible database of stored representations provided by designers from around the world. Given the considerable limitations with respect to how well the haptic system can inherently process static, raised-line representations of real objects and their spatial layout, we question the general value of relying on the use of such two-dimensional displays in educational venues. To serve as a useful communication tool, information in a display must be accessible both quickly and accurately. This has not reliably occurred with representational raised-line displays.

In contrast, other types of static, raised, two-dimensional displays that make less-precise spatial demands on haptic processing or that do not require explicit object identification may not be subject to the same limitations and can thus be quite useful. In one such example involving line graphs, congenitally blind participants haptically determined the features of one or more functions on a raised-line graph (e.g., shape, slope, converging versus diverging functions, the coordinates of a point of intersection). Presumably, they capitalized on the cutaneous and coarse proprioceptive information available during manual exploration (Lederman & Campbell, 1983). Under such circumstances, there is no need to resort to the extensive temporal integration that is necessary for producing a complete mental representation of the overall display.

Suggested Improvements

Adjustment of Two-Dimensional Details: Omission, Addition, and Distortion   Because the haptic system is relatively imprecise at processing raised two-dimensional geometric details, some map designers have proposed the judicious deletion, addition, and even distortion of critical features for proper map interpretation. They suggest that such map "editing" may facilitate easy access to and comprehension of information available in a specific tactile map (e.g., Eriksson et al., 2003). For example, details that are unnecessary to the required map function may be omitted. Important details that are relatively complex may be enlarged, whereas other parts of the map that are less significant may be reduced in magnitude to preserve a reasonable size of work space within which to explore manually. Although we do not know if these suggestions have been formally evaluated, the ideas seem plausible on the basis of what is now known about the nature of haptic processing. It is certainly the case that the visual system can recognize caricatures of faces, as opposed to photographs of those faces, quite easily despite the fact that the latter contain more "accurate information" (Gombrich, 1972, 1977).

Adding Three-Dimensional Structure   Structural information in the display may be enhanced by introducing full three-dimensional geometric variation wherever possible. In this way, the perceiver can capitalize on the valuable geometric cues made available by grasping, the exploratory procedure known as enclosure (Lederman & Klatzky, 1987), whereby the palm and fingers conform to the object's contours. As we have previously noted, the elimination of such information by rigidly splinting the fingers impairs haptic object recognition. Under such limited conditions of haptic exploration, the geometric information normally provided by the proprioceptive system is unavailable (Klatzky et al., 1993; see chapter 3).

A recent technological innovation known as a haptic interface will be described in detail in the subsequent section on haptic interfaces. These complex electromechanical systems provide force feedback to users as they explore objects and surfaces within a real or virtual work space by moving a probe. Most of the current systems deliver point-contact force feedback, which is experienced as resistance to movement over virtual objects and surfaces whose properties and locations are computationally modeled. The goal of such systems is to create a perceptual experience akin to feeling real objects and surfaces with a rigid probe.

Haptic interfaces offer the tangible graphic designer the possibility of creating virtual three-dimensional displays that may ultimately improve the identification of objects, their properties, and their spatial layout beyond what is currently achievable with static raised two-dimensional displays (Kilchenman-O'Malley & Goldfarb, 2002). Nevertheless, performance may still be somewhat limited, for reasons discussed in chapter 8 (Kirkpatrick & Douglas, 2002; Lederman & Klatzky, 2004).

A possible solution involves the use of multiple sensors and actuators in a display, which would provide haptic feedback distributed within or across the hand(s) (Burdea & Coiffet, 2003). Current multicontact systems (e.g., gloves) are costly and cumbersome to operate, thus reducing the effectiveness of the proprioceptive feedback normally available during hand movements (Bouzit, Popescu, Burdea, & Boian, 2002; Klatzky et al., 1993; Lederman & Klatzky, 2004). Nevertheless, we believe that systems that permit whole-hand exploration and grasping ultimately offer considerable promise for designing tangible graphic displays for the blind.

Adding Material Cues   As previously noted, research has highlighted the impressive effectiveness with which the haptic system processes variations in material features (e.g., Katz, 1925/1989; Lamb, 1983; Morley et al., 1983). The judicious addition of redundant variation in virtual material properties in a haptic interface may further enhance the efficacy of current raised-line two-dimensional graphic displays of objects (Holmes, Hughes, & Jansson, 1998; Klatzky et al., 1989; Lederman et al., 1996; C. L. Reed et al., 1990).

Reducing Tactile "Noise"   Like visual graphics, poorly designed raised two-dimensional graphic displays may appear to be highly cluttered. Any symbol in the display that is not a target becomes a distractor, or haptic "noise," and too many distractors sub-

stantially interfere with processing the target symbol. One method that has been proposed to reduce clutter on graphs (see figure 10.7) involves the use of an underlay, a sheet that presents valuable spatial reference information (e.g., a grid) beneath and physically separate from the main display (Lederman & Campbell, 1983). By freely exploring the graph with the preferred hand, the blind user can effectively obtain relatively simple spatial layout information about the raised-line functions in a reduced-noise environment (i.e., high signal/noise ratio). When more precise coordinate information is desired (e.g., about points of intersections), the grid underlay is used in conjunction with the graph above. Once again, participants use their preferred hand to explore the relatively uncluttered display on the top sheet (e.g., to locate a point of intersection

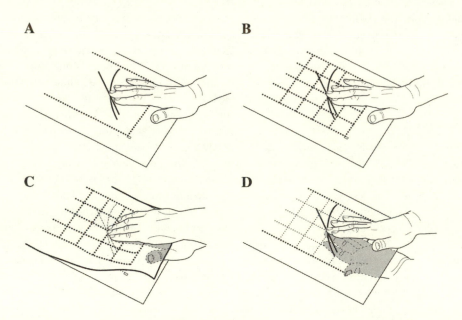

Figure 10.7. Four different modes of displaying a tangible line graph. (A) Graph only; (B) the graph and grid presented on the same surface; (C) separate grid overlay (top surface) and graph (bottom surface). The middle finger of the preferred hand (gray shows occluded portions) is used to find a desired point of intersection on the graph (bottom surface). The middle finger of the other hand (top surface) is then used to find and mark that point on the grid overlay (top surface), then to read off the coordinates. (D) Graph (top surface) with separate grid underlay attached to the base of the graph surface, its raised symbols facing down. The middle finger of the top (preferred) hand is used to find and mark a desired point of intersection on the graph. The middle finger of the other hand is then used to mark that point on the grid underlay (bottom surface; gray shows occluded portions) and then to read off the coordinates. As shown, the dark solid and dotted lines represent visible portions; the gray dashed lines and dots represent nonvisible contours occluded by the presence of the overlay or underlay. Four small squares were added to the display to mark the physical corners of the graph; in addition, Braille symbols (not shown here) were used to identify the X- and Y-axis values. Adapted from Lederman & Campbell, 1983, with permission of the authors.

between two line functions). They use the other hand to locate and mark the target position of the finger on the top sheet, feeling the latter through the underlay. Precise coordinate information is subsequently read off the grid underlay using the preferred hand. This system requires users to coordinate their hands with palms together, a two-handed configuration that is both normal and comfortable. It thus contrasts with the difficulty of coordinating the use of one hand on top of the other, as required when using an overlay haptically. Indeed, working with overlays is less effective than with underlays, to which blind users adapt quite effectively despite having no experience with this new technique.

Alternately, with the promising advent of computer-controlled force-feedback systems (discussed in detail below), the user should be able to add or omit background information easily as desired. Such electromechanical displays could also vibrate selected targets, thereby offering a new and potentially effective method for dynamically increasing signal-to-noise ratios.

Choosing Display Symbols   Much like those who design visual graphic displays, tangible graphic designers face several critical problems. The first—symbol tangibility, or the ability to differentiate effectively raised symbols by touch—critically affects the optimal selection of symbols for tangible graphic displays. The symbols (point, line, and area) cannot be too small because of limitations in tactile spatial acuity (see chapter 4). Unfortunately, the solution does not lie in increasing symbol size without limit because of the relatively small haptic work space in front of the body. Rather, to minimize clutter in the display, the smallest usable size should be selected based on what is now known about tactile spatial acuity thresholds.

The second problem concerns symbol meaningfulness. To reduce the time needed to identify a symbol from a legend, whenever possible it is desirable to use the symbol's physical structure to reflect its iconic meaning. An excellent example is provided by Schiff, Kaufer, and Mosak (1966), who designed a raised saw-toothed line to represent intended direction on a map. Schiff et al. avoided choosing a raised two-dimensional version of an arrow because a small arrowhead can be particularly difficult to resolve by hand. As shown in figure 10.8, moving

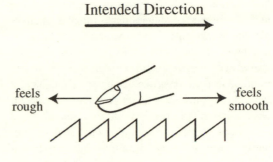

Directional Line

Figure 10.8. A raised-line symbol used to portray direction. The three-dimensional saw-toothed pattern feels rough when the finger moves across the line to the left and smooth when the finger moves across the same line to the right. The symbol in this figure is intended to represent direction to the right. Adapted from Schiff, Kaufer, & Mosak, 1966.

one's finger along the line to the right produces a smooth sensation, which represents the intended direction. In contrast, moving one's finger along the line to the left produces a rougher sensation, which is meant to represent the unintended direction.

Capitalizing on the value of three-dimensional contour for the haptic recognition of common objects (Klatzky et al., 1985), Lambert and Lederman (1989) designed a set of three-dimensional symbols (e.g., men's and women's washrooms, telephone, cafeteria) for use on tangible maps of building interiors. The three-dimensional structure of the objects being represented inherently suggested the meaning of the symbol (e.g., miniature telephone to represent a phone booth). The added three-dimensional structural cues proved effective in reducing the length of time taken to identify the symbol relative to other symbols that were tangible but without added meaning.

Symbol Standardization   The standardization of a fixed set of highly differentiable symbols is potentially desirable because it may reduce how much the visually impaired must learn about each new display. Not surprisingly, symbol standardization has been proposed a number of times in conjunction with the design of tangible graphic displays. Unfortunately, any such standardized symbol set must be usable in many different contexts. The need for a large number of standard symbols conflicts with the limited number of differentiable tangible symbols potentially available. As we have previously discussed in chapter 4, the tactile system has relatively poor spatial acuity; moreover, the user must be able to tactually detect the presence of any symbol against a noisy background. It is therefore doubtful that complete standardization is achievable, although blind individuals can spontaneously produce symbols similar to those used conventionally by sighted individuals when generating their own raised-line drawings (Kennedy, 1993).

Figure-Ground and Haptic Processing of Spatial Information   Both the visual and auditory systems demonstrate strong figure-ground organization together with a number of principles of perceptual grouping (Bregman, 1990; Goldstein, 2002). It is likely that similar or related processes are of critical importance to the effective design and use of tangible graphic displays. To date, however, we know of no comprehensive body of research that has systematically focused on how the haptic system organizes and spatially groups objects within a display.

One intriguing suggestion relates to figure-ground organization, a fundamental characteristic of early visual processing (Goldstein, 2002). In brief, blind children have demonstrated an ability to haptically separate figure from ground when interpreting a simple raised-line contour as the profile of a face (Kennedy & Domander, 1984). The results suggest that figure-ground relations depicted in a tangible graphic display affect the accuracy of haptic recognition of facial features in a manner that is similar to pictorial displays that require visual exploration. Understanding the processes that underlie haptic spatial organization in sighted, blindfolded, and blind individuals would seem critical for their epistemological value and for the effective design of tangible graphic displays for the blind.

Providing Multisensory Cues   A final potential design improvement is to provide additional redundant or complementary information to other functioning sensory systems in the form of a multimodal display. In the case of raised two-dimensional tangible graphic displays for the visually impaired, the auditory system is the most obvious choice. We note the success with which Yu and Brewster (2002) used two haptic interfaces, the high-cost PHANTOM

(SensAble Technologies) and the low-cost WingMan (Logitech) force-feedback mouse (for further details, see the section on haptic interfaces below) to render a two-dimensional bar chart either haptically or multimodally. The accuracy of unimodal and bimodal performance with each device was very similar for a number of performance measures. In addition, the bimodal condition was significantly superior to either unimodal condition. These encouraging results confirm that it is possible to use a cheaper haptic interface to produce a functional tangible graphic display. Pie charts should work well with such methods, but other more-complex forms of graphic display (e.g., multiple intersecting line graphs) will pose additional problems and therefore should be designed and evaluated separately.

Production Methods

It is important to keep in mind that the effectiveness of a tangible graphic display is significantly affected by the particular production method adopted. Until recently, tangible graphic displays have involved static presentations that are constructed in a variety of labor-intensive ways, which make further alterations difficult (Gill, 1982).

Some early methods included the following. One-of-a-kind masters have often been produced by spatially arranging and subsequently gluing various materials (e.g., string, wires, beads, sandpaper) to a base surface. These methods are time consuming, although they do allow the designer to capitalize on the haptic effectiveness with which material cues can be processed. Tangible graphics have also been created by manually producing an engraved, left-right reversed, female mold on thin metal foil. The metal sheet is then turned over to make accessible a raised display with the proper left-right spatial relations. Since the 1970s, it has also been possible to bypass the intermediate engraving stage by drawing directly on plastic "swellpaper," a material consisting of millions of tiny capsules filled with alcohol suspended in an emulsion paintlike substance that covers a backing paper. The graphic display is initially produced on the swellpaper and then exposed to heat using a specialized copy machine. The black areas on the paper expand with heat, leaving the remaining areas flat, to create a black, raised two-dimensional image that can be manually explored. The technique has become popular because original raised drawings and multiple copies may be easily produced with this technique. Unfortunately, the height cannot be precisely controlled. Flexible plastic copies of a male master (e.g., a metal foil, epoxy-resin master of the laminated female copy) have also been generated using a vacuum-forming machine. Although relatively easy to use, the vacuum-formed copies are even more imprecise than those produced with swellpaper.

In principle, the electromechanical generation of two-dimensional tangible graphic displays with large arrays of actuators is appealing; however, the cost of production and the limited reliability of such devices place such a system beyond the reach of most users. To reduce the size of the display, it has been suggested from time to time that the spatially extended physical display (e.g., map) be replaced with a digitally stored image. Much like the Optacon, the reading device for the blind described in chapter 4, the user could access the virtual display sequentially by moving a minia-ture camera over the surface of the map. A vibrotac-tile array of pins could present the portion sensed by the camera to the index finger of the other hand, updating the information displayed over time as the camera was moved across the virtual display. It would be even more desirable if the use of a second hand was entirely eliminated by combining a miniature camera and tactile array stimulator within a single device. An example of this technology is the VirTouch Mouse (http://www.skerpel.com/mouse.htm), a device that works similarly to the Optacon but that delivers infor-mation to two adjacent fingers via a pair of matrices of vibrating displays. Each display consists of 4×4 stim-ulating pins that rise and fall as the user moves the mouse across the display. It remains to be seen whether such a simple device used in conjunction with tangible graphic displays will produce a notable improvement in performance. Merely widening the tactile field of view to include two, as opposed to one, fingers does not improve the recognition of raised two-dimensional Braille characters, tactile maps, or depictions of common objects (Lappin & Foulke, 1973; Loomis et al., 1991). In contrast, distributing cutaneous and proprioceptive information across the whole hand does improve the identification of geo-metrically defined three-dimensional common objects (Jansson & Monaci, 2003, 2004; Klatzky et al., 1993), presumably because the palm and/or fingers can simultaneously mold to a considerable portion, if not all, of the object's contours.

To end this section, we would like to briefly men-tion the advent of a number of three-dimensional rapid prototyping technologies, such as stereolithog-raphy, which build plastic parts or objects a layer at a time by tracing a laser beam on the surface of a vat of liquid photopolymer. With the development of these new technologies, there now exist accurate and effi-cient ways of creating and copying three-dimensional structures, which offer promise in the construction of tangible graphic displays.

A System for Denominating Banknotes by Touch

Blind individuals are seriously limited in their ability to use banknotes in their daily financial transactions with individuals in the service industries, such as bank clerks and retail salespeople. Over the years, attempts have been made in many countries to develop an effective system of denominating ban-knotes for the blind. A report by the U.S. National Research Council (1995) examined the currency practices of 171 "issuing authorities" worldwide. Eight features, either singly or in some combination, were listed that intentionally or unintentionally could potentially help with banknote denomination by visually impaired users. These are variable-sized banknotes (e.g., Iceland, England), large numerals on banknotes (e.g., the Netherlands), variable-color banknotes for those with residual vision (all but the United States), special shaped patterns (e.g., Eng-land), specific engraved visible markings (e.g., Ger-many, the Netherlands, Malaysia), specific engraved invisible markings, such as randomly located dots (e.g., the Netherlands), watermark features (e.g., Japan), and machine-identifiable features (e.g., Canada). Because of traditions and issues of viabil-ity, no one method has been adopted as a worldwide standard.

One recent approach attempts to capitalize on our scientific understanding of how well the sense of touch is able to differentiate surface textures, as opposed to spatial patterns. A raised-texture symbol together with a banknote denomination code have been introduced on the new Canadian banknote series that has just been released (Lederman & Hamilton, 2002). In keeping with the research find-ings reported in chapters 4 and 5, the raised-dot symbol need only be identified as a "rough texture" patch as opposed to a specific microspatial pattern (e.g., Braille character), which additionally requires good spatial acuity and encoding relative to a spatial referent system.

The tactile code used to represent different banknote denominations capitalizes on redundant information available in the number of rough patches and in position or spacing cues, all easily detectable via manual exploration (figure 10.9). A $5 note has only one raised texture patch, which is located at the right side of the face of the banknote and thus at the far right end of the space allotted to the tactile code. The $10, $20, and $50 notes have two, three, and four texture patches separated by one, two, and three easily detectable gaps, respectively, therefore extending the added texture patch farther leftward in the tactile space. The $100 note consists of two texture patches in the extreme positions separated by one very large, smooth gap. In

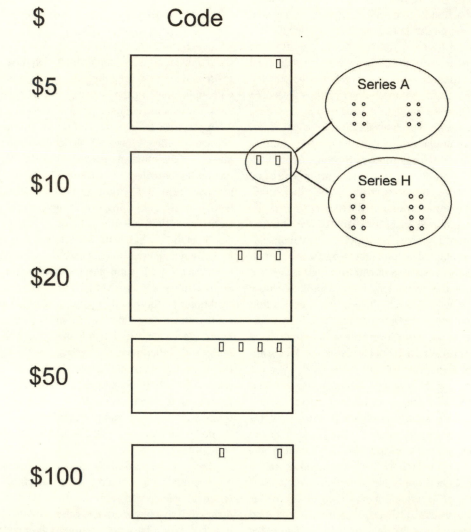

Figure 10.9. The denomination code and raised-texture symbol used to represent different Canadian banknote denominations during manual exploration. Series A and H are samples of a large number of different texture-symbol design sets that were evaluated in a banknote-denomination task performed by sighted blindfolded and blind participant groups. Both accuracy and response time were recorded. The design patterns varied in the number of rows and/or microgeometric features, such as dot size and interdot spacing. Reprinted from Lederman & Hamilton, 2002, with permission of the Human Factors and Ergonomics Society and the Bank of Canada.

keeping with research that has shown that people can haptically capitalize on redundant stimulus cues (Klatzky et al., 1989; Lederman et al., 1993; C. L. Reed et al., 1990), blind users successfully learned to use this code within a matter of minutes and with well over 90% accuracy.

Haptic Interfaces

Haptic interfaces are devices that present tactile and force feedback to a human operator who is interacting with a real or simulated object via a computer. They may enable a user to make contact with an object in the environment, to feel the object's properties (e.g., texture, compliance, or shape), or to manipulate the object directly (Biggs & Srinivasan, 2002b; Durlach & Mavor, 1994). These interfaces may be "masters" that control slave devices, such as robotic arms (telerobotics), or they may provide a means of interacting with a computer-generated virtual environment (e.g., model of the eye; Hunter, Jones, Sagar, Lafontaine, & Hunter, 1995). The interaction between the operator and the robot or virtual world is bidirectional, with either position or force being measured at the operator's hand and either force or position fed back via a computer to the operator. Usually, position (or velocity) is the input to the control loop and forces are fed back to the operator. Because there is bidirectional interaction between the user and the virtual world, this type of interface is fundamentally different from feedback systems using other modalities, such as vision and audition, as they are unidirectional.

Haptic displays are usually designed to be held or worn on the hand, and so the design and implementation of these interfaces is critically dependent on having a basic understanding of the human haptic system and how it functions. Characteristics of human performance, such as the absolute threshold for detecting a force or vibratory stimulus, changes in force amplitude or vibration frequency that are perceptible, or the stiffness required for a solid object to appear rigid, must be known in order for the haptic display to be useful and effective. Since the 1990s, the development of these devices has led to a resurgence of interest in sensory and motor studies of the human hand (Durlach & Mavor, 1994). It has also resulted in a highly interdisciplinary community of researchers that includes computer scientists involved in developing algorithms for haptic rendering, engineers designing and building haptic displays, and experimental psychologists and physiologists evaluating human performance when using these systems.

The haptic interfaces used in teleoperation or in virtual or simulated environments have matured from simple, ground-based, 2 degrees-of-freedom joysticks and cumbersome hand controllers to more dexterous interfaces that are body based and that present information to the fingertips or to the whole hand (Burdea, 1996; Burdea & Coiffet, 2003). The use of haptic devices for presenting the physical properties of objects, such as their mass or surface texture, becomes critical in those applications where the user actively manipulates the simulated or real world or where visual information is limited, such as in undersea environments, in space, or inside the human body. In these manipulation tasks, information provided haptically about an object's rigidity, mass, compliance, friction, or surface characteristics results in greater realism and assists task performance as reflected by faster completion times and decreased error rates (Hannaford & Wood, 1989; Howe & Kontarinis, 1992). The improvement in performance observed when force and tactile feedback are added to an interface is not limited to the manipulation of robotic devices, but also occurs with more simple systems, such as graphical user interfaces. For example, target acquisition has been found to be faster if tactile feedback is provided by a computer mouse, in addition to the normal visual feedback of cursor position on the computer screen (Akamatsu, MacKenzie, & Hasbroucq, 1995; Akamatsu, Sato, & MacKenzie, 1994).

Another more recent field of application of haptic interfaces is in environments in which a vast amount of information is presented visually, some of which may require an immediate response from an operator, but because of visual overload the possibility exists that such emergencies will not be detected (e.g., air traffic control, nuclear power plant regulation). An alternative mode of informing the operator of an emergency has considerable appeal in these contexts. This could be a thermal (Jones & Berris, 2003) or vibrotactile cue that functions as an alert (Hale & Stanney, 2004) and because of its novelty would be attended to, in much the same way as a vibrating cell phone or watch attracts the user's attention. Related to this application is the interest in assisting with data or scientific visualization by presenting some information haptically, for example, complex geometric forms (Pai & Reissell, 1996) or rheological properties, such as flow and deformation (Infed, Brown, Lee, Lawrence, Dougherty, & Pao, 1999). One early application of haptic interfaces to scientific visualization was in the study of molecular docking, in which the objective was to minimize the binding energy of two molecules by correctly positioning and orienting the smaller molecule with respect to the larger (Ouh-young, Beard, & Brooks, 1989). The Argonne mechanical arm was used to present the simulated forces of a molecular interaction between a drug molecule attempting to "dock" onto a protein molecule, and biochemists attempted to find the potential energy minimum in a six-dimensional space. It was found that, by using the force display, subjects were twice as fast at finding the zero force position than with a visual display alone. More recently, the PHANTOM haptic interface has been used to control the probe of an atomic force microscope (AFM). In this application, the operator pushes a nanoscale particle along a surface and can feel when the probe slips off the particle.

The quality of a haptic interface is generally interpreted in terms of the range of haptic sensations and percepts that can be presented to the user, such as hard surfaces, textures, geometry, surface deformations, and compliance, and the resolution and mechanical bandwidth of the force display (Lawrence, Pao, Salada, & Dougherty, 1996). These are a function of the components that comprise the interface (e.g., sensors, dampers, actuators) and the overall system's bandwidth, which in turn depends on computational and transmission delays. In terms of their physical configuration, a haptic interface can be a computer mouse (Akamatsu et al., 1994), a pen-based pantograph (Hayward, 1995), a stylus (Buttolo & Hannaford, 1995; Massie & Salisbury, 1994), a thimble (Massie & Salisbury, 1994), a glove (Burdea, Zhuang, Roskos, Silver, & Langrana, 1992; Caldwell, Kocak, & Anderson, 1995), or an exoskeleton (Bergamasco et al., 1994). A summary of some of these devices and their properties is presented in table 10.1. It is surprising to note, given this diversity of configurations of haptic interfaces, that relatively little research has been done on what is the optimal form or work space of an interface that is controlled by the hand. One early study in human manual control reported that it was easier to control a joystick-type device if it was held between the thumb and index finger as compared to being manipulated

solely by the thumb (Hammerton & Tickner, 1966). With the range of computer and robotic interface devices now available, including thimbles, styluses, gloves, trackballs, and exoskeletons, it would be fruitful to explore more fully the costs and benefits associated with these different interface designs.

The term haptic device is often used to describe two types of interfaces, namely, tactile and force-feedback displays, which differ in terms of the inputs or feedback that they provide to the user and in the hardware used in their fabrication (see Burdea & Coiffet, 2003). In terms of the definition of haptics provided in chapter 1, a tactile display is not a haptic device as it provides tactile but not kinesthetic inputs to the user. Kinesthetic inputs will, however, arise from the operator if the display is actively moved by hand and finger movements generated by the user. Force-feedback displays provide both tactile and kinesthetic feedback to the user. Tactile inputs arise from the contact between the device and the operator. Forces are actively produced by the display and are of sufficient magnitude that they can resist the movements of the user's hand and arm; the user's own movements constitute a second source of kinesthetic feedback, as in the case of tactile displays.

In general, force-feedback displays are powered by electric motors or other actuators and exert forces on the user. In contrast, tactile-feedback devices usually have pin arrays, vibrators, or, less commonly, air jets or bellows, that stimulate the skin. Force-feedback systems can provide coarse information to a human operator via a device such as a joystick, which is held within the palm of the hand, or more precise force information using a number of points of contact in the hand, for example, by an exoskeleton worn on the hand (Hasser & Massie, 1998). Tactile-feedback devices typically apply stimuli to the fingertip using arrays of pins that apply forces perpendicular to the skin and are actuated by piezoelectric crystals, shape memory alloy wires, solenoids, or pneumatics. Voice coil actuators and small electric motors have also been used to produce vibratory inputs to the skin.

The information about the environment that is conveyed by force-feedback and tactile-feedback devices differs. Force-feedback systems can provide information about the mechanical properties of objects, such as their compliance, weight, or inertia, in the real or simulated environment. These devices actively resist the user's movements and, provided they can generate sufficient force, can stop the

Table 10.1. Tactile and haptic devices for the hand

Name	Feedback	Actuators	Maximum force (N)	Weight (g)	Bandwidth
iFeel Mouse[a]	Vibrotactile	1	1.18 at 30 Hz	132	0–500
CyberTouch Glove[b]	Vibrotactile	6	1.2 at 125 Hz	142	0–125
WingMan 3D Joystick[a]	Force	2	3.3	NA	0–333
PHANTOM 1.5/6DOF[c]	Force	6	8.5	136 (apparent at tip)	15
PHANTOM 3.0/6DOF[c]	Force	6	22	44 (apparent at tip)	15
HapticMASTER[d]	Force	3	250	NA	10
CyberGrasp Glove[b]	Force	5	12	434	40
CyberForce[b]	Force	8	8.8	NA	NA

Adapted from Burdea and Coiffet (2003).
[a]Logitech Corporation.
[b]Immersion Technologies.
[c]SensAble Technologies.
[d]FCS Control Systems.

movements. The contact forces need to be large in order to simulate rigid objects in a virtual environment, but at the same time they should not be so large that they harm the user. In contrast, tactile-feedback systems can convey information about surface texture, geometry, and contact. These systems do not actively resist the user's contact motions and so in a virtual environment are unable to stop a user from moving through a virtual surface.

Tactile-Feedback Interfaces

A variety of tactile-feedback devices has been built for the hand, ranging from the relatively low-cost tactile computer mouse known as the iFeel Mouse (Logitech, 2004), which has a small actuator oriented perpendicular to the mouse base that can vibrate the outer shell of the mouse, to the considerably more expensive CyberTouch glove (Immersion Corporation, 2004), which has six vibrotactile actuators (electric motors with an eccentric mass mounted on the shaft), one mounted on the back of each finger and one on the palm, as illustrated in figure 10.10. These commercially available devices were clearly designed and built for different applications. The iFeel mouse was designed to assist in navigation on computer screens by providing an additional input to the user who typically relies on visual and auditory feedback to confirm that an action has been executed. The tac-

tile cues from the mouse do not require the user to look at the device and, with further developments, offer the possibility of having different tactile cues associated with different menu options or icons. Users are, however, constrained to keep their hands on the desk in contact with the mouse in order to receive feedback. In contrast, the CyberTouch glove was developed for use in virtual environment applications and allows significant motion of the hand. The configuration of the hand itself is recorded from sensors mounted over the finger joints and wrist and is then transmitted to the host computer. These measurements are used to drive the position and configuration of a virtual hand presented to the user on a visual display. The six vibrotactile actuators attached to the glove are activated whenever the virtual hand interacts with virtual objects, so they provide information about which fingers are in contact with the object. Each actuator can be individually programmed to vary the intensity of the vibration on the hand (Immersion Corporation, 2004). At present, these tactile displays deliver rather simple vibrotactile inputs to the hand at single frequencies that are within the range of maximal sensitivities (Bolanowski et al., 1988). The use of patterns of vibrotactile stimuli to convey information about objects or events in the environment holds promise and awaits more sophisticated control schemes that can control the characteristics of the inputs delivered to individual actuators.

Figure 10.10. CyberTouch (upper left), CyberGrasp (upper middle), and CyberForce (upper right) are tactile and haptic devices produced by Immersion Corporation. The CyberTouch is a tactile feedback glove that has six vibrotactile stimulators, one on each finger and one on the palm, that individually provide tactile feedback to the user. The CyberGrasp is a force-reflecting exoskeleton that provides force feedback to each finger. The CyberForce is a desktop force-feedback system that provides force feedback to each finger and the palm and measures the movement of the hand. The PHANTOM haptic devices are produced by SensAble Technologies. The PHANTOM 3/6DOF haptic device (lower left) provides force feedback in 6 degrees of freedom and has a range of motion equivalent to that of full arm movement around the shoulder joint. The PHANTOM 1.5/6DOF haptic device (lower right) also provides force feedback in 6 degrees of freedom, with a range of motion equivalent to a pivoting movement of the forearm around the elbow joint. Photographs are reproduced with permission of SensAble Technologies and Immersion Corporation. Copyright Immersion Corporation. All rights reserved.

The development of effective tactile array displays poses enormous technical challenges. Nevertheless, the results of psychophysical research offer promising scientific support for the value of displaying spatially distributed forces to the fingertips. For example, Lederman and Klatzky (1999) evaluated user performance on a battery of simple sensory and more complex perceptual tests when spatially distributed information was either presented to the distal pad of the thumb or index finger or eliminated by requiring participants to wear a rigid sheath molded to fit over the same region. The results indicated that two-point touch and pressure thresholds were both substantially impaired by the absence of spatially distributed finger deformation patterns, and detecting the orientation of a raised bar was at chance level. Roughness perception (as measured by magnitude estimation and two-alternative, forced-choice discrimination procedures) was only moderately impaired, suggesting that vibratory cues could still be used relatively effectively. Finally, detecting the

presence or absence of a simulated mass embedded within simulated tissue with 75% accuracy required the mass size to be approximately 44% larger (i.e., 6.6 mm) when the spatially distributed force cues were absent as opposed to present (i.e., 4.6 mm). Clearly, the availability of tactile cues on the fingertip improved performance across a broad spectrum of sensory and perceptual tasks.

Force-Feedback Interfaces

Force-feedback devices differ from tactile-feedback devices in that they are usually larger and heavier due to the force output requirements of the actuators. These devices may be attached to the hand and arm in the form of an exoskeleton or mounted on the floor or a desk, such as a joystick. The latter interfaces are termed earth-grounded, off-the-body systems and typically have fewer degrees of freedom than do exoskeletal devices. Several force-feedback devices are illustrated in figure 10.10. The most commonly used force-feedback devices are joysticks, which have a small number of degrees of freedom (usually 2–3), and produce relatively low forces but have a high mechanical bandwidth, defined in terms of the frequency of force/torque updates experienced by the user (Burdea & Coiffet, 2003). The WingMan Force 3-D joystick (Logitech, 2004) is a relatively simple force-feedback device with 3 degrees of freedom, 2 of which have force feedback that can produce a maximum force of 3.3 N. It was designed for use with computer games and so provides jolts or springlike forces that simulate collisions and inertial accelerations. A much more sophisticated force-feedback device, which is used in both industrial and research environments, is the PHANTOM, originally designed and built by Massie and Salisbury (1994) and now produced by SensAble Technologies. There are four different PHANTOM models, which vary with respect to their force-feedback work space, stiffness, and motor force. The premium models (see figure 10.10) provide 3 degrees of freedom of positional sensing and 6 degrees of freedom of force feedback. The interface has a serial feedback arm that ends with a stylus that is held in the hand (SensAble Technologies, 2004). Force-feedback devices, such as the PHANTOM, are used in industry and research for virtual assembly, virtual maintenance path planning (i.e., determining the optimal path for performing maintenance operations in accessible structures such as airplane engines), molecular modeling, and teleoperation.

The PHANTOM is a point-contact device and so assumes a point contact between the simulated probe and the virtual object. Point-contact displays are subject to the limitations previously raised concerning the absence of spatially distributed forces on the fingers (e.g., Lederman & Klatzky, 1999). In addition, as users are limited to a single point of contact when using the PHANTOM, unless two are used to grasp an object, each object must be explored serially in order to determine its geometry. We have previously discussed the substantial decline in haptic object recognition when using a rigid probe, which serves as a simple model for point-contact haptic interfaces. Such displays restrict the tactile (net force and vibratory cues) and kinesthetic information available to the user during manual exploration because only one extended "finger" is in effect being used. Such limited manual exploration imposes severe constraints on the temporal integration of sequentially extracted inputs, as previously discussed in chapter 5 and in the current chapter (see the section above on tangible graphics). Under these conditions of manual exploration, recognition accuracy declines markedly, and response times are unacceptably long.

The final type of device that is primarily used in virtual environment applications is force-feedback gloves that have both sensors to measure the position of the finger joints and actuators mounted over the fingers so that simulated forces can be controlled independently at each actuator. The CyberGrasp glove (Immersion Corporation, 2004) shown in figure 10.10 is one such system; it has 22 sensors to measure the position of the fingers and the wrist and 5 actuators mounted on the fingertips. The torques that are generated by these actuators are transmitted to the user's fingertips via a system of cables and a mechanical exoskeleton, which is worn on top of a CyberGlove that houses the position sensors. The exoskeleton both guides the cables and functions as a mechanical amplifier that increases the forces experienced at the fingertips. The forces applied are unidirectional and oppose flexion of the fingers. The maximum force that can be produced at each finger is 12 N within a work space of 1 m radius hemisphere (Immersion Corporation, 2004). Support for using force-feedback gloves in preference to point-contact devices comes from experimental studies that have demonstrated that whole-hand exploration is best for haptically recognizing real objects, in terms of both speed and accuracy (Lederman & Klatzky, 2004). This study also revealed impairments in performance when the finger was splinted, which highlights the importance of taking into consideration the potential sensory limitations associated with a rigid fixture such as an exoskeleton.

Although there have been considerable improvements in the design of haptic interfaces since the 1990s, limitations in hardware in providing the range, resolution, and bandwidth of forces have meant that the stimuli delivered by these systems are only a crude approximation to the natural interaction between the hand and the real environment. This does not mean that adding haptic information to a visual display lacks realism, since in the contexts in which haptic displays are used, there is no other source of information about the mechanical features of the environment. The degree of realism presented by the haptic display is often referred to as a "sense of presence." This concept has been interpreted to mean that the human operators receive sufficient information about the task environment that is displayed in a sufficiently natural way that they feel physically present at the remote (for teleoperation) or simulated (for virtual environments) site (Sheridan, 1992).

The technology used in developing haptic displays and virtual environments has enabled new questions to be asked about hand function. As described in chapters 6 and 8, instrumented gloves such as the CyberGlove (Immersion Corporation, 2004), which measure the position of each joint of the hand, have made it possible to analyze the temporal and spatial coordination between the fingers during multidigit grasping (Santello & Soechting, 1998). This work has demonstrated that there are consistent co-variations between the angular excursions of the fingers as the hand reaches to grasp an object and that the number of degrees of freedom that is independently controlled is much smaller than the number available (Schieber & Santello, 2004). The addition of robust force sensors to these gloves so that both the kinetics and kinematics of hand movements could be studied simultaneously would provide further valuable information about the coordinative control of hand movements. Glove-based haptic displays are also being used to evaluate the progress of therapy used to treat disorders affecting the hand (Burdea, 1996). As new materials are developed that can be used as both sensors and actuators, there will be additional opportunities to further our understanding of how the hand is used to sense and act on the environment.

Exploring Art by Touch

The Exploratorium is a scientific museum in San Francisco that was designed some years ago to encourage children and adults alike to gain valuable hands-on experience as a means of helping people to better understand the scientific phenomena on display. Indeed, many museums now include highly instructive hands-on presentations. However, the geodesic tactile dome, designed in 1971, is of particular interest to those who wish to experience the somewhat unsettling, always fascinating experience of exploring the sense of touch in total darkness. The

observer must crawl, climb, and squeeze through different rooms and openings whose surfaces are covered with various materials. The tactile dome offers an excellent opportunity for participants to become manually aware of the concrete world and its varied properties in a way that is denied when vision is concurrently available.

Those who visit art museums and galleries are usually prevented from touching the artistic works on display. Unfortunately, the constant wear and the acid deposited as a result of manual contact can cause significant damage to the art. The aesthetic experiences of those who are visually impaired are thus especially confined by restrictions on touching. However, even viewers with intact vision may benefit from haptic exploration. Over the years, occasional special art shows have invited the viewer to "please touch!" There now seems to be growing international interest in making art accessible to both nonsighted and sighted visitors through nonretinocentric means, including the sense of touch and other modalities. For example, a series of four workshops collectively entitled "Challenging Ocularcentricity in the Gallery" was sponsored in 2003 by the Economic and Scientific Research Council at the Tate Modern Gallery in London, England, to explore the possibilities of this uncommon mode of aesthetic experience.

In addition, Jansson, Bergamasco, and Frisoli (2003) describe a current European Union research project known as PURE-FORM, which is coordinated by PERCO in Pisa, Italy, and capitalizes on advances in haptic and robotic technologies. Its goal is to make it possible for blind individuals to visit a virtual museum, the Museum of PURE-FORM, in which they can manually explore virtual three-dimensional replicas of works of art via a haptic force-feedback display, as conceptually depicted in figure 10.11. Currently, a number of sculptures have been selected for digitization via three-dimensional laser scanning. Users will be able to select a work from the digital sculpture database and to explore it haptically using four of Lederman and Klatzky's (1987) exploratory procedures: contour following, enclosure, pressure, and lateral motion. The haptic displays include an arm exoskeleton, which delivers force feedback to the arm (figure 10.12A), and a hand exoskeleton attached to the distal end of the arm exoskeleton, which delivers forces to the tips of the thumb and index finger. The latter are inserted

Figure 10.11. The concept of PURE-FORM. When "visiting" the Museum of PURE-FORM, the human perceiver interacts haptically with a simulated, digitized three-dimensional sculpture (dotted form) via a haptic interface that delivers force feedback to the user's hand. The original object is shown on the right (solid form). Reprinted from Jansson, Bergamasco, & Frisoli, 2003, with the permission of Swets & Zeitlinger Publishers.

into thimbles mounted on the distal ends of the hand exoskeleton (figure 10.12B).

Rosalyn Driscoll, a visual artist, has been creating sculptures that appeal to the hands as well as to the eyes (Driscoll, 1995). She began this work by initially observing and learning from people with visual impairments as they explored works of art by touch. Since then, she has produced tactile/visual sculptures and exhibited them in art museums. When permitted, she also provided blindfolds and guide rails for viewers to explore art solely through their sense of touch. The scale of her sculptures is deliberately adapted to the human body. Thus, some pieces are small enough to be enclosed by the two hands; larger ones are contained within the haptic space of the arms' reach; finally, still larger ones define spaces through which the observer must climb or walk. Based on careful observation as to how the sense of touch normally functions, both on its own and when compared to vision, Driscoll has chosen to emphasize the importance of material properties and qualities in her work by creating sculptures out of a variety of durable, sensuous materials, such as wood, stone, metal, leather, and rope (e.g., figure 10.13). For Driscoll, active manual exploration of the contours, surfaces, and materials of her works plays a critical role in the aesthetic experience of the observer, thus capitalizing on what

we have learned in chapter 5 about the contributions of manual exploration to haptic perception. In addition, her sculptures tend toward abstract rather than representational imagery. This preference for abstract interpretation seems most appropriate, given the findings discussed in chapter 5 that highlight the considerable difficulty with which observers can perceive and identify the geometric properties of real objects and their spatial layout by hand.

Such an approach calls into question the validity of Révész's (1950) earlier rejection of haptic aesthetics in those born blind. His own comparisons of the visual and haptic systems in terms of normal function led him to conclude that the congenitally blind are incapable of having an aesthetic appreciation of sculpture: "The haptic process is not fit to fulfill the conditions for an aesthetic approach" (p. 169). Révész claimed that the blind observer merely attempts to recognize the form of the piece by touch. But since he limited his theoretical and practical discussions primarily to consideration of representational sculptures, it is not surprising that he arrived at this conclusion. Driscoll's approach would seem to offer a more promising avenue to the aesthetic appreciation and production of the plastic arts by both blind and sighted persons as mediated through the sense of touch.

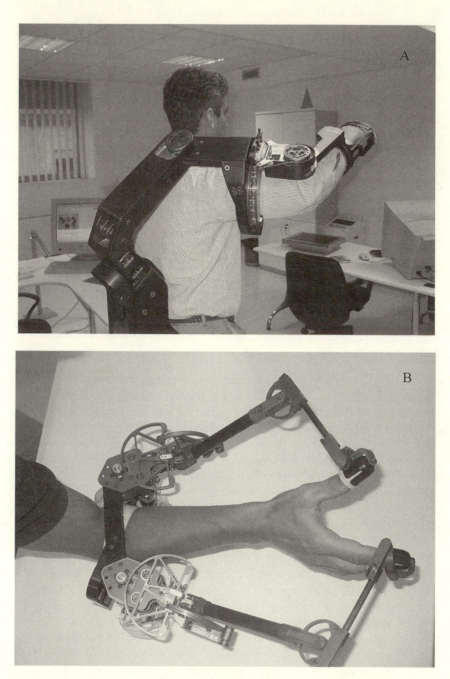

Figure 10.12. The arm (A) and hand (B) exoskeletons worn by the user to explore a virtual sculpture. Reprinted from Jansson, Bergamasco, & Frisoli, 2003, with permission of Swets & Zeitlinger Publishers.

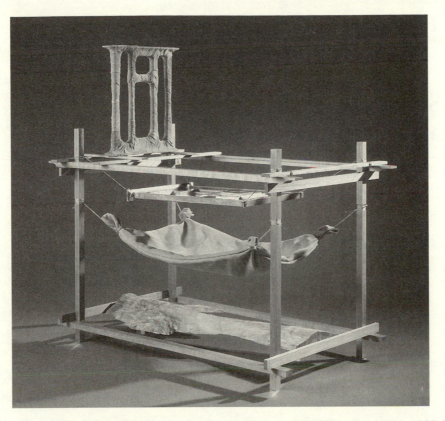

Figure 10.13. *Anatomy*, 1998. Wood, copper, and leather construction (147.32 × 86.36 × 137.16 cm) by sculptor Rosalyn Driscoll. Photograph reprinted with the permission of the artist and the photographer, David Stansbury.

Whether exploring art haptically or experiencing a computer-simulated virtual environment using a haptic display, it is clear that the active process of moving one's hand enhances the information derived and results in a more accurate image of the object. The importance of active hand movements to the identification of haptically perceived objects is further demonstrated by the impairments in haptic object perception seen in people with lesions in the parietal cortex. In these individuals, finger movements are less effectively controlled with the result that haptic information is not adequately synthesized into an object representation. The challenge for designers of robotic systems that are used for purposes of teleoperation or hand rehabilitation is to design a device that can mimic purposive manual exploration and provide information in a timely and natural fashion.

11

Summary, Conclusions, and Future Directions

Summary

The framework we created for conceptualizing hand function broadly differentiates four categories along a sensorimotor continuum that ranges from tactile sensing to non-prehensile skilled movements (see figure 1.1). Together, these four categories have permitted us to encompass the extensive range of activities considered in this review. They have also provided a structure for analyzing the factors that influence human manual performance. Any number of categories could have been used to distinguish different aspects of hand function, but four seemed to be the minimum number of categories that worked effectively.

At the sensory end of the continuum, the hand together with the central nervous system is capable of processing a variety of tactile inputs in the absence of any finger movements (i.e., tactile sensing). Based on this sensory information, people experience a wealth of subjective manual sensations that vary in their quality, intensity, and acuity. It is also possible to perceive coarse variations in the properties of objects, particularly when there is relative motion between

the skin and surface of the object. However, active haptic sensing offers many more cues about the external world. By selecting the appropriate exploratory procedure(s) (Lederman & Klatzky, 1987), people can choose not only the type(s) of information but also the appropriate level(s) of precision they require in order to achieve a particular goal. When either tactile or active haptic sensing is involved, the hand functions as the primary sensing organ for touch. In contrast, when the hand acts as a tool, its function is principally motoric, despite the demonstrated need for sensory information to achieve successful performance. This close interplay between the sensory and motor functions of the hand influences how we reach to grasp an object and then manipulate it. In the absence of sensory feedback from cutaneous mechanoreceptors, reaching movements take longer, and once contact is made with the object the forces produced at the fingertips cannot be controlled precisely and so the object often slips between the fingers (Gentilucci et al., 1997; Johansson & Cole, 1994). In contrast to the rather slow and deliberate hand movements that are made to explore objects in the environment, the movements made when the fingers interact with keyboards are rapid and efficient. These non-prehensile skilled movements are usually characterized in terms of their speed and consecutiveness. However, analyses of finger movements during typing and piano playing reveal that they are not strictly independent and serial and that they too rely on sensory feedback from muscles and skin to prevent errors in movement trajectories (Gordon & Soechting, 1995).

It is important to keep in mind that, in all of these activities, from tactile sensing to non-prehensile movements, the hand serves dual functions as a movable sensory organ and as a vehicle for motor activity. Indeed, in many situations, the hand must play both roles. For example, if we are hungry, we may reach into a bowl to select a piece of fruit. First, we squeeze it gently to determine its ripeness; if satisfied that it is indeed edible, we then pick it up and bring it to our mouth. It is even possible to perform these two sensory and motor functions simultaneously.

Conclusions

Since the 1980s, interest in studying the sensory and motor performance of the human hand has resurged in part stimulated by the development of dexterous robotic hands and haptic interfaces. For those interested in understanding the sensory neurophysiological and behavioral bases of somatosensory processing, the study of hand function is critical. Not only is the hand an important body site for studying tactile sensing, but it is undoubtedly paramount for active haptic sensing.

Tactile sensing with the hand is strongly influenced by the properties of the peripheral somatosensory system, particularly with respect to the structure, density, distribution, and sensitivities of the different mechanoreceptor and thermoreceptor populations in glabrous and hairy skin. In the case of tactile pattern recognition, performance appears to be set by such peripheral factors (Johnson & Phillips, 1981).

In contrast, when judging concrete objects, their properties, and the spaces they occupy, humans are most effective when they actively control their manual exploration. Systematic hand-movement patterns or exploratory procedures (Lederman & Klatzky, 1987) make the sensory information available. However, these must be selected and executed as appropriate to (1) the task (e.g., object recognition, object identification, or the perception of a specific object property, such as texture or shape); (2) the type of object (e.g., its size relative to the exploring hand or finger, whether it is defined by single or multiple attributes); and (3) the mode of exploration (e.g., bare hand, probe). What current research highlights is the fact that to understand fully the bases of sensory hand function, it is important to recognize that manual exploration necessarily constrains the information that is made available and thus, ultimately, one's performance. A random choice of hand movement pattern(s) does not guarantee success. Rather, EP selection serves as a funnel that determines the amount, quality, and resolution of the sensory information available. To use a simplistic example, if the only way to judge the weight of an object that is fixed to a supporting base is by rubbing the fingers back and forth across the object's surfaces with a lateral-motion EP, the observer will fail (Lederman & Klatzky, 1987). By molding the hand to whatever contours are accessible using an enclosure EP, performance will likely improve because that EP provides coarse information about object size, which is usually correlated with weight. Nevertheless, performance will not be optimal. To achieve this level, the observer must be able to lift (preferably heft) the object away from its supporting surface using an unsupported-holding EP, which provides the most precise information about weight.

For those interested in understanding the neurophysiological, biomechanical, and cognitive mechanisms involved in human movement control, the hand has several features that have made it an appealing site for study. First, the complexity of its anatomical structure with more than 20 degrees of freedom of movement poses an exciting challenge in terms of understanding how the central nervous system controls such a complex mechanical system. Second, given the diversity of actions that the hand is capable of performing, from reaching to grasp an object in the dark to playing a trill on a piano, what constraints are imposed by the central control mechanisms and the peripheral apparatus that enable such tasks to be carried out efficiently? These two issues have been addressed in a range of studies that have analyzed multidigit grasping and single-digit control (e.g., Mason et al., 2001; Santello & Soechting, 1998; Soechting & Flanders, 1997). It appears that the number of degrees of freedom that are independently controlled is considerably smaller than the number of mechanical degrees of freedom in the hand. Hence there is a simplification of the computational task during most manual activities. This is reflected in the synergistic action of the muscles of the hand during reaching-to-grasp movements when changes in finger joint angles are synchronized in time as the hand is preshaped to fit the object's dimensions (Gentilucci et al., 1991; Santello et al., 2002). As the object is grasped, the grip forces are synchronized in time across a variety of grasp conditions, so that they too are not independently controlled (Johansson & Cole, 1994). However, should it be necessary, the motor control system can rapidly adjust the trajectory of the fingers or forces produced by a single digit (Edin et al., 1992; Paulignan, Jeannerod, et al., 1991; Paulignan, MacKenzie, et al., 1991).

The analyses of keyboard skills, such as piano playing and typing, have contributed to our understanding of the mechanisms involved in the learning and retention of manual skills. This research has revealed what features of performance change as one progresses from being a novice to an expert and how, in these essentially serial motor tasks, so much of the planning and programming of finger movements occurs in parallel. The novice typist is primarily limited by cognitive constraints (i.e., processing text, identifying key positions, planning finger movements), whereas the main restriction affecting skilled performance is motoric. For the expert, the key variables that influence the speed with which letters are typed are the location of the letters on the keyboard and the fingers used to type them (Gentner, 1988). The speed with which the piano is played is explicitly represented in the musical score, but the force with which the keys are struck is actively controlled by pianists as it determines sound intensity and contributes to the dynamics of the musical performance (C. Palmer, 1997). For skilled pianists, it is their ability to use variations in the force or intensity of notes played that differentiates their performance from that of amateurs (Krampe & Ericsson, 1996; Shaffer, 1980).

Researchers involved in the design of robotic hands built for dexterous applications have carefully scrutinized these results on the sensory and motor capabilities of the human hand and used them to provide benchmarks for the design and evaluation of robotic hands (L. A. Jones, 1997; Venkataraman & Iberall, 1990). A major expectation of some early developers of dexterous robotic hands was that the robotic hand, with its anthropomorphic geometry, would attain performance levels and capabilities comparable to those of the human hand (Melchiorri & Vassura, 1993). This expectation neglected the formidable role played by the central nervous system in hand function. As the comparative anatomist Frederic Wood-Jones wrote in 1944: "It is not the hand that is perfect, but the whole nervous mechanism by which movements of the hand are evoked, co-ordinated and controlled" (p. 300). In this context, it is of interest to note that whereas sensing the internal positions (joint angles) and forces (joint torques) has been implemented in most dexterous robotic hands (e.g., Mason & Salisbury, 1985; McCammon & Jacobsen, 1990), imparting a sense of touch to these devices remains a significant challenge. A similar situation exists with the prosthetic hands used by people with upper-extremity amputations. A number of systems have been developed for providing tactile and proprioceptive feedback to prosthetic hands (e.g., Kyberd et al., 1995; Scott, Brittain, Caldwell, Cameron, & Dunfield, 1980), but these have met with limited clinical success due to problems with sensor technology, difficulties in affixing the sensors to the contact surface, and fragility of the sensors (Scott & Parker, 1988).

Human Hand Function has provided designers of robotic and prosthetic hands with current research information pertaining to what has been discovered about the sense of touch, from simple tactile sensation to the complex haptic perception of objects, their properties, and their spatial layout. The experimental studies reviewed in the book have also illustrated the importance of sensory information from the hand for the successful execution of many motor tasks. This research clearly indicates the value of incorporating effective tactile sensor technology in artificial hands, either prosthetic or robotic.

Future Directions

From a sensory perspective, it seems surprising that almost 120 years after the identification of various cutaneous mechanoreceptors, uncertainties still exist regarding the association between different cutaneous afferent fibers and their receptor endings. In the recent study by Paré et al. (2003) that is described in chapter 3, only one Ruffini corpuscle was found within the skin on the distal phalanx of three human cadaver hands. These results demonstrate the need for more detailed histological studies of human specimens, given the assumption that Ruffini corpuscles are the putative endings for a major class of cutaneous mechanoreceptor in the human hand. Such investigations are also required across the lifespan so that changes in the sensory and motor performance of the hand as it develops and matures can be analyzed with reference to the distribution and properties of cutaneous mechanoreceptors.

Recent research that uses the methods and models of genetics, molecular biology, and electrophysiology offers an exciting new multidisciplinary approach toward understanding the mechanisms of human tactile function. Molecular biological research is now focused on discovering the events that underlie the sense of touch (e.g., Goodman, Lumpkin, Ricci, Tracey, Kernan, & Nicolson, 2004). Goodman et al. have suggested that molecular identification of candidate mechanotransduction channels in touch may ultimately reveal how transduction channels are optimized to respond to tactile (and other forms of) mechanical energy and the extent to which accessory structures improve the response properties of mechanoreceptive cell types.

An understanding of the properties of sensory receptors in skin and muscles (chapter 3) is crucial to interpreting psychophysical data and has provided a framework for dealing with some of the application issues discussed in chapter 10. In many instances, questions at one level of analysis (e.g., what is the effect of mounting actuators in a haptic display on the dorsum of the fingers rather than on the palmar surface?) require an answer that depends on knowledge acquired at another level (i.e., there will be a loss in sensitivity due to the reduced density of mechanoreceptors on the dorsal surface and associated higher thresholds). In some areas of research, such as developmental studies of hand function, there is still insufficient information to make these connections between different levels (i.e., neurophysiological, perceptual, or behavioral) of analysis. For example, how much of the loss in manual sensory function in the elderly can be attributed to a decrease in peripheral mechanoreceptor populations and how much reflects changes in cortical processing? At the other end of the life spectrum, it would be of interest to know what maturational changes are occurring in children between the ages of 6 and 12 years that can account for the refinement in their manual skills. It is clear from studies of "elite performers," be they concert pianists, profes-sional jugglers, or expert typists, that practice is essential to the acquisition and maintenance of manual skills. These studies have also demonstrated the remarkable specificity of many manual skills, namely, that outstanding performance in one domain does not necessarily indicate superior ability in another, related field (Krampe & Ericsson, 1996; Salthouse, 1984).

There are a number of critical research questions related to tactile (chapter 4) and haptic (chapter 5) sensing that, if addressed in greater depth, would contribute to a fuller understanding and appreciation of human hand function. First, Bolanowksi, Gescheider, and their colleagues have proposed two multichannel models of tactile sensation involving the contribution of either four (glabrous skin) or three (hairy skin) psychophysical vibrotactile channels, each one associated with its own physiological substrate (Bolanowski et al., 1988; Bolanowski et al., 1994, respectively). We suggested in chapter 4 that the models apply more specifically to vibrotactile sensation than broadly to the entire range of tactile sensations. Are there other aspects of tactile sensation that can be explained by such multichannel models, as opposed to the feature-based models proposed by a number of investigators (e.g., Connor & Johnson, 1992; Dodson et al., 1998; Goodwin & Wheat, 1992; Johansson & Westling, 1987; LaMotte & Srinivasan, 1987a; Talbot et al., 1968)? A second important avenue of future research pertains to space-time interactions, which are surely important and necessary for a comprehensive understanding of tactile sensing by the hand. To date, this work has focused on spatiotemporal tactile illusions, such as apparent saltatory and continuous motion, and the tau and kappa illusions. However, we currently know little about the spatial and temporal parameters that affect susceptibility to these illusions when the hand is used. Designers of tactile displays may wish to capitalize on manual illusory effects in space and/or time that can be produced by effectively and discretely stimulating relatively few sites on the skin. A third critical issue pertains to the ways in which the sensory systems organize multiple sensory inputs from the hand, ranging from simple sensations to more complex cognitive experiences of the external environment. For example, as yet we know relatively little about the relevance of the Gestalt laws of organization for tactile and haptic sensing, despite their demonstrated importance for vision and audition. A fourth important research direction concerns the mechanisms by which one or more frames of reference affect the tactile perception of spatial patterns. This is a highly complex question that is currently far from well understood. A fifth research direction involves extending the early research on peripheral processing by the somatosensory system to include higher-level cognitive processes. This might include somatosensory attention, a topic that has been addressed most recently in terms of possible visual-tactile interactions (e.g., Spence, Pavani, & Driver, 2000; but see Sinclair, Kuo, & Burton, 2000). Another topic that has received relatively little study to date is that of somatosensory memory, surely a critical cognitive component of our manual experiences. The study of priming and the possible distinction between implicit and explicit memory with respect to manual contact would provide valuable additions to the limited research in this area (e.g., Srinivas, Greene, & Easton, 1997a, 1997b).

As our sixth and final research direction, we emphasize the fact that haptic perception of objects and their properties is usually performed in conjunction with other senses, particularly vision and/or audition. We concur with Goodwin and Wheat (2004), who suggest that a significant challenge for the future is to develop a global model of haptic exploration and manipulation that encompasses all of the senses. It is therefore critical to consider both how and how well people integrate multiple sources of sensory information that are simultaneously available. Furthermore, we need to understand better how perceptual discrepancies are resolved. Such issues have attracted a substantial amount of attention since the mid-1990s within both the cognitive science and cognitive neuroscience research communities (Calvert et al., 2004; Ernst & Banks, 2002).

Research on the motor functions of the hand has primarily focused on grasping and keyboard skills, as reviewed in chapters 6 and 7. Although the grasping studies described in chapter 6 have provided many valuable insights into the sensorimotor control of prehensile forces, the tasks used in most of these investigations are relatively simple from the perspective of the hands' capabilities. The most complex manipulative skills of the human hand are often considered to be evident in tasks such as microsurgery, watch making, and microelectronic assembly, all of which involve the use of tools and occupy a small work space (L. A. Jones, 1996; Starkes, Payk, Jennen,

& Leclair, 1993). As with all motor skills, repetition is essential to successful execution. In his book *Letters to a Young Doctor*, the surgeon-author Richard Selzer (1982) noted, "A facility in knot tying is gained only by tying ten thousand of them" (p. 49).

One area of future research that relates to the motor function of the hand is the analysis of highly skilled activities, such as microsurgery and microassembly, which have not been as extensively studied as reaching and grasping movements. A comprehensive analysis of these skills would provide a better understanding of human dexterity and the factors that influence its acquisition and retention. Many microsurgical procedures involve a degree of dexterity that is at the limit of human capability. Two factors are of particular importance in small-scale human manipulation: physiological tremor, which is a normal characteristic of the neuromuscular system, and the relatively high threshold for detecting forces. During microsurgical operations involving an operating microscope, movements with amplitudes between 150 and 200 μm can be made under visual guidance (Charles & Williams, 1989), even though they are not perceived kinesthetically. In this situation, the motor system can function effectively beyond the limits imposed by normal vision. Force feedback from the tool–tissue interface is usually absent during these procedures as the forces involved in manipulating tissues are below the human detection threshold. With the development of instrumented surgical tools that incorporate force and displacement sensors (L. A. Jones, 1996), it should become possible to quantify the movement and force control parameters that distinguish the performance of novice and experienced surgeons. These kinematic and kinetic measures may provide the bases for establishing criteria for evaluating microsurgical competence, in addition to the much more commonly used temporal indices (L. A. Jones, 1996; Starkes, Payk, & Hodges, 1998). The demanding precision requirements of microsurgery make it an interesting field in which to explore the dimensions of manual skill and its relation to proprioceptive and visual perceptual abilities (Grober et al., 2003).

A second area of future research that has received considerable attention but is still unresolved concerns the sensorimotor transformations involved when reaching or pointing with the arm to a target whose position is defined in a different coordinate system from that involved in encoding the position of the limb as it moves. When the target is presented visually, the transformation involves going from a retinocentric representation to an egocentric or body-centered representation, with the latter based on proprioceptive inputs. Flanders et al. (1992) proposed a two-stage process in which the retinocentric representation is transformed into a head-centered target representation and then into a shoulder-centered target representation. They derived this model from analyses of pointing errors that occurred when reaching under a fairly limited set of experimental conditions. More general models of sensorimotor transformations must be able to accommodate the diverse range of variables that have been shown to influence pointing and reaching movements. These include the initial position and orientation of the arm, the visual feedback conditions during movement execution (i.e., in the light or dark, with or without vision of the target or the initial position of the arm), the dynamics and kinetics of the movement, and the final configuration of the hand and arm as it reaches the target.

Third, we need a better understanding of the nature of the internal models of tasks and objects that have emerged as an important theoretical concept in motor control. In the context of grasping, these models are assumed to include both the dynamic and kinematic characteristics of the task and are assumed to be involved in the anticipatory control of grip forces. It has been proposed that, given the ease with which people can grasp objects with widely varying properties, there must be many internal models within the central nervous system. There may also be generic object models that are initially accessed based on visual and tactile information and then updated once contact is made with an object. The form and use of these models in human manual function warrant further investigation.

Finally, we wish to note the recent proliferation of neural-imaging techniques, which now make it possible to explore the complexities of how the brain processes and represents both sensory and motoric events. These range from simple internal tactile sensations (e.g., Francis, Kelly, Bowtell, Dunseath, Folger, & McGlone, 2000; Romo & Salinas, 2003) to the haptic cognition of external objects and their properties (e.g., James, Humphrey, Gati, Servos, Menon, & Goodale, 2002; Roland, O'Sullivan, & Kawashima, 1998; Stoeckel, Weder, Binkofski, Buccino, Shah, & Seitz, 2003). Motor activities such as those involved in prehension and non-prehensile skilled tasks (e.g., Ehrsson, Fagergren, Johansson, & Forssberg, 2003; Ehrsson, Kuhtz-Buschbeck, & Forssberg, 2002; Kim et al., 2004) have also been the focus of recent neuroimaging studies. We anticipate significant growth within this exciting field of neuroscience as imaging facilities become increasingly accessible to researchers.

Final Words

In conclusion, the sensorimotor continuum provides a valuable framework for conceptualizing the extensive range of activities in which humans engage as they explore, sense, perceive, and manipulate their environments by hand. From the simple task of knowing where an insect has landed on the hand to the highly complex challenge of playing Rachmaninov's Second Piano Concerto, the performer makes use of the unique sensory and motor attributes of the hand. The quotation by Charles Bell cited in the introduction seems a fitting way to draw our discussions to a close: "[W]e must confess that it is in the human hand that we have the consummation of all perfection as an instrument" (Bell, 1833, p. 207).

REFERENCES

Abravanel, E. (1972). Short-term memory for shape information processed intra- and intermodally at three ages. *Perceptual and Motor Skills, 35,* 419–425.

Adamovich, A. V., Berkinblit, M. B., Fookson, O., & Poizner, H. (1998). Pointing in 3D space to remembered targets: 1. Kinesthetic versus visual target representation. *Journal of Neurophysiology, 79,* 2833–2846.

Ager, C. L., Olivett, B. L., & Johnson, C. L. (1984). Grasp and strength in children 5 to 12 years old. *American Journal of Occupational Therapy, 38,* 107–113.

Aglioti, S., DeSouza, J. F. X., & Goodale, M. A. (1995). Size-contrast illusions deceive the eye but not the hand. *Current Biology, 5,* 679–685.

Agnew, P. J., & Maas, F. (1982). Hand function related to age and sex. *Archives of Physical Medicine and Rehabilitation, 63,* 269–271.

Akamatsu, M., MacKenzie, I. S., & Hasbroucq, T. (1995). A comparison of tactile, auditory, and visual feedback in a pointing task using a mouse-type device. *Ergonomics, 38,* 816–827.

Akamatsu, M., Sato, S., & MacKenzie, I. S. (1994). Multimodal mouse: A mouse-type device with tactile and force display. *Presence, 3,* 73–80.

Alcorn, S. (1932). The Tadoma method. *Volta Review, 34,* 195–198.

Alexander, R. M. (1992). *The human machine.* New York: Columbia University Press.

Amazeen, E. L., & Turvey, M. T. (1996). Weight perception and the haptic size-weight illusion are functions of the inertia tensor. *Journal of Experimental Psychology: Human Perception & Performance, 22,* 213–232.

American Academy of Orthopedic Surgeons. (1965). *Joint motion: Method of measuring and recording.* London: Livingstone.

American Medical Association. (1993). *Guides to the evaluation of permanent impairment* (4th ed.). Chicago: American Medical Association.

American Society for Surgery of the Hand. (1983). *The hand* (2d ed.). New York: Churchill Livingstone.

An, K. N., & Bejjani, F. J. (1990). Analysis of upper-extremity performance in athletes and musicians. *Hand Clinics, 6,* 393–403.

Andres, K. H., & Düring, M. von. (1973). Morphology of cutaneous receptors. In A. Iggo (Ed.), *Handbook of sensory physiology: Vol. 2. Somatosensory system* (pp. 3–28). New York: Springer.

Angelaki, D. E., & Soechting, J. F. (1993). Non-uniform temporal scaling of hand and finger kinematics during typing. *Experimental Brain Research, 95,* 319–329.

Annett, J., Annett, M., Hudson, P. T. W., & Turner, A. (1979). The control of movement in the preferred and non-preferred hands. *Quarterly Journal of Experimental Psychology, 31,* 641–652.

Anstis, S. M. (1964). Apparent size of holes felt with the tongue. *Nature, 203,* 792–793.

Appelle, S. (1972). Perception and discrimination as a function of stimulus orientation: The "oblique effect" in man and animals. *Psychological Bulletin, 78,* 266–278.

Appelle, S., & Gravetter, F. J. (1985). Effect of modality-specific experience on visual and haptic judgement of orientation. *Perception, 14,* 763–773.

Arbib, M. A., Iberall, T., & Lyons, D. (1985). Coordinated control programs for movements of the hand. In

A. W. Goodman & I. Darian-Smith (Eds.), *Hand function and the neocortex* (pp. 111–129). New York: Springer.

Armstrong, C. A., & Oldham, J. A. (1999). A comparison of dominant and non-dominant hand strengths. *Journal of Hand Surgery, 24B*, 421–425.

Armstrong, T. J., Foulke, J. A., Martin, B. J., Gerson, J., & Rempel, D. (1994). Investigation of applied forces in alphanumeric keyboard work. *American Industrial Hygiene Association Journal, 55*, 30–35.

American Radio Relay League. (2003, September 30). *National Association for Amateur Radio.* Available: http://www.arrl.org.

Asanuma, H., & Rosén, I. (1972). Topographical organization of cortical efferent zones projecting to distal forelimb muscles in the monkey. *Experimental Brain Research, 14*, 243–256.

Atkeson, C. G., & Hollerbach, J. M. (1985). Kinematic features of unrestrained vertical arm movements. *Journal of Neuroscience, 5*, 2318–2330.

Augurelle, A.-S., Smith, A. M., Lejeune, T., & Thonnard, J.-L. (2003). Importance of cutaneous feedback in maintaining a secure grip during manipulation of hand-held objects. *Journal of Neurophysiology, 89*, 665–671.

August, S., & Weiss, P. L. (1992). Objective and subjective approaches to the force and displacement characteristics of input devices used by the disabled. *Journal of Biomedical Engineering, 14*, 117–125.

Babkoff, H. (1978). Electrocutaneous psychophysical input-output functions and temporal integration. *Perception & Psychophysics, 23*, 251–257.

Bagesteiro, L. B., & Sainburg, R. L. (2002). Handedness: Dominant arm advantages in control of limb dynamics. *Journal of Neurophysiology, 88*, 2408–2421.

Ballesteros, S., Manga, D., & Reales, J. M. (1997). Haptic discrimination of bilateral symmetry in 2-dimensional and 3-dimensional unfamiliar displays. *Perception & Psychophysics, 59*, 37–50.

Ballesteros, S., Millar, S., & Reales, J. M. (1998). Symmetry in haptic and in visual shape perception. *Perception & Psychophysics, 60*, 389–404.

Bastian, C. (1887). "The muscular sense": Its nature and cortical localization. *Brain, 10*, 1–137.

Baud-Bovy, G., & Soechting, J. F. (2001). Two virtual fingers in the control of the tripod grasp. *Journal of Neurophysiology, 86*, 604–615.

Baud-Bovy, G., & Viviani, P. (2004). Amplitude and direction errors in kinesthetic pointing. *Experimental Brain Research, 157*, 197–214.

Bayley, N. (1969). *The Bayley scales of infant development.* New York: Psychological Corporation.

Bean, K. L. (1938). An experimental approach to the reading of music. *Psychological Monographs, 50*, 1–80.

Becker, J. D., & Mote, C. D., Jr. (1990). Identification of a frequency response model of joint rotation. *Journal of Biomechanical Engineering, 112*, 1–8.

Beek, P. J., & Lewbel, A. (1995). The science of juggling. *Scientific American, 273*, 92–97.

Beek, P. J., & van Santvoord, A. A. M. (1996). Dexterity in cascade juggling. In M. L. Latash & M. T. Turvey (Eds.), *Dexterity and its development* (pp. 377–391). Mahwah, NJ: Erlbaum.

Bejjani, F. J., & Halpern, N. (1989). Postural kinematics of trumpet playing. *Journal of Biomechanics, 23*, 439–446.

Békésy, G. (1939). Uber die Vibrationsempfindung (Study of vibration feelings). *Aleustische Zeitschrift, 4*, 315–334.

Bell, C. (1826). On the nervous circle which connects the voluntary muscles with the brain. *Philosophical Transactions of the Royal Society, 116*, 163–173.

Bell, C. (1833). *The hand: Its mechanism and vital endowments as evincing design: The Bridgewater treatises* (Vol. 4). London: William Pickering.

Bell-Krotoski, J. A. (2002). Sensory testing with the Semmes-Weinstein monofilaments. In E. J. Mackin, A. D. Callahan, T. M. Skirven, L. H. Schneider, & A. L. Osterman (Eds.), *Rehabilitation of the hand and upper extremity* (5th ed., pp. 194–213). St. Louis, MO: Mosby.

Bellugi, U., & Klima, E. (1976). Two faces of sign: Iconic and abstract. *Annals of the New York Academy of Sciences, 280*, 514–538.

Bender, M. B., Stacy, C., & Cohen, J. (1982). Agraphesthesia: A disorder of directional cutaneous kinesthesia or a disorientation in cutaneous space. *Journal of the Neurological Sciences, 53*, 531–555.

Benedetti, F. (1985). Processing of tactile spatial information with crossed fingers. *Journal of Experimental Psychology: Human Perception & Performance, 11*, 517–525.

Benedetti, F. (1986). Spatial organization of the diplesthetic and nondiplesthetic areas of the fingers. *Perception, 15*, 285–301.

Benedetti, F. (1988). Localization of tactile stimuli and body parts in space: Two dissociated perceptual experiences revealed by a lack of constancy in the presence of position sense and motor activity. *Journal of Experimental Psychology: Human Perception & Performance, 14*, 69–76.

Bennett, K. M. B., & Castiello, U. (1994). Reach to grasp: Changes with age. *Journal of Gerontology: Psychological Sciences, 49*, P1–P7.

Bensmaïa, S. J., & Hollins, M. (2003). The vibrations of texture. *Somatosensory & Motor Research, 20,* 33–43.

Benton, A. L., Hamsher, K. D., Varney, N. R., & Spreen, O. (1983). *Contributions to neuropsychological assessment: A clinical manual.* New York: Oxford University Press.

Bergamasco, M., Allotta, B., Bosio, L., Ferretti, L., Parrini, G., Prisco, G. M., Salsedo, F., & Sartini, G. (1994). An arm exoskeleton system for teleoperation and virtual environments applications. *Proceedings of the IEEE International Conference on Robotics and Automation, 2,* 1449–1454.

Berger, C., & Hatwell, Y. (1993). Dimensional and overall similarity classifications in haptics: A developmental study. *Cognitive Development, 8,* 495–516.

Berger, C., & Hatwell, Y. (1995). Developmental trends in haptic and visual free classifications: Influence of stimulus structure and exploration on decisional processes. *Journal of Experimental Child Psychology, 63,* 447–465.

Berger, C., & Hatwell, Y. (1996). Development of dimensional vs. global processing in haptics: The perceptual and decisional determinants of classification skills. *British Journal of Developmental Psychology, 13,* 143–162.

Bernstein, N. (1967). *The co-ordination and regulation of movements.* Oxford: Pergamon.

Bertelsen, A., & Capener, N. (1960). Fingers, compensation and King Canute. *Journal of Bone and Joint Surgery, 42B,* 390–392.

Bigelow, A. E. (1981). Children's tactile identification of miniaturized common objects. *Developmental Psychology, 17,* 111–114.

Biggs, S. J., & Srinivasan, M. A. (2002a). Tangential versus normal displacements of skin: Relative effectiveness for producing tactile sensations. In *IEEE Proceedings of the 10th International Symposium on Haptic Interfaces for Virtual Environment and Teleoperator Systems,* pp.121–127.

Biggs, S. J., & Srinivasan, M. A. (2002b). Haptic interfaces. In K. M. Stanney (Ed.), *Handbook of virtual environments: Design, implementation, and applications* (pp. 93–115). Mahwah, NJ: Erlbaum.

Binkofski, F., Kunesch, E., Classen, J., Seitz, R. J., & Freund, H.-J. (2001). Tactile apraxia: Unimodal apractic disorder of tactile object recognition associated with parietal lobe lesions. *Brain, 124,* 132–144.

Binkofski, F., Seitz, R. J., Arnold, S., Classen, J., Benecke, R., & Freund, H.-J. (1996). Thalamic metabolism and corticospinal tract integrity determine motor recovery in stroke. *Annals of Neurology, 39,* 460–470.

Biryukova, E. V., & Yourovskaya, V. Z. (1994). A model of human hand dynamics. In F. Schuind, K. N. An, W. P. Coney, & M. Garcia-Elias (Eds.), *Advances in the biomechanics of the hand and wrist* (pp. 107–122). New York: Plenum.

Birznieks, I., Jenmalm, P., Goodwin, A. W., & Johansson, R. S. (2001). Encoding of direction of fingertip forces by human tactile afferents. *Journal of Neuroscience, 21,* 8222–8237.

Bisley, J. W., Goodwin, A. W., & Wheat, H. E. (2000). Responses of slowly adapting type I afferents from the sides and end of the finger to stimuli applied to the central part of the fingerpad. *Journal of Neurophysiology, 84,* 57–64.

Blake D. T., Hsiao, S. S., & Johnson K. O. (1997). Neural coding mechanisms in tactile pattern recognition: The relative contributions of slowly and rapidly adapting mechanoreceptors to perceived roughness. *Journal of Neuroscience, 17,* 7480–7489.

Blake D. T., Johnson K. O., & Hsiao, S. S. (1997). Monkey cutaneous SA I and RA responses to raised and depressed scanned patterns: Effects of width, height, orientation, and a raised surround. *Journal of Neurophysiology, 78,* 2503–2517.

Bliss, J. (1978). Reading machines for the blind. In G. Gordon (Ed.), *Active touch: The mechanism of recognition of objects by manipulation* (pp. 243–248). Oxford: Pergamon.

Bliss, J. C., Katcher, M. H., Rogers, C. H., & Shepard, R. P. (1970). Optical-to-tactile image conversion for the blind. *IEEE Transactions on Man-Machine Systems, 11,* 58–65.

Blix, M. (1884). Experimentelle beiträge zur lösung der frage über die specifische energie der hautnerven (Experimental contribution toward determination of the case for specific nerve energies in the skin). *Zeitschrift für Biologie, 20,* 141–156.

Boatright, J. R., Kiebzak, G. M., O'Neil, D. M., & Peindl, R. D. (1997). Measurement of thumb abduction strength: Normative data and a comparison with grip and pinch strength. *Journal of Hand Surgery, 22A,* 843–848.

Bock, O., & Eckmiller, R. (1986). Goal-directed movements in absence of visual guidance: Evidence for amplitude rather than position control. *Experimental Brain Research, 62,* 451–458.

Boecker, H., Dagher, A., Ceballos-Baumann, O., Passingham, R. E., Samuel, M., Friston, K. J., Poline, J.-B., Dettmers, C., Conrad, B., & Brooks, D. J. (1998). Role of the human rostral supplementary motor area and the basal ganglia in motor sequence control: Investigations with $H_2\,^{15}O$ PET. *Journal of Neurophysiology, 79,* 1070–1080.

Bolanowski, S. J. (1981). *Intensity and frequency characteristics of Pacinian corpuscles*. Unpublished Ph.D. dissertation and Special Report ISR-S-20. Institute for Sensory Research, Syracuse University, Syracuse, NY.

Bolanowski, S. J., Gescheider, G. A., & Verrillo, R. T. (1994). Hairy skin: Psychophysical channels and their physiological substrates. *Somatosensory and Motor Research, 11*, 279–290.

Bolanowski, S. J., Gescheider, G. A., Verrillo, R. T., & Checkosky, C. M. (1988). Four channels mediate the mechanical aspects of touch. *Journal of the Acoustical Society of America, 84*, 1680–1694.

Bolanowski, S. J., Jr., & Verrillo, R. T. (1982). Temperature and criterion effects in a somatosensory subsystem: A neurophysiological and psychophysical study. *Journal of Neurophysiology, 48*, 836–855.

Bolton, C. F., Winkelmann, R. K., & Dyck, P. J. (1966). A quantitative study of Meissner's corpuscles in man. *Neurology, 16*, 1–9.

Boring, E. G. (1942). *Sensation and perception in the history of experimental psychology*. New York: Appleton-Century-Crofts.

Bornstein, R. A. (1986). Normative data on intermanual differences on three tests of motor performance. *Journal of Clinical Experimental Neuropsychology, 8*, 12–20.

Bouche, P., Cattelin, F., Saint-Jean, O., Leger, J. M., Queslati, S., Guez, D., Moulonguet, A., Brault, Y., Aquino, J. P. & Simunek, P. (1993). Clinical and electrophysiological study of the peripheral nervous system in the elderly. *Journal of Neurology, 240*, 263–268.

Bouzit, M., Popescu, G., Burdea, G., & Boian, R. (2002). The Rutgers Master II-ND force feedback glove. In *IEEE Proceedings of the 10th Symposium on Haptic Interfaces for Virtual Environment and Teleoperator Systems*, pp. 145–152.

Bower, T. G. R. (1972). Object perception in infants. *Perception, 1*, 15–30.

Brand, P. W. (1985). *Clinical mechanics of the hand*. St. Louis, MO: Mosby.

Bregman, A. S. (1990). *Auditory scene analysis: The perceptual organization of sound*. Cambridge, MA: Bradford Books/MIT Press.

Brink, E. E., & Mackel, R. (1987). Sensorimotor performance of the hand during peripheral nerve regeneration. *Journal of the Neurological Sciences, 77*, 249–266.

Brisben, A. J., Hsiao, S. S., & Johnson, K. O. (1999). Detection of vibration transmitted through an object grasped in the hand. *Journal of Neurophysiology, 81*, 1548–1558.

Brock, D. (1993). *A sensor based strategy for automatic grasping*. Unpublished doctoral dissertation, Massachusetts Institute of Technology, Cambridge, MA.

Brodie, E. E., & Ross, H. E. (1984). Sensorimotor mechanisms in weight discrimination. *Perception & Psychophysics, 36*, 477–481.

Brodie, E. E., & Ross, H. E. (1985). Jiggling a lifted weight does aid discrimination. *American Journal of Psychology, 98*, 469–471.

Brown, K. W., & Gottfried, A. W. (1986). Cross-modal transfer of shape in early infancy: Is there reliable evidence? In L. P. Lipsitt & C. Rovee-Collier (Eds.), *Advances in infancy research* (Vol. 4, pp. 163–170). Norwood, NJ: Ablex.

Brown, J. K., Omar, T., & O'Regan, M. (1997). Brain development and the development of tone and movement. In K. J. Connolly & H. Forssberg (Eds.), *Neurophysiology and neuropsychology of motor development* (pp. 1–41). London: MacKeith.

Bruce, M. F. (1980). The relations of tactile thresholds to histology in the fingers of elderly people. *Journal of Neurology, Neurosurgery and Psychiatry, 43*, 730–734.

Bryan, W. L., & Harter, N. (1899). Studies on the telegraphic language: The acquisition of a hierarchy of habits. *Psychological Review, 6*, 345–375.

Bryant, P. E., Jones, P., Claxton, V., & Perkins, G. M. (1972). Recognition of shapes across modalities by infants. *Nature, 240*, 303–304.

Bryant, P. E., & Raz, I. (1975). Visual and tactual perception of shape by young children. *Developmental Psychology, 11*, 525–526.

Buchholz, B., Armstrong, T. J., & Goldstein, S. A. (1992). Anthropometric data for describing the kinematics of the human hand. *Ergonomics, 35*, 261–273.

Buchholz, B., Frederick, L. J., & Armstrong, T. J. (1988). An investigation of human palmar skin friction and the effects of materials, pinch force and moisture. *Ergonomics, 31*, 317–325.

Buford, W. L., & Thompson, D. E. (1987). A system for three-dimensional interactive simulation of hand biomechanics. *IEEE Transactions on Biomedical Engineering, 34*, 444–453.

Bujas, Z. (1938). La sensibilité au froid en function du temps (Sensitivity to cold as a function of time). *L'Anneé Psychologique, 38*, 140.

Buonomano, D. V., & Merzenich, M. M. (1998). Cortical plasticity: From synapses to maps. *Annual Reviews of Neuroscience, 21*, 49–86.

Burdea, G. (1996). *Force and touch feedback for virtual reality*. New York: Wiley.

Burdea, G., & Coiffet, P. (2003). *Virtual reality technology* (2d ed.). Hoboken, NJ: Wiley-Interscience.

Burdea, G., Zhuang, J., Roskos, E., Silver, D., & Langrana, N. (1992). A portable dextrous master with force feedback. *Presence, 1,* 18–27.

Burgess, P. R., Mei, J., Tuckett, R. P., Horch, K. W., Ballinger, C. M., & Poulos, D. A. (1983). The neural signal for indentation depth. I. Changing indentations. *Journal of Neuroscience, 3,* 1572–1585.

Burke, D., Gandevia, S. C., & Macefield, G. (1988). Responses to passive movement of receptors in joint, skin and muscle of the human hand. *Journal of Physiology, 402,* 347–361.

Burstedt, M. K. O., Edin, B. B., & Johansson, R. S. (1997). Coordination of fingertip forces during human manipulation can emerge from independent neural networks controlling each digit. *Experimental Brain Research, 117,* 67–79.

Burton, H. (2002). Cerebral cortical regions devoted to the somatosensory system: Results from brain imaging studies in humans. In R. J. Nelson (Ed.), *The somatosensory system: Deciphering the brain's own body image* (pp. 27–72). Boca Raton, FL: CRC.

Bushnell, E. W. (1978). *Cross-modal object recognition in infancy.* Paper presented at the annual meeting of the American Psychological Association, Toronto, Canada.

Bushnell, E. W. (1982). Visual-tactual knowledge in 8-, 9½- and 11-month old infants. *Infant Behavior and Development, 5,* 63–75.

Bushnell, E. W., & Baxt, C. (1999). Children's haptic and cross-modal recognition with familiar and unfamiliar objects. *Journal of Experimental Psychology: Human Perception & Performance, 25,* 1867–1881.

Bushnell, E. W., & Boudreau, P. (1991). The development of haptic perception during infancy. In M. Heller and W. Schiff (Eds.), *The psychology of touch* (pp. 139–161). Hillsdale, NJ: Erlbaum.

Bushnell, E. W., & Boudreau, P. (1993). Motor development and the mind: The potential role of motor abilities as a determinant of aspects of perceptual development. *Child Development, 64,* 1005–1021.

Bushnell, E. W., Shaw, L., & Strauss, D. (1985). Relationship between visual and tactual exploration by 6-month-olds. *Developmental Psychology, 21,* 591–600.

Bushnell, E. W., & Weinberger, N. (1987). Infants' detection of visual-tactual discrepancies: Asymmetries that indicate a directive role of visual information. *Journal of Experimental Psychology: Human Perception and Performance, 13,* 601–608.

Bushnell, E. W., Weinberger, N., & Sasseville, A. (1989, April). *Interactions between vision and touch during infancy: The development of cooperative relations and specializations.* Paper presented at the biennial meeting of the Society for Research in Child Development, Kansas City, MO.

Butsch, R. L. C. (1932). Eye movements and the eye-hand span in typewriting. *Journal of Educational Psychology, 23,* 104–121.

Buttolo, P., & Hannaford, B. (1995). Pen-based force display for precision manipulation in virtual environments. In *Proceedings of IEEE Virtual Reality Annual International Symposium,* pp. 217–224.

Cadoret, G., & Smith, A. M. (1996). Friction, not texture, dictates grip forces used during object manipulation. *Journal of Neurophysiology, 75,* 1963–1969.

Cain, W. S. (1973). Spatial discrimination of cutaneous warmth. *American Journal of Psychology, 86,* 169–181.

Caldwell, D., Kocak, O., & Anderson, U. (1995). Multi-armed dextrous manipulator operation using glove/exoskeleton control and sensory feedback. *Proceedings of the IEEE/RSJ International Conference on Intelligent Robots and Systems, 2,* 567–572.

Calford, M. B., & Tweedale, R. (1991). Immediate expansion of receptive fields of neurons in area 3b of Macaque monkeys after digit denervation. *Somatosensory and Motor Research, 8,* 249–260.

Callahan, A. D. (2002). Sensibility assessment for nerve lesions-in-continuity and nerve lacerations. In E. J. Mackin, A. D. Callahan, T. M. Skirven, L. H. Schneider, & A. L. Osterman (Eds.), *Rehabilitation of the hand and upper extremity* (5th ed., pp. 214–239). St. Louis, MO: Mosby.

Calvert, G., Stein, B., & Spence, C. (2004). *The handbook of multisensory processes.* Cambridge, MA: MIT Press.

Campbell, R. A. (1963). Detection of a noise signal of varying duration. *Journal of the Acoustical Society of America, 35,* 1732–1737.

Caparro, A. J., Verrillo, R. T., & Zwislocki, J. J. (1979). Psychophysical evidence for a triplex system of cutaneous mechanoreception. *Sensory Processes, 3,* 334–352.

Carlton, L. G. (1979). Control processes in the production of discrete aiming responses. *Journal of Human Movement Studies, 5,* 115–124.

Carmeli, E., Patish, H., & Coleman, R. (2003). The aging hand. *Journal of Gerontology: Medical Sciences, 58A,* 146–152.

Carnahan, H., Vandervoort, A. A., & Swanson, L. R. (1998). The influence of aging and target motion on the control of prehension. *Experimental Aging Research, 24,* 289–306.

Carter-Wilson, M. (1991). Sensory re-education. In R. H. Gelberman (Ed.), *Operative nerve repair and reconstruction* (Vol. 1, pp. 827–844). Philadelphia: Lippincott.

Cascio, C. J., & Sathian, K. (2001). Temporal cues contribute to tactile perception of roughness. *Journal of Neuroscience, 21,* 5289–5296.

Caselli, R. J. (1993). Ventrolateral and dorsomedial somatosensory association cortex damage produces distinct somesthetic syndromes in humans. *Neurology, 43,* 762–771.

Casla, M., Blanco, F., & Travieso, D. (1999). Haptic perception of geometric illusions by persons who are totally congenitally blind. *Journal of Visual Impairment and Blindness, 93,* 583–588.

Cauna, N. (1965). The effects of aging on the receptor organs of the human dermis. In W. Montagna (Ed.), *Advances in biology of the skin* (Vol. 6, pp. 63–96). New York: Pergamon.

Cesari, P., & Newell, K. M. (2000). Body-scaled transitions in human grip configurations. *Journal of Experimental Psychology: Human Perception and Performance, 26,* 1657–1668.

Chambers, M. R., Andres, K. H., Düring, M. von, & Iggo, A. (1972). The structure and function of the slowly adapting type II mechanoreceptor in hairy skin. *Quarterly Journal of Experimental Physiology, 57,* 417–445.

Chao, E. Y. S., An, K.-N., Cooney, W. P., & Linscheid, R. L. (1989). *Biomechanics of the hand.* Singapore: World Scientific.

Chapman, C. E. (1994). Active versus passive touch: Factors influencing the transmission of somatosensory signals to primary somatosensory cortex. *Canadian Journal of Physiology and Pharmacology, 72,* 558–570.

Chapman, C. E., & Wiesendanger, M. (1982). Recovery of function following unilateral lesions of the bulbar pyramid in the monkey. *Electroencephalography and Clinical Neurophysiology, 53,* 374–387.

Charles, S., & Williams, R. (1989). Measurement of hand dynamics in a microsurgery environment: Preliminary data in the design of a bimanual telemicro-operation test bed. *Proceedings of the NASA Conference on Space Telerobotics, 1,* 109–118.

Charpentier, A. (1891). Analyse expérimentale de quelques éléments de la sensation de poids (Experimental study of some aspects of weight perception). *Archives de Physiologie Normales et Pathologiques, 3,* 122–135.

Chieffi, S., & Gentilucci, M. (1993). Coordination between the transport and the grasp components during prehension movements. *Experimental Brain Research, 94,* 471–477.

Cholewiak, R. W., & Collins, A. A. (2000). The generation of vibrotactile patterns on a linear array: Influences of body site, time, and presentation mode. *Perception & Psychophysics, 62,* 1220–1235.

Chua, R., & Elliot, D. (1993). Visual regulation of manual aiming. *Human Movement Science, 12,* 365–401.

Churchill, A., Hopkins, B., Rönnqvist, L., & Vogt, S. (2000). Vision of the hand and environmental context in human prehension. *Experimental Brain Research, 134,* 81–89.

Cisek, P., & Kalaska, J. F. (2004). Neural correlates of mental rehearsal in dorsal premotor cortex. *Nature, 431,* 993–996.

Cisek, P., & Kalaska, J. F. (2005). Neural correlates of reaching decisions in dorsal premotor cortex: Specification of multiple direction choices and final selection of action. *Neuron, 45,* 801–814.

Clark, F. J., Grigg, P., & Chapin, J. W. (1989). The contribution of articular receptors to proprioception with the fingers in humans. *Journal of Neurophysiology, 61,* 186–193.

Clarkson, P., & Pelly, A. (1962). *The general and plastic surgery of the hand.* Oxford: Blackwell.

Classen, J., Liepert, J., Wise, S. P., Hallett, M., & Cohen, L. G. (1998). Rapid plasticity of human cortical movement representation induced by practice. *Journal of Neurophysiology, 79,* 1117–1123.

Cohen, J., Hansel, C. E., & Sylvester, J. D. (1953). Interdependence of judgments of space, time, and movement. *Acta Psychologica, 11,* 360–372.

Cohen, J., Hansel, C. E., & Sylvester, J. D. (1954). Interdependence of temporal and auditory judgments. *Nature, 174,* 642–645.

Cohen, L. G., Bandinelli, S., Findley, T. W., & Hallett, M. (1991). Motor reorganization after upper limb amputation in man. A study with focal magnetic stimulation. *Brain, 114,* 615–627.

Cohen, R. H., & Vierck, C. J., Jr. (1993). Population estimates for responses of cutaneous mechanoreceptors to a vertically indenting probe on the glabrous skin of monkeys. *Experimental Brain Research, 94,* 105–119.

Cole, K. (1991). Grasp force control in older adults. *Journal of Motor Behavior, 23,* 251–258.

Cole, K., & Beck, C. L. (1994). The stability of precision grip force in older adults. *Journal of Motor Behavior, 26,* 171–177.

Cole, K. J., Rotella, D. L., & Harper, J. G. (1998). Tactile impairments cannot explain the effect of age on a grasp and lift task. *Experimental Brain Research, 121,* 263–269.

Cole, K. J., Rotella, D. L., & Harper, J. G. (1999). Mechanisms for age-related changes of fingertip forces during precision gripping and lifting in adults. *Journal of Neuroscience, 19,* 3238–3247.

Colgate, J. E., & Brown, J. M. (1994). Factors affecting the Z-width of a haptic interface. *Proceedings of the*

IEEE International Conference on Robotics and Automation, 4, 3205–3210.

Collins, D. F., & Prochazka, A. (1996). Movement illusions evoked by ensemble cutaneous input from the dorsum of the human hand. *Journal of Physiology, 496,* 857–871.

Connolly, J. D., & Goodale, M. A. (1999). The role of visual feedback of hand position in the control of manual prehension. *Experimental Brain Research, 125,* 281–286.

Connolly, K., Brown, K., & Bassett, C. (1968). Developmental changes in some components of a motor skill. *British Journal of Psychology, 59,* 305–314.

Connolly, K. J., & Elliot, J. (1972). The evolution and ontogeny of hand function. In N. Blurton-Jones (Ed.), *Etiological studies of child behavior* (pp. 329–383). Cambridge: Cambridge University Press.

Connor, C. E., Hsiao, S. S., Phillips, J. R., & Johnson, K. O. (1990). Tactile roughness: Neural codes that account for psychophysical magnitude estimates. *Journal of Neuroscience, 10,* 3823–3836.

Connor, C. E., & Johnson, K. O. (1992). Neural coding of tactile texture: Comparison of spatial and temporal mechanisms for roughness perception. *Journal of Neuroscience, 12,* 3414–3426.

Cooke, J. D., Brown, S. H., & Cunningham, D. A. (1989). Kinematics of arm movements in elderly humans. *Neurobiology of Aging, 10,* 159–165.

Cooney, W. P., Lucca, M. J., Chao, E. Y. S., & Linscheid, R. L. (1981). The kinesiology of the thumb trapeziometacarpal joint. *Journal of Bone and Joint Surgery, 63A,* 1371–1381.

Corballis, M. C. (1991). *The lopsided ape.* New York: Oxford University Press.

Corcoran, D. (1977). The phenomena of the disembodied eye or is it a matter of personal geography? *Perception, 6,* 247–253.

Corkin, S., Milner, B., & Rasmussen, T. (1970). Somatosensory thresholds: Contrasting effects of postcentral-gyrus and posterior parietal-lobe excisions. *Archives of Neurology, 23,* 41–58.

Corkin, S., Milner, B., & Taylor, L. (1973). Bilateral sensory loss after unilateral cerebral lesion in man. *Transactions of the American Neurological Association, 98,* 25–29.

Craig, J. C. (1972). Difference threshold for intensity of tactile stimuli. *Perception & Psychophysics, 11,* 150–152.

Craig, J. C. (1974). Vibrotactile difference thresholds for intensity and the effect of a masking stimulus. *Perception & Psychophysics, 15,* 123–127.

Craig, J. C. (1980). Modes of vibrotactile pattern perception. *Journal of Experimental Psychology: Human Perception & Performance, 6,* 151–166.

Craig, J. C. (1981). Tactile letter recognition: Pattern duration and modes of pattern generation. *Perception & Psychophysics, 30,* 540–546.

Craig, J. C. (1982a). Vibrotactile masking: A comparison of energy and pattern maskers. *Perception & Psychophysics, 31,* 523–529.

Craig, J. C. (1982b). Temporal integration of vibrotactile patterns. *Perception & Psychophysics, 32,* 219–229.

Craig, J. C. (1983a). Some factors affecting tactile pattern recognition. *International Journal of Neuroscience, 19,* 47–57.

Craig, J. C. (1983b). The role of onset in perception of sequentially presented vibrotactile patterns. *Perception & Psychophysics, 34,* 421–432.

Craig, J. C. (1985). Attending to two fingers: Two hands are better than one. *Perception & Psychophysics, 38,* 496–511.

Craig, J. C. (1999). Grating orientation as a measure of tactile spatial acuity. *Somatosensory and Motor Research, 16,* 197–206.

Craig, J. C., & Johnson, K. O. (2000). The two-point threshold: Not a measure of tactile spatial resolution. *Current Directions in Psychological Science, 9,* 29–32.

Craig, J. C., & Lyle, K. B. (2001). A comparison of tactile spatial sensitivity on the palm and fingerpad. *Perception & Psychophysics, 63,* 337–347.

Craig, J. C., & Xu, B. (1990). Temporal order and tactile patterns. *Perception & Psychophysics, 47,* 22–34.

Cross, M. J., & McCloskey, D. I. (1973). Position sense following surgical removal of joints in man. *Brain Research, 55,* 443–445.

Cross, D. V., Tursky, B., & Lodge, M. (1975). Magnitude scales and somatic evoked potentials to percutaneous electrical stimulation. *Physiology & Behaviour, 3,* 947–953.

Crossman, E. R. F. W., & Goodeve, P. J. (1983). Feedback control of hand-movement and Fitts' law. *Quarterly Journal of Experimental Psychology, 35A,* 251–278.

Cutkosky, M. R. (1989). On grasp choice, grasp models, and the design of hands for manufacturing tasks. *IEEE Transactions on Robotics and Automation, 5,* 269–279.

Cutkosky, M. R., & Howe, R. D. (1990). Human grasp choice and robotic grasp analysis. In S. T. Venkataraman & T. Iberall (Eds.), *Dextrous robot hands* (pp. 5–31). New York: Springer-Verlag.

Cutkosky, M. R., & Wright, P. K. (1986). Modelling manufacturing grips and correlations with design of robotic hands. Proceedings of the *IEEE International Conference on Robotics and Automation, 3,* 1533–1539.

Dandekar, K., Raju, B. I., & Srinivasan, M. A. (2003). 3-D finite-element models of human and monkey fingertips to investigate the mechanics of tactile sense. *Journal of Biomechanical Engineering, 125,* 682–691.

D'Angiulli, A., Kennedy, J. M., & Heller, M. A. (1988). Blind children recognizing tactile pictures respond like sighted children given guidance in exploration. *Scandinavian Journal of Psychology, 39,* 187–190.

Dannenbaum, R. M., & Jones, L. A. (1993). The assessment and treatment of patients who have sensory loss following cortical lesions. *Journal of Hand Therapy, 6,* 130–138.

Daprati, E., & Gentilucci, M. (1997). Grasping an illusion. *Neuropsychologia, 35,* 1577–1582.

Darian-Smith, I. (1984). Thermal sensibility. In I. Darian-Smith (Ed.), *Handbook of physiology: The nervous system: Vol. 3. Sensory processes* (pp. 879–913). Bethesda, MD: American Physiological Society.

Darian-Smith, I., & Johnson, K. O. (1977). Thermal sensibility and thermoreceptors. *Journal of Investigative Dermatology, 69,* 146–153.

Darian-Smith, I., & Kenins, P. (1980). Innervation density of mechanoreceptive fibers supplying glabrous skin of the monkey's index finger. *Journal of Physiology, 309,* 146–155.

Davidoff, R. A. (1990). The pyramidal tract. *Neurology, 40,* 332–339.

Davidson, P. R., & Wolpert, D. M. (2004). Internal models underlying grasp can be additively combined. *Experimental Brain Research, 155,* 334–340.

Davidson, P. W. (1972). Haptic judgments of curvature by blind and sighted humans. *Journal of Experimental Psychology, 93,* 43–55.

Day, R. H., & Wong, T. S. (1971). Radial and tangential movement directions as determinants of the haptic illusion in an L-figure. *Journal of Experimental Psychology, 87,* 19–22.

Dearborn, G. V. N. (1910). *Moto-sensory development: Observations on the first three years of a child.* Baltimore, MD: Warwick & York.

Dellon, A. L. (1978). The moving two-point discrimination test: Clinical evaluation of the quickly adapting fiber/receptor system. *Journal of Hand Surgery, 3A,* 474–481.

Dellon, A. L. (1981). *Evaluation of sensibility and re-education of sensation in the hand.* Baltimore, MD: Williams & Wilkins.

Dellon, A. L., & Kallman, C. H. (1983). Evaluation of functional sensation in the hand. *Journal of Hand Surgery, 8,* 865–870.

Dennerlein, J. T., Mote, C. D., & Rempel, D. M. (1998). Control strategies for finger movement during touch typing: The role of the extrinsic muscles during a keystroke. *Experimental Brain Research, 121,* 1–6.

Derouesné, C., Mas, J. L., Bolgert, F., & Castaigne, P. (1984). Pure sensory stroke caused by a small cortical infarct in the middle cerebral artery territory. *Stroke, 15,* 660–662.

Desmedt, J. E., Noel, P., Debecker, J., & Nameche, J. (1973). Maturation of afferent conduction velocity as studied by sensory nerve potentials and by cerebral evoked potentials. In J. E. Desmedt (Ed.), *New developments in electromyography and clinical neurophysiology* (Vol. 2, pp. 52–63). Basel: Karger.

Desrosiers, J., Hebert, R., Bravo, G., & Rochette, A. (1999). Age-related changes in upper extremity performance of elderly people: A longitudinal study. *Experimental Gerontology, 34,* 393–405.

Deuchar, M. (1999). Spoken language and sign language. In A. Lock & C. R. Peters (Eds.), *Handbook of human symbolic function* (pp. 553–570). Oxford: Blackwell.

Devanandan M. S., Ghosh S., & John, K. T. (1983). A quantitative study of muscle spindles and tendon organs in some intrinsic muscles of the hand in the bonnet monkey (Macaca radiata). *Anatomical Record, 207,* 263–266.

Dodrill, C. B. (1979). Sex differences on the Halstead-Reitan neuropsychological battery and on other neuropsychological measures. *Journal of Clinical Psychology, 35,* 236–241.

Dodson, M. J., Goodwin, A. W., Browning, A. S., & Gehring, H. M. (1998). Peripheral neural mechanisms determining the orientation of cylinders grasped by the digits. *Journal of Neuroscience, 18,* 521–530.

Donoghue, J. P., Leibovic, S., & Sanes, J. N. (1992). Organization of the forelimb area in squirrel monkey cortex: Representation of digit, wrist and elbow muscles. *Experimental Brain Research, 89,* 1–19.

Driscoll, R. (1995). Touching art. *Contact Quarterly,* Summer/Fall, 42–48.

Durlach, N. I., & Mavor, A. S. (Eds.). (1994). *Virtual reality: Scientific and technical challenges.* Washington, DC: National Academy.

Dvorak, A. (1943). There is a better typewriter keyboard. *National Business Education Quarterly, 12,* 51–58.

Edin, B. B. (1990). Finger joint movement sensitivity of non-cutaneous mechanoreceptor afferents in human radial nerve. *Experimental Brain Research, 82,* 417–422.

Edin, B. B. (1992). Quantitative analysis of static strain sensitivity in human mechanoreceptors from hairy skin. *Journal of Neurophysiology, 67,* 1105–1113.

Edin, B. B., & Abbs, J. H. (1991). Finger movement responses of cutaneous mechanoreceptors in the dorsal skin of the human hand. *Journal of Neurophysiology, 65,* 657–670.

Edin, B. B., & Johansson, N. (1995). Skin strain patterns provide kinaesthetic information to the human central nervous system. *Journal of Physiology, 487,* 243–251.

Edin, B. B., & Vallbo, A. B. (1990). Dynamic response of human muscle spindle afferents to stretch. *Journal of Neurophysiology, 63,* 1297–1306.

Edin, B. B., Westling, G., & Johansson, R. S. (1992). Independent control of human finger-tip forces at individual digits during precision lifting. *Journal of Physiology, 450,* 547–564.

Ehrsson, H. H., Fagergren, A., Johansson, R. S., & Forssberg, H. (2003). Evidence for the involvement of the posterior parietal cortex in coordination of fingertip forces for grasp stability in manipulation. *Journal of Neurophysiology, 90,* 2978–2986.

Ehrsson, H. H., Kuhtz-Buschbeck, J. P., & Forssberg, H. (2002). Brain regions controlling nonsynergistic versus synergistic movement of the digits: A functional magnetic resonance imaging study. *Journal of Neuroscience, 22,* 5074–5080.

Elbert, T., Pantev, C., Wienbruch, C., Rockstroh, B., & Taub, E. (1995). Increased cortical representation of the fingers of the left hand in string players. *Science, 270,* 305–307.

Elliot, J. M., & Connolly, K. J. (1984). A classification of manipulative hand movements. *Developmental Medicine and Child Neurology, 26,* 283–296.

Elliott, D., Carson, R. G., Goodman, D., & Chua, R. (1991). Discrete vs. continuous visual control of manual aiming. *Human Movement Science, 10,* 393–418.

Elliott, D., & Chua, R. (1996). Manual asymmetries in goal-directed movement. In D. Elliott & E. A. Roy (Eds.), *Manual asymmetries in motor performance* (pp. 143–158). Boca Raton, FL: CRC.

Elliott, D., Helsen, W. F., & Chua, R. (2001). A century later: Woodworth's (1899) two-component model of goal-directed aiming. *Psychological Bulletin, 127,* 342–357.

Ellis, R. R., Flanagan, J. R., & Lederman, S. J. (1999). The influence of visual illusions on grasp position. *Experimental Brain Research, 125,* 109–114.

Ellis, R. R., & Lederman, S. J. (1993). The role of haptic vs. visual volume cues in the size-weight illusion. *Perception & Psychophysics, 53,* 315–324.

Ellis, R. R., & Lederman, S. J. (1998). The golf-ball illusion: Evidence for top-down processing in weight perception. *Perception, 27,* 193–201.

Ellis, R. R., & Lederman, S. J. (1999). The material-weight illusion revisited. *Perception & Psychophysics, 61,* 1564–1576.

Emmorey, K., Grabowski, T., McCullough, S., Damasio, H., Ponto, L. L. B., Hichwa, R. D., & Bellugi, U. (2003). Neural systems underlying lexical retrieval for sign language. *Neuropsychologia, 41,* 85–95.

Engel, K. C., Flanders, M., & Soechting, J. F. (1997). Anticipatory and sequential motor control in piano playing. *Experimental Brain Research, 113,* 189–199.

Ericsson, K. A., Krampe, R. T., & Tesch-Römer, C. (1993). The role of deliberate practice in the acquisition of expert performance. *Psychological Review, 100,* 363–406.

Eriksson, Y., Jansson, G., & Strucel, M. (2003). *Tactile maps: Guidelines for the production of maps for the visually impaired.* Enskede, Sweden: Swedish Braille Authority.

Ernst, M. O., & Banks, M. S. (2002). Humans integrate visual and haptic information in a statistically optimal fashion. *Nature, 415,* 429–433.

Essick, G. K., Franzen, O., & Whitsel, B. L. (1988). Discrimination and scaling of velocity of stimulus motion across the skin. *Somatosensory and Motor Research, 6,* 21–40.

Evans, A. L., Harrison, L. M., & Stephens, J. A. (1990). Maturation of the cutaneomuscular reflex recorded from the first dorsal interosseous muscle in man. *Journal of Physiology, 428,* 425–440.

Fagard, J. (1998). Changes in grasping skills and the emergence of bimanual coordination during the first year of life. In K. Connelly (Ed.), *The psychobiology of the hand* (pp. 123–143). London: MacKeith.

Fasse, E. D., Hogan, N., Kay, B. A., & Mussa-Ivaldi, F. A. (2000). Haptic interaction with virtual objects: Spatial perception and motor control. *Biological Cybernetics, 82,* 69–83.

FCS Control Systems. (2004, June 16). Available: http://www.fcs-cs.com.

Fechner, G. (1966). *Elements of psychophysics* (H. E. Adler, Trans., D. H. Howes & E. G. Boring, Eds.). New York: Holt, Rinehart & Winston. (Original work published 1860)

Feldman, A. G. (1974). Change in the length of the muscle as a consequence of a shift in equilibrium in the muscle-load system. *Biofizika, 19,* 535–538.

Ferrell, W. R. (1980). The adequacy of stretch receptors in the cat knee joint for signalling joint angle throughout a full range of movement. *Journal of Physiology, 299,* 85–99.

Ferrell, W. R., & Milne, S. E. (1989). Factors affecting the accuracy of position matching at the proximal interphalangeal joint in human subjects. *Journal of Physiology, 411,* 575–583.

Ferrier, D. (1886). *The functions of the brain* (2d ed.). New York: Smith Elder.

Fetters, L., & Todd, J. (1987). Quantitative assessment of infant reaching movements. *Journal of Motor Behavior, 19*, 147–166.

Fish, J., & Soechting, J. F. (1992). Synergistic finger movements in a skilled motor task. *Experimental Brain Research, 91*, 327–334.

Fitts, P. M. (1954). The information capacity of the human motor system in controlling the amplitude of movement. *Journal of Experimental Psychology, 47*, 381–391.

Fitts, P. M. (1964). Perceptual-motor skill learning. In A. W. Melton (Ed.), *Categories of human learning* (pp. 244–285). New York: Academic.

Flanagan, J. R., & Beltzner, M. A. (2000). Independence of perceptual and sensorimotor predictions in the size-weight illusion. *Nature Neuroscience, 3*, 737–741.

Flanagan, J. R., & Wing, A. M. (1995). The stability of precision grip forces during cyclic arm movements with a hand-held load. *Experimental Brain Research, 105*, 455–464.

Flanagan, J. R., & Wing, A. M. (1997). The role of internal models in motion planning and control: Evidence from grip force adjustments during movements of hand-held loads. *Journal of Neuroscience, 17*, 1519–1528.

Flanders, M., Helms Tillery, S. I., & Soechting, J. F. (1992). Early stages in a sensorimotor transformation. *Behavioral and Brain Sciences, 15*, 309–362.

Flanders, M., & Soechting, J. F. (1990). Parcellation of sensorimotor transformations for arm movements. *Journal of Neuroscience, 10*, 2420–2427.

Flanders, M., & Soechting, J. F. (1992). Kinematics of typing: Parallel control of the two hands. *Journal of Neurophysiology, 67*, 1264–1274.

Flash, T., & Hogan, N. (1985). The coordination of arm movements: An experimentally confirmed mathematical model. *Journal of Neuroscience, 5*, 1688–1703.

Flatt, A. E., & Burmeister, L. (1979). A comparison of hand growth in elementary schoolchildren in Czechoslovakia and the United States. *Developmental Medicine and Child Neurology, 21*, 515–524.

Florence, S. L., Wall, J. T., & Kaas, J. H. (1989). Somatotopic organization of inputs from the hand to the spinal gray and cuneate nucleus of monkeys with observations on the cuneate nucleus of humans. *Journal of Comparative Neurology, 286*, 48–70.

Folio, M. R., & Fewell, R. R. (1983). *Peabody developmental motor scales and activity cards*. Allen, TX: Teaching Resources.

Forssberg, H. (1998). The neurophysiology of manual skill development. In K. J. Connolly (Ed.), *The psy-chobiology of the hand* (pp. 97–122). London: MacKeith.

Forssberg, H., Eliasson, A. C., Kinoshita, H., Johansson, R. S., & Westling, G. (1991). Development of human precision grip: I. Basic coordination of force. *Experimental Brain Research, 85*, 451–457.

Forssberg, H., Eliasson, A. C., Kinoshita, H., Westling, G., & Johansson, R. S. (1995). Development of human precision grip: IV. Tactile adaptation of isometric finger forces to the frictional condition. *Experimental Brain Research, 104*, 323–330.

Forssberg, H., Eliasson, A. C., Redon-Zouitenn, C., Mercuri, E., & Dubowitz, L. (1999). Impaired grip-lift synergy in children with unilateral brain lesions. *Brain, 122*, 1157–1168.

Forssberg, H., Kinoshita, H., Eliasson, A. C., Johansson, R. S., Westling, G., & Gordon, A. M. (1992). Development of human precision grip: II. Anticipatory control of isometric forces targeted for object's weight. *Experimental Brain Research, 90*, 393–398.

Foucher, G., & Chabaud, M. (1998). The bipolar lengthening technique: A modified partial toe transfer for thumb reconstruction. *Plastic and Reconstructive Surgery, 102*, 1981–1987.

Francis, S. T., Kelly, E. F., Bowtell, R., Dunseath, W. J., Folger, S. E., & McGlone, F. (2000). fMRI of the responses to vibratory stimulation of digit tips. *NeuroImage, 11*, 188–202.

Frey, M. von. (1896). Untersuchungen über die sinnesfunctionen der menschlichen haut: Druckempfindung und schmerz (Investigation on the sensory function of human skin: Sensitivity to pressure and pain). *Abhandlungen der Sächsischen Akademie der Wissenschaften zu Leipzig, 23*, 175–266.

Frey, M. von. (1914). Studien über den Kraftsinn (Studies of the muscle sense). *Zeitschrift für Biologie, 63*, 129–154.

Fung, Y. C. (1993). *Biomechanics: Mechanical properties of living tissue*. New York: Springer-Verlag.

Galganski, M. E., Fuglevand, A. J., & Enoka, R. M. (1993). Reduced control of motor output in a human hand muscle of elderly subjects during submaximal contractions. *Journal of Neurophysiology, 69*, 2108–2115.

Gandevia, S. C. (1996). Kinesthesia: Roles for afferent signals and motor commands. In L. B. Rowell & J. T. Shepherd (Eds.), *Handbook of physiology: Sec. 12. Exercise regulation and integration of multiple systems* (pp. 128–172). New York: Oxford University Press.

Gandevia, S. C., & McCloskey, D. I. (1976). Joint sense, muscle sense, and their combination as position sense, measured at the distal interphalangeal joint of the middle finger. *Journal of Physiology, 260*, 387–407.

Gandevia, S. C., & McCloskey, D. I. (1977). Effects of related sensory inputs on motor performances in man studied through changes in perceived heaviness. *Journal of Physiology, 272,* 653–672.

Gandevia, S. C., & Phegan, C. M. L. (1999). Perceptual distortions of the human body image produced by local anesthesia, pain and cutaneous stimulation. *Journal of Physiology, 514,* 609–616.

Gandevia, S. C., Wilson, L., Cordo, P. J., & Burke, D. (1994). Fusimotor reflexes in relaxed forearm muscle produced by cutaneous afferents from the human hand. *Journal of Physiology, 479,* 499–508.

Gardner, E. P., Martin, J. M., & Jessell, T. M. (2000). The bodily senses. In E. R. Kandel, J. H. Schwartz, & T. M. Jessell (Eds.), *Principles of neural science* (4th ed., pp. 430–450). New York: McGraw-Hill.

Gardner, E. P., & Sklar, B. (1994). Discrimination of the direction of motion on the human hand: A psychophysical study of stimulation parameters. *Journal of Neurophysiology, 71,* 2414–2429.

Gardner, R. A., & Broman, M. (1979). The Purdue pegboard: Normative data on 1334 school children. *Journal of Clinical Child Psychology, 1,* 156–162.

Garnett, R., & Stephens, J. A. (1980). The reflex responses of single motor units in human first dorsal interosseous muscle following cutaneous afferent stimulation. *Journal of Physiology, 303,* 351–364.

Garraghty, P. E., & Kaas, J. H. (1991). Functional reorganization in adult monkey thalamus after peripheral nerve injury. *NeuroReport, 2,* 747–750.

Garrett, J. W. (1971). The adult human hand: Some anthropometric and biomechanical considerations. *Human Factors, 13,* 117–131.

Geldard, F. A. (1972). *The human senses* (2d ed.). New York: Wiley.

Geldard, F. A. (1975). *Sensory saltation: Metastability in the perceptual world.* Hillsdale, NJ: Erlbaum.

Geldard, F. A., & Sherrick, C. E. (1972). The cutaneous "rabbit": A perceptual illusion. *Science, 178,* 178–179.

Geldard, F. A., & Sherrick, C. E. (1983). The cutaneous salutatory area and its presumed neural basis. *Perception & Psychophysics, 33,* 299–304.

Gellis, M., & Pool, R. (1977). Two point discrimination distances in the normal hand and forearm. *Plastic and Reconstructive Surgery, 59,* 57–63.

Gentaz, E., & Hatwell, Y. (1995). The haptic "oblique effect" in children's and adults' perception of orientation. *Perception, 24,* 631–646.

Gentaz, E., & Hatwell, Y. (2004). Geometrical haptic illusions: The role of exploration in the Müller-Lyer, vertical-horizontal, and Delboeuf illusions. *Psychonomic Bulletin Review, 11,* 31–40.

Gentilucci, M., Caselli, L., & Secchi, C. (2003). Finger control in the tripod grasp. *Experimental Brain Research, 149,* 351–360.

Gentilucci, M., Castiello, U., Corradini, M. L., Scarpa, M., Umilta, C., & Rizzolatti, G. (1991). Influence of different types of grasping on the transport component of prehension movements. *Neuropsychologia, 29,* 361–378.

Gentilucci, M., Chieffi, S., Scarpa, M., & Castiello, U. (1992). Temporal coupling between transport and grasp components during prehension movements: Effect of visual perturbation. *Behavioral Brain Research, 47,* 71–82.

Gentilucci, M., Daprati, E., Toni, I., Chieffi, S., & Saetti, M. C. (1995). Unconscious updating of grasp motor program. *Experimental Brain Research, 105,* 291–303.

Gentilucci, M., Toni, I., Daprati, E., & Gangitano, M. (1997). Tactile input of the hand and the control of reaching to grasp movements. *Experimental Brain Research, 114,* 130–137.

Gentner, D. R. (1983). The acquisition of typewriting skill. *Acta Psychologica, 54,* 233–248.

Gentner, D. R. (1987). Timing of skilled motor performance: Tests of the proportional duration model. *Psychological Review, 94,* 255–276.

Gentner, D. R. (1988). Expertise in typewriting. In M. T. H. Chi, R. Glaser, & M. J. Farr (Eds.), *The nature of expertise* (pp. 1–21). Hillsdale, NJ: Erlbaum.

Georgopoulos, A. P. (1986). On reaching. *Annual Review of Neuroscience, 9,* 147–170.

Georgopoulos, A. P., Kalaska, J. F., & Massey, J. T. (1981). Spatial trajectories and reaction times of aimed movements: Effects of practice, uncertainty, and change in target location. *Journal of Neurophysiology, 46,* 725–743.

Gerard, M. J., Armstrong, T. J., Franzblau, T. J., Martin, B. J., & Rempel, D. M. (1999). The effects of keyswitch stiffness on typing force, finger electromyography, and subjective discomfort. *American Industrial Hygiene Association Journal, 60,* 762–769.

Gescheider, G. A. (1974). Effects of signal probability on vibrotactile signal recognition. *Perceptual & Motor Skills, 38,* 15–23.

Gescheider, G. A. (1997). *Psychophysics: The fundamentals* (3d ed.). Mahwah, NJ: Erlbaum.

Gescheider, G. A., Beiles, E. J., Checkosky, C. M., Bolanowski, S. J., & Verrillo, R. T. (1994). The effects of aging on information-processing channels in the sense of touch: II. Temporal summation in the P channel. *Somatosensory & Motor Research, 11,* 359–365.

Gescheider, G. A., Berryhill, M. E., Verrillo, R. T., & Bolanowski, S. J. (1999). Vibrotactile temporal summation: Probability summation or neural integration? *Somatosensory & Motor Research, 16,* 229–242.

Gescheider, G. A., Bolanowski, S. J., Hall, K. L., Hoffman, K. E., & Verrillo, R. T. (1994). The effects of aging on information-processing channels in the sense of touch: I. Absolute sensitivity. *Somatosensory and Motor Research, 11,* 345–357.

Gescheider, G. A., Bolanowski, S. J., Pope, J., & Verrillo, R. T. (2002). A four-channel analysis of the tactile sensitivity of the fingertip: Frequency selectivity, spatial summation, and temporal summation. *Somatosensory & Motor Research, 19,* 114–124.

Gescheider, G. A., Edwards, R. R., Lackner, E. A., Bolanowski, S. J., & Verrillo, R. T. (1996). The effects of aging on information-processing channels in the sense of touch: III. Differential sensitivity to changes in stimulus intensity. *Somatosensory & Motor Research, 13,* 73–80.

Gescheider, G. A., Hoffman, K. E., Harrison, M. A., Travis, M. L., & Bolanowski, S. J. (1994). The effects of masking on vibrotactile temporal summation in the detection of sinusoidal and noise signals. *Journal of the Acoustical Society of America, 95,* 1006–1016.

Gescheider, G. A., Sklar, B. F., Van Doren, C. L., & Verrillo, R. T. (1985). Vibrotactile forward masking: Psychophysical evidence for a triplex theory of cutaneous mechanoreception. *Journal of the Acoustical Society of America, 78,* 534–543.

Gescheider, G. A., Verrillo, R. T., & Van Doren, C. L. (1982). Prediction of vibrotactile masking functions. *Journal of the Acoustical Society of America, 72,* 1421–1426.

Gesell, A. (1928). *Infancy and human growth.* New York: Macmillan.

Geyer, S., Ledberg, A., Schleicher, A., Kinomura, S., Schormann, T., Bürgel, U., Klingberg, T., Larsson, J., Zilles, K., & Roland, P. E. (1996). Two different areas within the primary motor cortex of man. *Nature, 382,* 805–807.

Gibson, J. J. (1962). Observations on active touch. *Psychological Review, 69,* 477–490.

Gibson, J. J. (1966). *The senses considered as perceptual systems.* Boston: Houghton Mifflin.

Gibson, J. J., & Walker, A. (1984). Development of knowledge of visual-tactual affordances of substance. *Child Development, 55,* 453–460.

Gill, J. (1982). Production of tangible graphic displays. In W. Schiff & E. Foulke (Eds.), *Tactual perception: A sourcebook* (pp. 405–416). New York: Cambridge University Press.

Goldscheider, A. (1886). Zur dualität des temperatursinns (On the duality of the temperature senses). *Archiv für die gesammte Physiologie, 39,* 96–120.

Goldscheider, A. (1889). Untersuchungen über den Muskelsinn (Investigations of the muscle sense). *Archiv für Anatomie und Physiologie, 3,* 369–502.

Goldstein, E. B. (2002). *Sensation and perception* (6th ed.). Belmont, CA: Wadsworth.

Gombrich, E. H. (1972). *Art, perception and reality.* Baltimore, MD: Johns Hopkins University Press.

Gombrich, E. H. (1977). *Art and illusion: A study in the psychology of pictorial representation* (5th ed.). Oxford: Phaidon.

Goodman, M. B., Lumpkin, E. A., Ricci, A., Tracey, W. D., Kernan, M., & Nicolson, T. (2004). Molecules and mechanisms of mechanotransduction. *Journal of Neuroscience, 24,* 9220–9222.

Goodwin, A. W., Jenmalm, P., & Johansson, R. S. (1998). Control of grip force when tilting objects: Effect of curvature of grasped surfaces and applied tangential torque. *Journal of Neuroscience, 18,* 10724–10734.

Goodwin, A. W., John, K. T., & Marceglia, A. H. (1991). Tactile discrimination of curvature by humans using only cutaneous information from the fingerpads. *Experimental Brain Research, 86,* 663–672.

Goodwin, A. W., & Wheat, H. E. (1992). Magnitude estimation of contact force when objects with different shapes are applied passively to the fingerpad. *Somatosensory and Motor Research, 9,* 339–344.

Goodwin, A. W., & Wheat, H. E. (2004). Sensory signals in neural populations underlying tactile perception and manipulation. *Annual Review of Neuroscience, 27,* 53–77.

Gordon, A. M., & Forssberg, H. (1997). Development of neural mechanisms underlying grasping in children. In K. J. Connolly & H. Forssberg (Eds.), *Neurophysiology and neuropsychology of motor development* (pp. 214–231). London: MacKeith.

Gordon, A. M., Forssberg, H., Johansson, R. S., Eliasson, A. C., & Westling, G. (1992). Development of human precision grip: III. Integration of visual cues during programming of isometric forces. *Experimental Brain Research, 90,* 399–403.

Gordon, A. M., Forssberg, H., Johansson, R. S., & Westling, G. (1991a). Visual size cues in the programming of manipulative forces during precision grip. *Experimental Brain Research, 83,* 477–482.

Gordon, A. M., Forssberg, H., Johansson, R. S., & Westling, G. (1991b). The integration of haptically acquired size information in the programming of precision grip. *Experimental Brain Research, 83,* 483–488.

Gordon, A. M., & Soechting, J. F. (1995). Use of tactile afferent information in sequential finger movements. *Experimental Brain Research, 107,* 281–292.

Gordon, A. M., Westling, G., Cole, K. J., & Johansson, R. S. (1993). Memory representations underlying motor commands used during manipulation of common and novel objects. *Journal of Neurophysiology, 69*, 1789–1796.

Gordon, G. (Ed.). (1978). *Active touch: The mechanism of recognition of objects by manipulation: A multidisciplinary approach.* London: Pergamon.

Gordon, I. E., & Morison, V. (1982). The haptic perception of curvature. *Perception & Psychophysics, 31*, 446–450.

Gordon, J., Ghilardi, M. F., & Ghez, C. (1995). Impairments of reaching movements in patients without proprioception: 1. Spatial errors. *Journal of Neurophysiology, 73*, 347–360.

Granit, R. (1981). Comments on history of motor control. In V. B. Brooks (Ed.), *Handbook of physiology: The nervous system, motor control* (Vol. 2, part 1, pp. 1–16). Bethesda, MD: American Physiological Society.

Green, B. G. (1976). Vibrotactile temporal summation: Effect of frequency. *Sensory Processes, 1*, 138–149.

Green, B. G. (1977). Localization of thermal sensation: An illusion and synthetic heat. *Perception & Psychophysics, 22*, 331–337.

Green, B. G., & Zaharchuk, R. (2001). Spatial variation in sensitivity as a factor in measurements of spatial summation of warmth and cold. *Somatosensory & Motor Research, 18*, 181–190.

Greenspan, J. D., & Bolanowski, S. J. (1996). The psychophysics of tactile perception and its peripheral physiological basis. In L. Kruger (Ed.), *Pain and touch* (pp. 25–103). San Diego, CA: Academic.

Greenspan, J. D., & Kenshalo, D. R., Sr. (1985). The primate as a model for the human temperature-sensing system: 2. Area of skin receiving thermal stimulation (spatial summation). *Somatosensory Research, 2*, 315–324.

Grill, S. E., & Hallett, M. (1995). Velocity sensitivity of human muscle spindle afferents and slowly adapting type II cutaneous mechanoreceptors. *Journal of Physiology, 489*, 593–602.

Grober, E. D., Hamstra, S. J., Wanzel, K. R., Reznick, R. K., Matsumoto, E. D., Sidhu, R. S., & Jarvi, K. A. (2003). Validation of novel and objective measures of microsurgical skill: Hand-motion analysis and stereoscopic acuity. *Microsurgery, 23*, 317–322.

Grudin, J. T. (1983). Error patterns in skilled and novice transcription typing. In W. E. Cooper (Ed.), *Cognitive aspects of skilled typewriting* (pp. 121–144). New York: Springer-Verlag.

Grunwald, A. P. (1965). A Braille reading machine. *Science, 154*, 144–146.

Hackel, M. E., Wolfe, G. A., Bang, S. M., & Canfield, J. S. (1992). Changes in hand function in the aging adult as determined by the Jebsen test of hand function. *Physical Therapy, 72*, 373–377.

Häger-Ross, C., & Johansson, R. S. (1996). Nondigital afferent input in reactive control of fingertip forces during precision grip. *Experimental Brain Research, 110*, 131–141.

Häger-Ross, C., & Schieber, M. H. (2000). Quantifying the independence of human finger movements: Comparisons of digits, hands, and movement frequencies. *Journal of Neuroscience, 20*, 8542–8550.

Haggard, P., & Wing, A. M. (1991). Remote responses to perturbation in human prehension. *Neuroscience Letters, 122*, 103–108.

Haggard, P., & Wing, A. (1997). On the hand transport component of prehensile movements. *Journal of Motor Behavior, 29*, 282–287.

Hahn, J. F. (1958). Cutaneous vibratory thresholds for square-wave electrical pulses. *Science, 127*, 879–880.

Hajian, A. Z., & Howe, R. D. (1997). Identification of the mechanical impedance at the human finger tip. *Journal of Biomechanical Engineering, 119*, 109–114.

Hajnis, K. (1969). The dynamics of hand growth since the birth till 18 years of age. *Panminerva Medica, 11*, 123–132.

Hale, K. S., & Stanney, K. M. (2004). Deriving haptic design guidelines from human physiological, psychological, and neurological foundations. *IEEE Computer Graphics and Applications, 24*, 33–39.

Hall, E. J., Flament, D., Fraser, C., & Lemon, R. N. (1990). Non-invasive brain stimulation reveals reorganized cortical outputs in amputees. *Neuroscience Letters, 116*, 379–386.

Hall, L. A., & McCloskey, D. I. (1983). Detections of movements imposed on finger, elbow and shoulder joints. *Journal of Physiology, 335*, 519–533.

Hallin, R. G., Wiesenfeld, Z., & Lindblom, U. (1981). Neurophysiological studies on patients with sutured median nerves: Faulty sensory localization after nerve regeneration and its physiological correlates. *Experimental Neurology, 73*, 90–106.

Halverson, H. M. (1931). An experimental study of prehension in infants by means of systematic cinema records. *Genetic Psychology Monographs, 10*, 107–286.

Hämäläinen, H., Hiltunen, J., & Titievskaja, I. (2002). Activation of somatosensory cortical areas varies with attentional state: An fMRI study. *Behavioral & Brain Research, 135*, 159–165.

Hämäläinen, H., & Järvilehto, T. (1981). Peripheral neural basis of tactile sensations in man: I. Effect of frequency and probe area on sensations elicited by single mechanical pulses on hairy and glabrous skin of the hand. *Brain Research, 219*, 1–12.

Hammerton, M., & Tickner, A. H. (1966). An investigation into the comparative suitability of forearm, hand and thumb controls on acquisition tasks. *Ergonomics, 9,* 125–130.

Hannaford, B., & Wood, L. (1989). Performance evaluation of a 6 axis high fidelity generalized force reflecting teleoperator. *Proceedings of NASA Conference on Space Telerobotics, 2,* 87–96.

Hanten, W. P., Chen, W. Y., Austin, A. A., Brooks, R. E., Carter, H. C., Law, C. A., Morgan, M. K., Sanders, D. J., Swan, C. A., & Vanderslice, A. L. (1999). Maximum grip strength in normal subjects from 20 to 64 years of age. *Journal of Hand Therapy, 12,* 193–200.

Harju, E. (2002). Cold and warmth perception mapped for age, gender, and body area. *Somatosensory and Motor Research, 19,* 61–73.

Harper, R., & Stevens, S. S. (1964). Subjective hardness of compliant materials. *Quarterly Journal of Experimental Psychology, 16,* 214–215.

Harrington, T., & Merzenich, M. (1970). Neural coding in the sense of touch: Human sensations of skin indentation compared with the responses of slowly adapting mechanoreceptive afferents innervating the hairy skin of monkeys. *Experimental Brain Research, 10,* 251–264.

Hasser, C. J., & Massie, T. H. (1998). The haptic illusion. In C. Dodsworth, Jr. (Ed.), *Digital illusion: Entertaining the future with high technology* (pp. 287–310). Reading, MA: Addison-Wesley.

Havenith, G., van de Linde, E. J. G., & Heus, R. (1992). Pain, thermal sensation and cooling rates of hands while touching cold materials. *European Journal of Applied Physiology, 65,* 43–51.

Hawkes, G. R. (1961). Information transmitted via electrical cutaneous stimulus duration. *Journal of Psychology, 51,* 293–298.

Hay, L. (1981). The effect of amplitude and accuracy requirements on movement time in children. *Journal of Motor Behavior, 13,* 177–186.

Hayward, V. (1995). Toward a seven axis haptic device. *Proceedings of the IEEE/RSJ International Conference on Intelligent Robots and Systems, 3,* 133–139.

Head, H. (1920). *Studies in neurology.* Oxford: Oxford University Press.

Heller, M. A. (1989). Texture perception in sighted and blind observers. *Perception & Psychophysics, 45,* 49–54.

Heller, M. A. (2000). *Touch, representation, and blindness.* New York: Oxford University Press.

Heller, M. A., Calcaterra, J. A., Burson, L. L., & Tyler, L. A. (1996). Tactual picture identification by blind and sighted people: Effects of providing categorical information. *Perception & Psychophysics, 58,* 310–323.

Heller, M. A., & Joyner, T. D. (1993). Mechanisms in the haptic horizontal-vertical illusion: Evidence from sighted and blind subjects. *Perception & Psychophysics, 53,* 422–428.

Helmholtz, H. von. (1925). *Treatise on physiological optics* (Vol. 3, J. P. C. Southall, Ed. and Trans.). Menasha, WI: Optical Society of America. (Original work published 1866)

Helson, H., & King, S. M. (1931). The Tau effect: An example of psychological relativity. *Journal of Experimental Psychology, 46,* 483.

Henriques, D. Y., Flanders, M., & Soechting, J. F. (2004). Haptic synthesis of shapes and sequences. *Journal of Neurophysiology, 91,* 1808–1821.

Henriques, D. Y., & Soechting, J. F. (2003). Bias and sensitivity in the haptic perception of geometry. *Experimental Brain Research, 150,* 95–108.

Hepp-Reymond, M.-C., Trouche, E., & Wiesendanger, M. (1974). Effects of unilateral and bilateral pyramidotomy on a conditioned rapid precision grip in monkeys (macaca fascicularis). *Experimental Brain Research, 21,* 519–527.

Hepp-Reymond, M.-C., & Wiesendanger, M. (1972). Unilateral pyramidotomy in monkeys: Effect on force and speed of a conditioned precision grip. *Brain Research, 36,* 117–131.

Hermsdörfer, J., Hagl, E., & Nowak, D. A. (2004). Deficits of anticipatory grip force control after damage to peripheral and central sensorimotor systems. *Human Movement Science, 23,* 643–662.

Hermsdörfer, J., Marquardt, C., Philipp, J., Zierdt, A., Nowak, D., Glasauer, S., & Mai, N. (2000). Moving weightless objects: Grip force control during microgravity. *Experimental Brain Research, 132,* 52–64.

Higashiyama, A., & Hayashi, M. (1993). Localization of electrocutaneous stimuli on the fingers and forearm: Effects of electrode configuration and body axis. *Perception & Psychophysics, 54,* 108–120.

Higashiyama, A., & Tashiro, T. (1983). Temporal and spatial integration for electrocutaneous stimulation. *Perception & Psychophysics, 33,* 437–442.

Higashiyama, A., & Tashiro, T. (1988). Temporal integration of double electrical pulses. *Perception & Psychophysics, 43,* 172–178.

Higashiyama, A., & Tashiro, T. (1989). Magnitude estimates for electrical pulses: Evidence for two neural systems. *Perception & Psychophysics, 45,* 537–549.

Higashiyama, A., & Tashiro, T. (1990). Electrocutaneous spatial integration at threshold: The effects of electrode size. *Perception & Psychophysics, 48,* 389–397.

Higashiyama, A., & Tashiro, T. (1993a). Electrocutaneous spatial integration at suprathreshold levels:

An additive neural model. *Journal of Experimental Psychology: Human Perception & Performance, 19,* 912–923.

Higashiyama, A., & Tashiro, T. (1993b). Localization of electrocutaneous stimuli on the fingers and forearm: Effects of electrode configuration and body axis. *Perception & Psychophysics, 54,* 108–120.

Hikosaka, O., Sakai, K., Miyauchi, S., Takino, R., Sasaki, Y., & Pütz, B. (1996). Activation of human presupplementary motor area in learning of sequential procedures: A functional MRI study. *Journal of Neurophysiology, 76,* 617–621.

Hikosaka, O., Tanaka, M., Sakamoto, M., & Iwamura, Y. (1985). Deficits in manipulative behaviors induced by local injections of muscimol in the first somatosensory cortex of the conscious monkey. *Brain Research, 325,* 375–380.

Ho, H., & Jones, L. A. (2004). Material identification using real and simulated thermal cues. *Proceedings of the 26th Annual International Conference of the IEEE Engineering in Medicine and Biology Society, 4,* 2462–2465.

Ho, H., & Jones, L. A. (in press). Contribution of thermal cues to material discrimination and localization. *Perception & Psychophysics, 68.*

Hofsten, C. von. (1979). Development of visually directed reaching: The approach phase. *Journal of Human Movement Studies, 5,* 160–178.

Hofsten, C. von. (1982). Eye-hand coordination in the newborn. *Developmental Psychology, 18,* 450–461.

Hofsten, C. von. (1984). Developmental changes in the organization of prereaching movements. *Developmental Psychology, 20,* 378–388.

Hofsten, C. von. (1991). Structuring of early reaching movements: A longitudinal study. *Journal of Motor Behavior, 23,* 280–292.

Hofsten, C. von., & Fazel-Zandy, S. (1984). Development of visually guided hand orientation in reaching. *Journal of Experimental Child Psychology, 38,* 208–219.

Hofsten, C. von., & Rönnqvist, L. (1988). Preparation for grasping an object: A developmental study. *Journal of Experimental Psychology: Human Perception and Performance, 14,* 610–621.

Hofsten, C. von., & Rösblad, B. (1988). The integration of sensory information in the development of precise manual pointing. *Neuropsychologia, 26,* 805–821.

Hollerbach, J. M., & Atkeson, C. G. (1986). Characterization of joint-interpolated arm movements. *Experimental Brain Research, 15*(Suppl.), 41–54.

Hollingworth, H. L. (1909). The inaccuracy of movement. *Archives of Psychology,* No. 13. New York: Johnson Associates.

Hollins, M., Bensmaïa, S., Karlof, K., & Young, F. (2000). Individual differences in perceptual space for tactile textures: Evidence from multidimensional scaling. *Perception & Psychophysics, 62,* 1534–1544.

Hollins, M., Bensmaïa, S., & Risner, R. (1998). The duplex theory of tactile texture perception. In Grondin, S., & Lacouture (Eds.), Fechner Day 98. *Proceedings of the Fourteenth Annual Meeting of the International Society for Psychophysics,* (pp.115–120). Quebec, Canada: International Society for Psychophysics.

Hollins, M., Bensmaïa, S. J., & Washburn, S. (2001). Vibrotactile adaptation impairs discrimination of fine, but not coarse, textures. *Somatosensory & Motor Research, 18,* 253–262.

Hollins, M., Faldowski, R., Rao, S., & Young, F. (1993). Perceptual dimensions of tactile surface texture: A multidimensional scaling analysis. *Perception & Psychophysics, 54,* 697–705.

Hollins, M., & Favorov, O. (1994). The tactile movement aftereffect. *Somatosensory & Motor Research, 11,* 153–162.

Hollins, M., & Goble, A. (1988). Perception of the length of voluntary movements. *Somatosensory Research, 5,* 335–348.

Hollins, M., & Risner, S. R. (2000). Evidence for the duplex theory of tactile texture perception. *Perception & Psychophysics, 62,* 695–705.

Holmes, E., Hughes, B., & Jansson, G. (1998). Haptic perception of texture gradients. *Perception, 27,* 993–1008.

Holmes, G. (1927). Disorders of sensation produced by cortical lesions. *Brain, 50,* 413–427.

Holst, E. von. (1954). Relations between the central nervous system and the peripheral organs. *British Journal of Animal Behaviour, 2,* 89–94.

Hoop, N. (1971). Haptic perception in preschool children. *American Journal of Occupational Therapy, 25,* 340–344.

Howard, I. P. (1982). *Human visual orientation.* New York: Wiley.

Howard, W. S., & Kumar, V. (1996). On the stability of grasped objects. *IEEE Transactions on Robotics and Automation, 12,* 904–917.

Howe, R., & Kontarinis, D. (1992). Task performance with a dextrous teleoperated hand system. *Proceedings of the SPIE International Society for Optical Engineering, 1833,* 199–207.

Hsiao, S. S., Johnson, K. O., & Twombly, I. A. (1993). Roughness coding in the somatosensory system. *Acta Psychologica, 84,* 53–67.

Hulliger, M. (1984). The mammalian muscle spindle and its central control. *Reviews in Physiology, Biochemistry and Pharmacology, 101,* 2–107.

Hulliger, M., Nordh, E., & Vallbo, A. B. (1982). The absence of position response in spindle afferent units from human finger muscles during accurate position holding. *Journal of Physiology, 322,* 167–179.

Hunter, I. W., Jones, L. A., Sagar, M. A., Lafontaine, S. R., & Hunter, P. J. (1995). Ophthalmic microsurgical robot and associated virtual environment. *Computers in Biology and Medicine, 25,* 173–182.

Iberall, T. (1997). Human prehension and dexterous robot hands. *International Journal of Robotics Research, 16,* 285–299.

Iberall, T., & MacKenzie, C. L. (1990). Opposition space and human prehension. In S. T. Venkataraman & T. Iberall (Eds.), *Dextrous robot hands* (pp. 32–54). New York: Springer-Verlag.

Iggo, A., & Muir, A. R. (1969). The structure and function of a slowly adapting touch corpuscle in hairy skin. *Journal of Physiology, 200,* 763–796.

Immersion Corporation. (2004, June 14). Available: http://www.immersion.com.

Infed, F., Brown, S. W., Lee, C. D., Lawrence, D. A., Dougherty, A. M., & Pao, L. Y. (1999). Combined visual/haptic rendering modes for scientific visualization. *Proceedings of ASME Dynamic Systems and Control Division, 67,* 93–99.

Inhoff, A. W. (1991). Word frequency during copy typing. *Journal of Experimental Psychology: Human Perception and Performance, 17,* 478–487.

Inhoff, A. W., Briihl, D., Bohemier, G., & Wang, J. (1992). Eye-hand span and coding of text during copytyping. *Journal of Experimental Psychology: Learning, Memory, & Cognition, 18,* 298–306.

Ino, S., Shimizu, S., Odagawa, T., Sato, M., Takahashi, M., Izumi, T., & Ifukube, T. (1993). A tactile display for presenting quality of materials by changing the temperature of skin surface. *IEEE International Workshop on Robot and Human Communication,* 220–224.

Iriki, A., Tanaka, M., & Iwamura, Y. (1996). Coding of modified body schema during tool use by macaque postcentral neurons. *NeuroReport, 7,* 2325–2330.

Jagacinski, R. J., Repperger, D. W., Moran, M. S., Ward, S. L., & Glass, B. (1980). Fitts' law and the microstructure of rapid discrete movements. *Journal of Experimental Psychology: Human Perception and Performance, 6,* 309–320.

Jain, N., Catania, M. V., & Kaas, J. H. (1997). Deactivation and reactivation of somatosensory cortex after dorsal spinal cord injury. *Nature, 386,* 495–498.

Jakobson, L. S., & Goodale, M. A. (1991). Factors affecting higher-order movement planning: A kinematic analysis of human prehension. *Experimental Brain Research, 86,* 199–208.

James, T. W., Humphrey, G. K., Gati, J. S., Servos, P., Menon, R. S., & Goodale, M. A. (2002). Haptic study of three-dimensional objects activates extrastriate visual areas. *Neuropsychologia, 40,* 1706–1714.

Jami, L. (1992). Golgi tendon organs in mammalian skeletal muscle: Functional properties and central actions. *Physiological Reviews, 72,* 623–666.

Jäncke, L., Schlaug, G., & Steinmetz, H. (1997). Hand skill asymmetry in professional musicians. *Brain & Cognition, 34,* 424–432.

Jansson, G., Bergamasco, M., & Frisoli, A. (2003). A new option for the visually impaired to experience 3D art at museums: Manual exploration of virtual copies. *Visual Impairment Research, 5,* 1–12.

Jansson, G., & Monaci, L. (2003). Exploring tactile maps with one or two fingers. *Cartographic Journal, 40,* 269–271.

Jansson, G., & Monaci, L. (2004). Haptic identification of objects with different numbers of fingers. In S. Ballesteros & M. A. Heller (Eds.), *Touch, blindness and neuroscience* (pp. 203–213). Madrid, Spain: UNED.

Järvilehto, T., Hämäläinen, H., & Laurinen, P. (1976). Characteristics of single mechanoreceptive fibers innervating hairy skin of the human hand. *Experimental Brain Research, 25,* 45–61.

Järvilehto, T., Hämäläinen, H., & Soininen, K. (1981). Peripheral neural basis of tactile sensations in man: II. Characteristics of human mechanoreceptors in the hairy skin and correlations of their activity with tactile sensations. *Brain Research, 219,* 13–27.

Jeannerod, M. (1984). The timing of natural prehension movements. *Journal of Motor Behavior, 16,* 235–254.

Jeannerod, M. (1988). *The neural and behavioral organization of goal-directed movements.* Oxford: Clarendon.

Jeannerod, M., Michel, F., & Prablanc, C. (1984). The control of hand movements in a case of hemi-anaesthesia following a parietal lesion. *Brain, 107,* 899–920.

Jebsen, R. H., Taylor, N., Trieschmann, R. B., Trotter, M. J., & Howard, L. A. (1969). An objective and standardized test of hand function. *Archives of Physical Medicine and Rehabilitation, 50,* 311–319.

Jenkins, W. M., Merzenich, M. M., Ochs, M. T., Allard, T., & Guic-Robles, E. (1990). Functional reorganization of primary somatosensory cortex in adult owl monkeys after behaviorally controlled tactile stimulation. *Journal of Neurophysiology, 63,* 82–104.

Jenmalm, P., Dahlstedt, S., & Johansson, R. S. (2000). Visual and tactile information about object-curvature control fingertip forces and grasp kinematics in human dexterous manipulation. *Journal of Neurophysiology, 84,* 2984–2997.

Jenmalm, P., Goodwin, A. W., & Johansson, R. S. (1998). Control of grasp stability when humans lift objects with different surface curvatures. *Journal of Neurophysiology, 79*, 1643–1652.

Jenmalm, P., & Johansson, R. S. (1997). Visual and somatosensory information about object shape control manipulative fingertip forces. *Journal of Neuroscience, 17*, 4486–4499.

Jerde, T. E., Soechting, J. F., & Flanders, M. (2003). Coarticulation in fluent fingerspelling. *Journal of Neuroscience, 23*, 2383–2393.

Jessel, A. (1987). *Cutaneous spatial resolution and haptic recognition of common objects in elderly and young adults.* Unpublished honors bachelor's thesis, Queen's University, Kingston, Ontario, Canada.

Jindrich, D. L., Zhou, Y., Becker, T., & Dennerlein, J. T. (2003). Non-linear viscoelastic models predict fingertip pulp force-displacement characteristics during voluntary tapping. *Journal of Biomechanics, 36*, 497–503.

Johansson, R. S., & Cole, K. J. (1994). Grasp stability during manipulative actions. *Canadian Journal of Physiology and Pharmacology, 72*, 511–524.

Johansson, R. S., Häger, C., & Riso, R. (1992). Somatosensory control of precision grip during unpredictable pulling loads. *Experimental Brain Research, 89*, 192–203.

Johansson, R. S., & LaMotte, R. H. (1983). Tactile detection thresholds for a single asperity on an otherwise smooth surface. *Somatosensory Research, 1*, 21–31.

Johansson, R. S., Landström, U., & Lundström, R. (1982). Responses of mechanoreceptive afferent units in the glabrous skin of the human hand to sinusoidal skin displacements. *Brain Research, 244*, 17–25.

Johansson, R. S., Riso, R., Häger, C., & Bäckström, L. (1992). Somatosensory control of precision grip during unpredictable pulling loads. 1. Changes in load force amplitude. *Experimental Brain Research, 89*, 181–191.

Johansson, R. S., & Vallbo, A. B. (1979). Tactile sensibility of the human hand: Relative and absolute densities of four types of mechanoreceptive units in glabrous skin. *Journal of Neurophysiology, 286*, 283–300.

Johansson, R. S., & Vallbo, A. B. (1983). Tactile sensory coding in the glabrous skin of the human hand. *Trends in Neurosciences, 6*, 27–32.

Johansson, R. S., & Westling, G. (1984). Roles of glabrous skin receptors and sensorimotor memory in automatic control of precision grip when lifting rougher or more slippery objects. *Experimental Brain Research, 56*, 550–564.

Johansson, R. S., & Westling, G. (1987). Signals in tactile afferents from the fingers eliciting adaptive motor responses during precision grip. *Experimental Brain Research, 66*, 141–154.

Johansson, R. S., & Westling, G. (1988). Programmed and triggered actions to rapid load changes during precision grip. *Experimental Brain Research, 71*, 72–86.

Johansson, R. S., Westling, G., Backström, A., & Flanagan, J. R. (2001). Eye-hand coordination in object manipulation. *Journal of Neuroscience, 21*, 6917–6932.

Johnson, K. O. (2001). The roles and functions of cutaneous mechanoreceptors. *Current Opinions in Neurobiology, 11*, 455–461.

Johnson, K. O., Darian-Smith, I., & LaMotte, C. (1973). Peripheral neural determinants of temperature discrimination in man: A correlative study of responses to cooling skin. *Journal of Neurophysiology, 36*, 347–370.

Johnson, K. O., Darian-Smith, I., LaMotte, C., Johnson, B., & Oldfield, S. (1979). Coding of incremental changes in skin temperature by a population of warm fibers in the monkey: Correlation with intensity discrimination in man. *Journal of Neurophysiology, 42*, 1332–1353.

Johnson, K. O., & Lamb, G. D. (1981). Neural mechanisms of spatial tactile discrimination: Neural patterns evoked by Braille-like dot patterns in the monkey. *Journal of Physiology, 310*, 117–144.

Johnson, K. O., & Phillips, J. R. (1981). Tactile spatial resolution: 1. Two-point discrimination, gap detection, grating resolution, and letter recognition. *Journal of Neurophysiology, 46*, 1177–1191.

Johnson, K. O., Van Boven, R. W., & Hsiao, S. S. (1994). The perception of two points is not the spatial resolution threshold. In J. Boivie, P. Hansson, & U. Lindblom (Eds.), *Touch, temperature, and pain in health and disease: Mechanisms and assessments* (pp. 389–404). Seattle, WA: IASP.

Johnson, K. O., Yoshioka, T., & Vega-Bermudez, F. (2000). Tactile functions of mechanoreceptive afferents innervating the hand. *Journal of Clinical Neurophysiology, 17*, 539–558.

Jones, B. (1981). The developmental significance of cross-modal matching. In R. D. Walk & H. L. Pick, Jr. (Eds.), *Intersensory perception and sensory integration* (pp. 109–136). New York: Plenum.

Jones, E. G. (1985). *The thalamus.* New York: Plenum.

Jones, E. G. (2000). Cortical and subcortical contributions to activity-dependent plasticity in primate somatosensory cortex. *Annual Reviews of Neuroscience, 23*, 1–37.

Jones, L. A. (1986). The perception of force and weight: Theory and research. *Psychological Bulletin, 100*, 29–42.

Jones, L. A. (1989). The assessment of hand function: A critical review of techniques. *Journal of Hand Surgery, 14A*, 211–228.

Jones, L. A. (1996). Proprioception and its contribution to manual dexterity. In A. M. Wing, P. Haggard, & J. R. Flanagan (Eds.), *Hand and brain: The neurophysiology and psychology of hand movements* (pp. 349–362). San Diego, CA: Academic.

Jones, L. A. (1997). Dextrous hands: Human, prosthetic and robotic. *Presence: Teleoperators, & Virtual Environments, 6*, 29–56.

Jones, L. A. (1998). Manual dexterity. In K. J. Connolly (Ed.), *The psychobiology of the hand* (pp. 46–62). London: MacKeith.

Jones, L. A., & Berris, M. (2003). Material discrimination and thermal perception. In *IEEE Proceedings of the 11th International Symposium on Haptic Interfaces for Virtual Environment and Teleoperator Systems*, pp. 171–178.

Jones, L. A., & Hunter I. W. (1983). Effect of fatigue on force sensation. *Experimental Neurology, 81*, 640–650.

Jones, L. A., & Hunter, I. W. (1992). Changes in pinch force with bidirectional load forces. *Journal of Motor Behavior, 24*, 157–164.

Jones, M. C. (1926). The development of early patterns in young children. *Pedagogical Seminary, 33*, 537–585.

Kaas, J. H. (1991). Plasticity of sensory and motor maps in adult mammals. *Annual Reviews of Neuroscience, 14*, 137–167.

Kaas, J. H., Nelson, R. J., Sur, M., & Merzenich, M. M. (1981). Organization of somatosensory cortex in primates. In F. O. Schmitt, F. G. Worden, G. Adelman, & S. G. Dennis (Eds.), *The organization of the cerebral cortex: Proceedings of a neurosciences research program colloquium* (pp. 237–261). Cambridge, MA: MIT Press.

Kakei, S., Hoffman, D. S., Strick, P. L. (2001). Direction of action is represented in the ventral premotor cortex. *Nature Neuroscience, 4*, 1020–1025.

Kandel, E. R., Schwartz, J. H., & Jessell, T. M. (1991). *Principles of neural science.* New York: Elsevier

Kapandji, I. A. (1970). *The physiology of the joints: Vol. 1. Upper limb* (2d ed., L. H. Honore, Trans.). Edinburgh: E & S Livingstone.

Kappers, A. M. (1999). Large systematic deviations in the haptic perception of parallelity *Perception, 28,* 1001–1012.

Kappers, A. M., & Koenderink, J. J. (1999). Haptic perception of spatial relations. *Perception, 28,* 781–795.

Karni, A., Meyer, G., Jezzard, P., Adams, M. M., Turner, R., & Ungerlieder, L. G. (1995). Functional MRI evidence for adult motor cortex plasticity during motor skill learning. *Nature, 377*, 155–158.

Katz, D. (1937). Studies on test baking: III. The human factor in test baking: A psychological study. *Cereal Chemistry, 14*, 382–396.

Katz, D. (1989). *The world of touch [Der aufbau der tastwelt]* (L. Krueger, Trans.). Mahwah, NJ: Erlbaum. (Original work published 1925)

Kawai, S. (2002a). Heaviness perception: I. Constant involvement of haptically perceived size in weight discrimination. *Experimental Brain Research, 147*, 16–22.

Kawai, S. (2002b). Heaviness perception: II. Contributions of object weight, haptic size, and density to the accurate perception of heaviness or lightness. *Experimental Brain Research, 147*, 23–28.

Keele, S. W. (1968). Movement control in skilled motor performance. *Psychological Bulletin, 70*, 387–403.

Keele, S. W., & Posner, M. I. (1968). Processing of visual feedback in rapid movements. *Journal of Experimental Psychology, 77*, 155–158.

Kelso, J. A. S. (1977). Motor control mechanisms underlying human movement reproduction. *Journal of Experimental Psychology: Human Perception and Performance, 3*, 529–543.

Kelso, J. A. S. (1984). Phase transitions and critical behavior in human bimanual coordination. *American Journal of Physiology, 15*, R1000–1004.

Kelso, J. A. S. (1995). *Dynamic patterns: The self-organization of brain and behavior.* Cambridge, MA: MIT Press.

Kelso, J. A. S., Holt, K. G., & Flatt, A. E. (1980). The role of proprioception in the perception and control of human movement: Toward a theoretical reassessment. *Perception & Psychophysics, 28*, 45–52.

Kelso, J. A. S., Southard, D. L., & Goodman, D. (1979). On the coordination of two-handed movements. *Journal of Experimental Psychology: Human Perception and Performance, 5*, 229–238.

Kemble, J. V. (1987). Man's hand in evolution. *Journal of Hand Surgery, 12B*, 396–399.

Kendon, A. (1993). Human gesture. In K. R. Gibson & T. Ingold (Eds.), *Tools, language, and cognition in human evolution* (pp. 43–62). Cambridge: Cambridge University Press.

Kennedy, J. M. (1993). *Drawing and the blind: Pictures to touch.* New Haven, CT: Yale University Press.

Kennedy, J. M., & Domander, R. (1984). Pictorial foreground/background reversal reduces tactual recognition by blind subjects. *Journal of Visual Impairment and Blindness, 78*, 215–216.

Kenshalo, D. R., Sr. (1986). Somesthetic sensitivity in young and elderly humans. *Journal of Gerontology, 41*, 732–742.

Kenshalo, D. R., Sr., Decker, T., & Hamilton, A. (1967). Comparisons of spatial summation on the forehead, forearm, and back produced by radiant and conducted heat. *Journal of Comparative and Physiological Psychology, 63*, 510–515.

Kenshalo, D. R., Holmes, C. E., & Wood, P. B. (1968). Warm and cool thresholds as a function of rate of stimulus temperature change. *Perception & Psychophysics, 3*, 81–84.

Kew, J. J. M., Halligan, P. W., Marshall, J. C., Passingham, R. E., Rothwell, J. C., Ridding, M. C., Marsden, C. D., & Brooks, D. J. (1997). Abnormal access of axial vibrotactile input to deafferented somatosensory cortex in human upper limb amputees. *Journal of Neurophysiology, 77*, 2753–2764.

Khater-Boidin, J., & Duron, B. (1991). Postnatal development of descending motor pathways studied in man by percutaneous stimulation of the motor cortex and the spinal cord. *International Journal of Developmental Neuroscience, 9*, 15–26.

Kilbreath, S. L., & Gandevia, S. C. (1991). Independent digit control: Failure to partition perceived heaviness of weights lifted by digits of the human hand. *Journal of Physiology, 442*, 585–599.

Kilbreath, S. L., & Gandevia, S. C. (1992). Independent control of the digits: Changes in perceived heaviness over a wide range of force. *Experimental Brain Research, 91*, 539–542.

Kilbreath, S. L., & Gandevia, S. C. (1994). Limited independent flexion of the thumb and fingers in human subjects. *Journal of Physiology, 479*, 487–497.

Kilbreath, S. L., Gorman, R. B., Raymond, J., & Gandevia, S. C. (2002). Distribution of the forces produced by motor unit activity in the human flexor digitorum profundus. *Journal of Physiology, 543*, 289–296.

Kilchenman-O'Malley, M., & Goldfarb, M. (2002). Comparison of human haptic size identification and discrimination performance in real and simulated environments. In *IEEE Proceedings of the 10th Annual Symposium on Haptic Interfaces for Virtual Environment and Teleoperator Systems*, pp. 10–17.

Kilgour, A., & Lederman, S. J. (2002). Face recognition by hand. *Perception & Psychophysics, 64*, 339–352.

Kim, D.-E., Shin, M.-J., Lee, K.-M., Chu, K., Woo, S. H., Kim, Y. R., Song, E.-C., Lee, J.-W., Park, S.-H., & Roh, J.-K.(2004). Musical training-induced functional reorganization of the adult brain: Functional magnetic resonance imaging and transcranial magnetic stimulation study on amateur string players. *Human Brain Mapping, 23*, 188–199.

Kimura, D., & Vanderwolf, C. H. (1970). The relation between hand preference and the performance of individual finger movements by left and right hands. *Brain, 93*, 769–774.

King, P. M. (1997). Sensory function assessment: A pilot comparison study of touch pressure threshold with texture and tactile discrimination. *Journal of Hand Therapy, 10*, 24–28.

Kinoshita, H., & Francis, P. R. (1996). A comparison of prehension force control in young and elderly individuals. *European Journal of Applied Physiology, 74*, 450–460.

Kinoshita, H., Kawai, S., & Ikuta, K. (1995). Contributions and co-ordination of individual fingers in multiple finger prehension. *Ergonomics, 38*, 1212–1230.

Kinoshita, H., Kawai, S., Ikuta, K., & Teraoka, T. (1996). Individual finger forces acting on a grasped object during shaking actions. *Ergonomics, 39*, 243–256.

Kinoshita, H., Murase, T., & Bandou, T. (1996). Grip posture and forces during holding cylindrical objects with circular grips. *Ergonomics, 39*, 1163–1176.

Kirkpatrick, A. E., & Douglas, S. A. (2002). Application-based evaluation of haptic interfaces. In *IEEE Proceedings of the 10th International Symposium on Haptic Interfaces for Virtual Environment and Teleoperator Systems*, pp. 32–39.

Kirman, J. H. (1974). Tactile apparent movement: The effects of number of stimulators. *Journal of Experimental Psychology, 103*, 1175–1180.

Klapp, S. T. (1975). Feedback versus motor programming in the control of aimed movements. *Journal of Experimental Psychology: Human Perception and Performance, 104*, 147–153.

Klatzky, R. L., & Lederman, S. J. (1993). Toward a computational model of constraint-driven exploration and haptic object identification. *Perception, 22*, 597–621.

Klatzky, R. L., & Lederman, S. J. (1999). Tactile roughness perception with a rigid link from surface to skin. *Perception & Psychophysics, 61*, 591–607.

Klatzky, R. L., & Lederman, S. J. (2002). Perceiving texture through a probe. In M. L. McLaughlin, J. P. Hespanha, & G. S. Sukhatme (Eds.), *Touch in virtual environments* (pp. 180–193). Upper Saddle River, NJ: Prentice Hall.

Klatzky, R. L., & Lederman, S. J. (2003a). The haptic identification of everyday life objects. In Y. Hatwell, A. Streri, & E. Gentaz (Eds.), *Touching for knowing: Cognitive psychology of haptic manual perception* (pp. 105–121). Amsterdam: Benjamins.

Klatzky, R. L., & Lederman, S. J. (2003b). Representing spatial location and layout from sparse kinaesthetic contacts. *Journal of Experimental Psychology: Human Perception & Performance, 29,* 310–325.

Klatzky, R. L., & Lederman, S. J. (2003c). Touch. In I. B. Weiner (Series Ed.) & A. F. Healy & R. W. Proctor (Vol. Eds.), *Handbook of psychology: Vol. 4. Experimental psychology* (pp. 147–176). New York: Wiley.

Klatzky, R. L., & Lederman, S. J. (in press). Object recognition by touch. In J. Rieser, D. Ashmead, F. Ebner, & A. Corn (Eds.), *Blindness, brain plasticity and spatial function.*Mahwah, NJ: Erlbaum.

Klatzky, R. L., Lederman, S. J., & Balakrishnan, J. (1991). Task-driven extraction of object contour by human haptics: I. *Robotica, 9,* 43–51.

Klatzky, R. L., Lederman, S. J., Hamilton, C., Grindley, M., & Swendsen, R. (2003). Feeling textures through a probe: Effects of probe/plate geometry and exploratory factors. *Perception & Psychophysics, 65,* 613–631.

Klatzky, R. L., Lederman, S. J., Hamilton, C., & Ramsey, G. I. (1999). Perceiving roughness via a rigid probe: Effects of exploration speed. *Proceedings of the ASME Dynamic Systems and Control Division, 67,* 27–34.

Klatzky, R. L., Lederman, S.J., & Mankinen, J. (2005). Manual exploration for function testing and object perception in 5-year-olds. *Infant Behavior and Development, 28,* 240–249.

Klatzky, R. L., Lederman, S. J., & Metzger, V. (1985). Identifying objects by touch: An "expert system." *Perception & Psychophysics, 37,* 299–302.

Klatzky, R. L., Lederman, S. J., & Reed, C. (1987). There's more to touch than meets the eye: Relative salience of object dimensions for touch with and without vision. *Journal of Experimental Psychology: General, 116,* 356–369.

Klatzky, R. L., Lederman, S. J., & Reed, C. (1989). Haptic integration of object properties: Texture, hardness, and planar contour. *Journal of Experimental Psychology: Human Perception & Performance, 15,* 45–57.

Klatzky, R. L., Loomis, J. M., Lederman, S. J., Wake, H., & Fujita, N. (1993). Haptic identification of objects and pictures of objects. *Perception & Psychophysics, 54,* 170–178.

Klima, E. S., & Bellugi, U. (1979). *The signs of language.* Cambridge, MA: Harvard University Press.

Knibestöl, M. (1973). Stimulus-response functions of rapidly adapting mechanoreceptors in human glabrous skin area. *Journal of Physiology, 232,* 427–452.

Knibestöl, M. (1975). Stimulus-response functions of slowly adapting mechanoreceptors in the human glabrous skin area. *Journal of Physiology, 245,* 63–80.

Knibestöl, M., & Vallbo, A. (1970). Single unit analysis of mechanoreceptor activity from the human glabrous skin. *Acta Physiologica Scandinavica, 80,* 178–195.

Knibestöl, M., & Vallbo, A. B. (1980). Intensity of sensation related to activity of slowly adapting mechanoreceptive units in the human hand. *Journal of Physiology, 300,* 251–267.

Krakauer, J., & Ghez, C. (2000). Voluntary movement. In E. R. Kandel, J. H. Schwartz, & T. M. Jessell (Eds.), *Principles of neural science* (4th ed., pp. 756–781). New York: McGraw-Hill.

Krampe, R. T., & Ericsson, K. A. (1996). Maintaining excellence: Deliberate practice and elite performance in young and older pianists. *Journal of Experimental Psychology: General, 125,* 331–359.

Krech, D., & Crutchfield, R. S. (1958). *Elements of psychology.* New York: Knopf.

Kuhtz-Buschbeck, J. P., Stolze, H., Jöhnk, K., Boczek-Funcke, A., & Illert, M. (1998). Development of prehension movements in children: A kinematic study. *Experimental Brain Research, 122,* 424–432.

Kumamoto, K., Senuma, H., Ebara, S., & Matsuura, T. (1993). Distribution of Pacinian corpuscles in the hand of the monkey, Macaca fuscata. *Journal of Anatomy, 183,* 149–154.

Kurth, R., Villringer, K., Curio, G., Wolf, K. J., Krause, T., Repenthin, J., Schwiemann, J., Deuchert, M., & Villringer, A. (2000). fMRI shows multiple somatotopic digit representations in human primary somatosensory cortex. *NeuroReport, 11,* 1487–1491.

Kyberd, P. J., Holland, O. E., Chappell, P. H., Smith, S., Tregidgo, R., Bagwell, P. J., & Snaith, M. (1995). MARCUS: A two-degree-of-freedom hand prosthesis with hierarchical grip control. *IEEE Transactions on Rehabilitation Engineering, 3,* 70–76.

Lacquaniti, F., & Soechting, J. F. (1982). Invariant characteristics of a pointing movement in man. *Journal of Neuroscience, 1,* 710–720.

Lacquaniti, F., Soechting, J. F., & Terzuolo, C. A. (1982). Some factors pertinent to the organization and control of arm movements. *Brain Research, 252,* 394–397.

Lakatos, S., & Marks, L. E. (1999). Haptic form perception: Relative salience of local and global features. *Perception & Psychophysics, 61,* 895–908.

Lakatos, S., & Shepard, R. N. (1997). Constraints common to apparent motion in visual, tactile, and auditory space. *Journal of Experimental Psychology: Human Perception & Performance, 23,* 1050–1060.

Lamb, G. D. (1983). Tactile discrimination of textured surfaces: Psychophysical performance measurements in humans. *Journal of Physiology, 38,* 551–565.

Lambert, L., & Lederman, S. (1989). An evaluation of the legibility and meaningfulness of potential map symbols. *Journal of Visual Impairment & Blindness, 83*, 397–403.

LaMotte, R. H. (2000). Softness discrimination with a tool. *Journal of Neurophysiology, 83*, 1777–1786.

LaMotte, R. H., & Srinivasan, M. A. (1987a). Tactile discrimination of shape: Responses of slowly adapting mechanoreceptive afferents to a step stroked across the monkey fingerpad. *Journal of Neuroscience, 7*, 1655–1671.

LaMotte, R. H., & Srinivasan, M. A. (1987b). Tactile discrimination of shape: Responses of rapidly adapting mechanoreceptive afferents to a step stroked across the monkey fingerpad. *Journal of Neuroscience, 7*, 1672–1681.

LaMotte, R. H., & Srinivasan, M. A. (1991). Surface microgeometry: Tactile perception and neural encoding. In O. Franzen & J. Westman (Eds.), *Information processing in the somatosensory system* (pp. 49–58). London: Macmillan.

Landau, B. (1990). *Spatial representation of objects in the young blind child.* Paper presented at the International Conference on Infant Studies, Montreal, Canada.

Langolf, G. D., Chaffin, D. B., & Foulke, J. A. (1976). An investigation of Fitts' law using a wide range of movement amplitudes. *Journal of Motor Behavior, 8*, 113–128.

Lansdell, H., & Donnelly, E. F. (1977). Factor analysis of the Wechsler Adult Intelligence Scale subtests and the Halstead-Reitan category and tapping tests. *Journal of Consulting and Clinical Psychology, 45*, 412–416.

Lansky, L. M., Feinstein, H., & Peterson, J. M. (1988). Demography of handedness in two samples of randomly selected adults (N = 2083). *Neuropsychologia, 26*, 465–477.

Lappin, J. S., & Foulke, E. (1973). Expanding the tactual field of view. *Perception & Psychophysics, 14*, 237–241.

Larochelle, S. (1984). Some aspects of movements in skilled typewriting. In H. Bouma & D. G. Bouwis (Eds.), *Attention & performance,* (Vol. 10, pp. 43–54). Hillsdale, NJ: Erlbaum.

Larsson, L., & Ansved, T. (1995). Effects of ageing on the motor unit. *Progress in Neurobiology, 45*, 397–458.

Larsson, L., Sjödin, B., & Karlsson, J. (1978). Histochemical and biochemical changes in human skeletal muscle with age in sedentary males, age 22–65 years. *Acta Physiologica Scandinavica, 103*, 31–39.

Lawrence, D. A., Pao, L. Y., Salada, M. A., & Dougherty, A. M. (1996). Quantitative experimental analysis of transparency and stability in haptic interfaces. *Proceedings of ASME Dynamic Systems and Control Division, 59*, 441–449.

Lawrence, D. G., & Kuypers, H. G. J. M. (1968). The functional organization of the motor system in the monkey: Part 1. The effects of bilateral pyramidal lesions. *Brain, 91*, 1–14.

Lazarus, J. C., & Haynes, J. M. (1997). Isometric pinch force control and learning in older adults. *Experimental Aging Research, 23*, 179–200.

Leakey, R. E., & Lewin, R. (1977). *Origins.* New York: Dutton.

Lechelt, E. C. (1975). Temporal numerosity discrimination: Intermodal comparisons revisited. *British Journal of Psychology, 66*, 101–108.

Lechelt, E. (1988). Spatial asymmetries in tactile discrimination of line orientation: A comparison of the sighted, visually impaired, and blind. *Perception, 17*, 579–585.

Lechelt, E. C., Eliuk, J., & Tanne, G. (1976). Perceptual orientation asymmetries: A comparison of visual and haptic space. *Perception & Psychophysics, 20*, 463–469.

Lederman, S. J. (1974). Tactile roughness of grooved surfaces: The touching process and effects of macro- and micro-surface structure. *Perception & Psychophysics, 16*, 385–395.

Lederman, S. J. (1976). The "callus-thenics" of touching. *Canadian Journal of Psychology, 30*, 82–89.

Lederman, S. J. (1981). The perception of surface roughness by active and passive touch. *Bulletin of the Psychonomic Society, 18*, 253–255.

Lederman, S. J. (1983). Tactual roughness perception: Spatial and temporal determinants. *Canadian Journal of Psychology, 37*, 498–511.

Lederman, S. J., & Campbell, J. (1983). Tangible graphics for the blind. *Human Factors, 24*, 85–100.

Lederman, S. J., & Hamilton, C. (2002). Using tactile features to help functionally blind individuals denominate banknotes. *Human Factors, 44*, 413–428.

Lederman, S. J., Jones, B., & Segalowitz, S. J. (1984). Lateral symmetry in the tactual perception of roughness. *Canadian Journal of Psychology, 38*, 599–609.

Lederman, S. J., & Klatzky, R. L. (1987). Hand movements: A window into haptic object recognition. *Cognitive Psychology, 19*, 342–368.

Lederman, S. J., & Klatzky, R. L. (1990). Haptic classification of common objects: Knowledge-driven exploration. *Cognitive Psychology, 22*, 421–459.

Lederman, S. J., & Klatzky, R. L. (1997). Relative availability of surface and object properties during early haptic processing. *Journal of Experimental Psychology: Human Perception and Performance, 23*, 1680–1707.

Lederman, S. J., & Klatzky, R. L. (1999). Sensing and displaying spatially distributed fingertip forces in haptic interfaces for teleoperator and virtual environment systems. *Presence, Teleoperators, & Virtual Environments, 8,* 86–103.

Lederman, S. J., & Klatzky, R. L. (2001). Feeling surfaces and objects remotely. In R. J. Nelson (Ed.), *The somatosensory sensory system: Deciphering the brain's own body image* (pp. 103–120). Boca Raton, FL: CRC.

Lederman, S. J., & Klatzky, R. L. (2004). Haptic identification of common objects: Effects of constraining the manual exploration process. *Perception & Psychophysics, 66,* 618–628.

Lederman, S. J., Klatzky, R. L., & Balakrishnan, J. (1991). Task-driven extraction of object contour by human haptics: II. *Robotica, 9,* 179–188.

Lederman, S. J., Klatzky, R. L., & Barber, P. O. (1985). Spatial and movement-based heuristics for encoding pattern information through touch. *Journal of Experimental Psychology: General, 114,* 33–49.

Lederman, S. J., Klatzky, R. L., Chataway, C., & Summers, C. (1990). Visual mediation and the haptic recognition of two-dimensional pictures of common objects. *Perception & Psychophysics, 47,* 54–64.

Lederman, S. J., Klatzky, R. L., Hamilton, C., & Ramsay, G. I. (1999). Perceiving roughness via a rigid probe: Effects of exploration speed and mode of touch. *Haptics-e: The Electronic Journal of Haptics Research, 1,* 1–20.

Lederman, S. J., Klatzky, R. L., & Reed, C. (1993). Constraints on the haptic integration of spatially shared object dimensions. *Perception, 22,* 723–743.

Lederman, S. J., Loomis, J. M., & Williams, D. (1982). The role of vibration in the tactual perception of roughness. *Perception & Psychophysics, 32,* 109–116.

Lederman, S. J., Summers, C., & Klatzky, R. L. (1996). Cognitive salience of haptic object properties: Role of modality-encoding bias. *Perception, 25,* 983–998.

Lederman, S. J., & Taylor, M. M. (1969). Perception of interpolated position and orientation by vision and active touch. *Perception & Psychophysics, 6,* 153–159.

Lederman, S. J., & Taylor, M. M. (1972). Fingertip force, surface geometry and the perception of roughness by active touch. *Perception & Psychophysics, 12,* 401–408.

Lederman, S. J., & Wing, A. M. (2003). Perceptual judgement, grasp point selection and object symmetry. *Experimental Brain Research, 152,* 156–165.

LeGros-Clark, W. E. (1959). *The antecedents of man.* Edinburgh: Edinburgh University Press.

Lemon, R. N. (1999). Neural control of dexterity: What has been achieved? *Experimental Brain Research, 128,* 6–12.

Lemon, R. N., & Griffiths, J. (2005). Comparing the function of the corticospinal system in different species: Organizational differences for motor specialization? *Muscle & Nerve, 32,* 261–279.

Leonard, J. A. (1959). Tactual choice reactions: 1. *Quarterly Journal of Experimental Psychology, 11,* 76–83.

Lerner, E. A., & Craig, J. C. (2002). The prevalence of tactile motion aftereffects. *Somatosensory and Motor Research, 19,* 24–29.

Levin, S., Pearsall, G., & Ruderman, R. J. (1978). Von Frey's method of measuring pressure sensibility in the hand: An engineering analysis of the Weinstein-Semmes pressure aesthesiometer. *Journal of Hand Surgery, 3,* 211–216.

Lewes, G. H. (1879). Motor-feelings and the muscular sense. *Brain, 1,* 14–28.

Lhote, M., & Streri, A. (1998). Haptic memory and handedness in 2-month-old infants. *Laterality, 3,* 173–192.

Li, S., Danion, F., Latash, M. L., Li, Z.-M., & Zatsiorsky, V. M. (2001). Bilateral deficit and symmetry in finger-force production during two-hand multifinger tasks. *Experimental Brain Research, 141,* 530–540.

Li, Z.-M., Latash, M., & Zatsiorsky, V. M. (1998). Force sharing among fingers as a model of the redundancy problem. *Experimental Brain Research, 119,* 276–286.

Lindblom, U. (1965). Properties of touch receptors in distal glabrous skin of the monkey. *Journal of Neurophysiology, 28,* 966–985.

Lockman, J. J., & McHale, J. P. (1989). Object manipulation in infancy: Developmental and contextual determinants. In J. J. Lockman & N. L. Hazen (Eds.), *Action in social context: Perspectives on early development* (pp. 129–167). New York: Plenum.

Lockman, J. J., & Wright, M. H. (1988). *A longitudinal study of banging.* Paper presented at the International Conference on Infant Studies, Washington, DC.

Löfvenberg, J., & Johansson, R. S. (1984). Regional differences and interindividual variability in sensitivity in the glabrous skin of the human hand. *Brain Research, 301,* 65–72.

Logan, F. A. (1999). Errors in copy typewriting. *Journal of Experimental Psychology: Human Perception and Performance, 25,* 1760–1773.

Logitech. (2004, June 14). Available: http://www.logitech.com.

Long, C., Conrad, P. W., Hall, E. A., & Furler, S. L. (1970). Intrinsic-extrinsic muscle control of the hand in power grip and precision handling. *Journal of Bone and Joint Surgery, 52A*, 853–867.

Loomis, J. M. (1981a). On the tangibility of letters and Braille. *Perception & Psychophysics, 29*, 37–46.

Loomis, J. M. (1981b). Tactile pattern perception. *Perception, 10*, 5–27.

Loomis, J. M. (1982). Analysis of tactile and visual confusion matrices. *Perception & Psychophysics, 29*, 37–46.

Loomis, J. M. (1990). A model of character recognition and legibility. *Journal of Experimental Psychology: Human Perception and Performance, 16*, 106–120.

Loomis, J., Klatzky, R., & Lederman, S. J. (1991). Similarities of tactual and visual picture recognition with limited field of view. *Perception, 20*, 167–177.

Loomis, J. M., & Lederman, S. J. (1986). Tactual perception. In K. Boff, L. Kaufman, & J. Thomas (Eds.), *Handbook of perception and human performance* (pp. 31-1–31-41). New York: Wiley.

Louis, D. S., Greene, T. L., Jacobson, K. E., Rasmussen, C., Kolowich, P., & Goldstein, S. A. (1984). Evaluation of normal values for stationary and moving two-point discrimination in the hand. *Journal of Hand Surgery, 9A*, 552–555.

Louw, S., Kappers, A. M. L., & Koenderink, J. J. (2000). Haptic detection thresholds of Gaussian profiles over the whole range of spatial scales. *Experimental Brain Research, 132*, 369–374.

Lovelace, E. A., & Aikens, J. E. (1990). Vision, kinesthesis, and control of hand movement by young and old adults. *Perceptual and Motor Skills, 70*, 1131–1137.

Luce, D. (1963). Detection and recognition. In R. D. Luce, B. R. Bush, & E. Galanter, (Eds.), *Handbook of mathematical psychology* (Vol. 1, pp. 103–189). New York: Wiley.

Lundborg, G. (1993). Peripheral nerve injuries: Pathophysiology and strategies for treatment. *Journal of Hand Therapy, 6*, 179–188.

Lynn, B., & Perl, E. R. (1996). Afferent mechanisms of pain. In L. Kruger (Ed.), *Pain and touch* (pp. 213–241). San Diego, CA: Academic.

Macefield, G., Gandevia, S. C., & Burke, D. (1990). Perceptual responses to microstimulation of single afferents innervating joints, muscles and skin of the human hand. *Journal of Physiology, 429*, 113–129.

Macefield, V. G., Häger-Ross, C., & Johansson, R. S. (1996). Control of grip force during restraint of an object held between finger and thumb: Responses of cutaneous afferents from the digits. *Experimental Brain Research, 108*, 155–171.

Macefield, V. G., & Johansson, R. S. (1996). Control of grip force during restraint of an object held between finger and thumb: Responses of muscle and joint afferents from the digits. *Experimental Brain Research, 108*, 172–184.

MacKenzie, C. L., & Iberall, T. (1994). *The grasping hand.* Amsterdam: North-Holland.

MacKenzie, C. L., Marteniuk, R. G., Dugas, C., Liske, D., & Eickmeir, B. (1987). Three dimensional movement trajectories in Fitts' law: Implications for control. *Quarterly Journal of Experimental Psychology, 39*, 629–647.

MacKenzie, C. L., & Van Eerd, D. L. (1990). Rhythmic precision in the performance of piano scales: Motor psychophysics and motor programming. In M. Jeannerod (Ed.), *Attention and performance: Vol. 13. Motor representation and control* (pp. 375–408). Hillsdale, NJ: Erlbaum.

Mackinnon, S. E., & Dellon, A. L. (1988). *Surgery of the peripheral nerve.* New York: Thieme Medical.

Magee, L., & Kennedy, J. M. (1980). Exploring pictures tactually. *Nature, 283*, 287–288.

Mallon, W. J., Brown, H. R., & Nunley, J. A. (1991). Digital ranges of motion: Normal values in young adults. *Journal of Hand Surgery, 16A*, 882–887.

Manoel, E. de J., & Connolly, K. J. (1998). The development of manual dexterity in young children. In K. J. Connolly (Ed.), *The psychobiology of the hand* (pp. 177–198). London: MacKeith.

Marchetti, F. M., & Lederman, S. J. (1983). The haptic radial-tangential effect: Two tests of Wong's (1977) "moments-of-inertia" hypothesis. *Bulletin of the Psychonomic Society, 21*, 43–46.

Marks, L. (1979). Summation of vibrotactile intensity: An analog to auditory critical bands. *Sensory Processes, 3*, 188–203.

Marteniuk, R. G., Jeannerod, M., Athenes, S., & Dugas, C. (1987). Constraints on human arm movement trajectories. *Canadian Journal of Psychology, 41*, 365–378.

Marteniuk, R. G., Leavitt, J. L., MacKenzie, C. L., & Athenes, S. (1990). Functional relationships between grasp and transport components in a prehension task. *Human Movement Science, 9*, 149–176.

Martin, B. J., Armstrong, T. J., Foulke, J. A., Natarajan, S., Klinenberg, E., Serina, E., & Rempel, D., (1996). Keyboard reaction force and finger flexor electromyograms during computer keyboard work. *Human Factors, 38*, 654–664.

Marzke, M. W. (1992). Evolutionary development of the human thumb. *Hand Clinics, 8*, 1–8.

Marzke, M. W. (1997). Precision grips, hand morphology, and tools. *American Journal of Physical Anthropology, 102*, 91–110.

Marzke, M. W., & Marzke, R. F. (2000). Evolution of the human hand: Approaches to acquiring, analyzing and interpreting the anatomical evidence. *Journal of Anatomy, 197,* 121–140.

Marzke, M. W., & Wullstein, K. L. (1996). Chimpanzee and human grips: A new classification with a focus on evolutionary morphology. *International Journal of Primatology, 17,* 117–139.

Mason, C. R., Gomez, J. E., & Ebner, T. J. (2001). Hand synergies during reach-to-grasp. *Journal of Neurophysiology, 86,* 2896–2910.

Mason, M. T., & Salisbury, J. K. (1985). *Robot hands and the mechanics of manipulation.* Cambridge, MA: MIT Press.

Massie, T. H., & Salisbury, J. K. (1994). The PHANTOM haptic interface: A device for probing virtual objects. *Proceedings of ASME Dynamic Systems and Control Division, 55,* 295–301.

Mathiowetz, V., Kashman, N., Volland, G., Weber, K., Dowe, M., & Rogers, S. (1985). Grip and pinch strength: Normative data for adults. *Archives of Physical Medicine and Rehabilitation, 66,* 69–72.

Mathiowetz, V., Rogers, S. L., Dowe-Keval, M., Donahoe, L., & Rennells, C. (1986). The Purdue pegboard: Norms for 14- to 19-year-olds. *American Journal of Occupational Therapy, 40,* 174–179.

Mathiowetz, V., Wiemer, D. M., & Federman, S. M. (1986). Grip and pinch strength: Norms for 6- to 19-year-olds. *American Journal of Occupational Therapy, 40,* 705–711.

Matsuzaka, Y., Aizawa, H., & Tanji, J. (1992). A motor area rostral to the supplementary motor area (presupplementary motor area) in the monkey: Neuronal activity during a learned motor task. *Journal of Neurophysiology, 76,* 2327–2342.

Matthews, P. B. C. (1984). The contrasting stretch reflex responses of the long and short flexor muscles in the human thumb. *Journal of Physiology, 348,* 545–558.

McCall, R.B. (1974). Exploratory manipulation and play in the human infant. *Monographs of the Society for Research in Child Development, 39,* number 155.

McCammon, I. D., & Jacobsen, S. C. (1990). Tactile sensing and control for the Utah/MIT hand. In S. T. Venkataraman & T. Iberall (Eds.), *Dextrous robot hands* (pp. 239–266). New York: Springer-Verlag

McCloskey, D. I. (1981). Corollary discharges: Motor commands and perception. In V. B. Brooks (Ed.), *Handbook of physiology: Sec. 1. The nervous system* (Vol. 2, pp.1415–1445). Bethesda, MD: American Physiological Society.

McComas, A. J. (1996). *Skeletal muscle: Form and function.* Champaign, IL: Human Kinetics.

McKeon, R. (1941). *The basic works of Aristotle.* New York: Random House.

McManus, C. (2002). *Right hand, left hand: The origins of asymmetry in brains, bodies, atoms and cultures.* Cambridge, MA: Harvard University Press.

McPherson, G. E. (1995). The assessment of musical performance: Development and validation of five new measures. *Psychology of Music, 23,* 142–161.

Mechsner, F., Kerzel, D., Knoblich, G., & Prinz, W. (2001). Perceptual basis of bimanual coordination. *Nature, 414,* 69–73.

Meftah, E. M., Belingard, L., & Chapman, C. E. (2000). Relative effects of the spatial and temporal characteristics of scanned surfaces on human perception of tactile roughness using passive touch. *Experimental Brain Research, 132,* 351–361.

Meissner, M., & Philpott, S. B. (1975). The sign language of sawmill workers in British Columbia. *Sign Language Studies, 6,* 291–308.

Melchiorri, C., & Vassura, G. (1993). Mechanical and control issues for integration of an arm-hand robotic system. In R. Chatila & G. Hirzinger (Eds.), *Experimental Robotics II: The Second International Symposium,* (pp. 136–152). London: Springer-Verlag.

Melzack, R., & Bromage, P. R. (1973). Experimental phantom limbs. *Experimental Neurology, 39,* 261–269.

Merzenich, M. M., & Harrington, T. (1969). The sense of flutter-vibration evoked by stimulation of the hairy skin of primates: Comparison of human sensory capacity with the responses of mechanoreceptive afferents innervating the hairy skin of monkeys. *Experimental Brain Research, 9,* 236–260.

Meyer, D. E., Abrams, R. A., Kornblum, S., Wright, C. E., & Smith, J. E. K. (1988). Optimality in human motor performance: Ideal control of rapid aimed movements. *Psychological Review, 95,* 340–370.

Miall, R. C., & Haggard, P. N. (1995). The curvature of human arm movements in the absence of visual experience. *Experimental Brain Research, 103,* 421–428.

Millar, S. (1985). Movement cues and body orientation in recall of locations of blind and sighted children. *Quarterly Journal of Experimental Psychology, 37A,* 257–279.

Millar, S., & Al-Attar, Z. (2002). The Müller-Lyer illusion in touch and vision: Implications for multisensory processes. *Perception & Psychophysics, 64,* 353–365.

Miller, L. J. (1982). *Miller assessment for preschoolers.* Littleton, CO: Foundation for Knowledge in Development.

Mills, A. W. (1958). On the minimum audible angle. *Journal of the Acoustical Society of America, 30*, 237–246.

Milner, A. D., & Goodale, M. A. (1995). *The visual brain in action*. Oxford: Oxford University Press.

Mima, T., Ikeda, A., Terada, K., Yazawa, S., Mikuni, N., Kunieda, T., Taki, W., Kimura, J., & Shibasaki, H. (1997). Modality-specific organization for cutaneous and proprioceptive sense in human primary sensory cortex studied by chronic epicortical recording. *Electroencephalography and Clinical Neurophysiology, 104*, 103–107.

Ministry of Pensions (1947). *Report of the Inter-Departmental Committee on the Assessment of Disablement*. London: Her Majesty's Stationery Office.

Moberg, E. (1958). Objective methods for determining the functional value of sensibility in the hand. *Journal of Bone and Joint Surgery, 40B*, 454–476.

Moberg, E. (1962). Criticism and study of methods for examining sensibility in the hand. *Neurology, 12*, 8–19.

Molina, M., & Jouen, F. (2003). Haptic intramodal comparison of texture in human neonates. *Developmental Psychobiology, 42*, 378–385.

Montagna, W. (1965). Morphology of the aging skin: The cutaneous appendages. In W. Montagna (Ed.), *Advances in biology of the skin* (Vol. 6, pp. 1–16). New York: Pergamon.

Montagna, W., Kligman, A. M., & Carlisle, K. S. (1992). *Atlas of normal human skin*. New York: Springer-Verlag.

Mon-Williams, M., & McIntosh, R. D. (2000). A test between two hypotheses and a possible third way for the control of prehension. *Experimental Brain Research, 134*, 268–273.

Monzée, J., Lamarre, Y., & Smith, A. M. (2003). The effects of digital anesthesia on force control using a precision grip. *Journal of Neurophysiology, 89*, 672–683.

Moran, C. A., & Callahan, A. D. (1984). Sensibility measurement and management. In C. A. Moran (Ed.), *Hand rehabilitation* (pp. 45–68). New York: Churchill Livingstone.

Morange-Majoux, F., Cougnot, P., & Bloch, H. (1997). Hand tactual exploration of textures in infants from 4 to 6 months. *Early Development and Parenting, 6*, 127–135.

Morasso, P. (1981). Spatial control of arm movements. *Experimental Brain Research, 42*, 223–227.

Morgan, M., Phillips, J. G., Bradshaw, J. L., Mattingley, J. B., Iansek, R., & Bradshaw, J. A. (1994). Age-related motor slowness: Simply strategic? *Journal of Gerontology: Medical Sciences, 49*, M133–M139.

Morley, J. W., Goodwin, A. W., & Darian-Smith, I. (1983). Tactile discrimination of gratings. *Experimental Brain Research, 49*, 291–299.

Morrongiello, B. A., Humphrey, G. K., Timney, B., Choi, J., & Rocca, P. T. (1994). Tactual object exploration and recognition in blind and sighted children. *Perception, 23*, 833–848.

Mott, F. W., & Sherrington, C. S. (1895). Experiments upon the influence of sensory nerves upon movement and nutrition of the limbs. Preliminary communication. *Proceedings of the Royal Society, Series B, 57*, 481–488.

Mounod, P., & Bower, T. G. R. (1974). Conservation of weight in infants. *Cognition, 3*, 29–40.

Mountcastle, V. B. (1967). The problem of sensing and the neural coding of sensory events. In G. C. Quarton, T. Melnecheck, & F. O. Smith (Eds.), *The neurosciences* (pp. 393–407). New York: Rockefeller University Press.

Mountcastle, V. B. (1975). The view from within: Pathways to the study of perception. *Johns Hopkins Medical Journal, 136*, 109–131.

Müller, F., & Dichgans, J. (1994). Dyscoordination of pinch and lift forces during grasp in patients with cerebellar lesions. *Experimental Brain Research, 101*, 485–492.

Müller, K., Hömberg, V., & Lenard, H.-G. (1991). Magnetoelectrical stimulation of motor cortex and nerve roots in children: Maturation of corticomotoneural projections. *Electroencephalography and Clinical Neurophysiology, 81*, 63–70.

Munk, H. (1909). *Uber die Functionen von Hirn und Rückenmark. Gesammelte Mitteilungen* [About the function of the brain and spinal cord: Collected works and notes]. Berlin: Hirschwald.

Nagasaki, H., Itoh, H., Maruyama, H., & Hashizume, K. (1989). Characteristic difficulty in rhythmic movement with aging and its relation to Parkinson's disease. *Experimental Aging Research, 14*, 171–176.

Nakada, M. (1993). Localization of a constant-touch and moving touch stimulus in the hand: A preliminary study. *Journal of Hand Therapy, 6*, 23–28.

Nakazawa, N., Ikeura, R., & Inooka, H. (2000). Characteristics of human fingertips in the shearing direction. *Biological Cybernetics, 82*, 207–214.

Napier, J. R. (1956). The prehensile movements of the human hand. *Journal of Bone and Joint Surgery, 38B*, 902–913.

Napier, J. R. (1961). Prehensibility and opposability in the hands of primates. *Symposium of the Zoological Society of London, 5*, 115–132.

Napier, J. R. (1976). *The human hand*. Burlington, NC: Carolina Biological Supply.

Napier, J. R. (1993). *Hands* (revised by R. H. Tuttle). Princeton, NJ: Princeton University Press.

Napier, J. R., & Napier, P. H. (1967). *A handbook of living primates*. London: Academic.

Napier, J. R., & Napier, P. H. (1985). *The natural history of the primates*. Cambridge, MA: MIT Press.

Natsoulas, T. (1966). Locus and orientation of the perceiver (EGO) under variable, constant, and no perspective instructions. *Journal of Personality and Social Psychology, 3*, 190–196.

Newell, F. N., Ernst, M. O., Tjan, B. S., & Bulthoff, H. (2001). Viewpoint dependence in visual and haptic object recognition. *Psychological Science, 12*, 37–42.

Newell, K. M., & McDonald, P. V. (1997). The development of grip patterns in infancy. In K. J. Connolly & H. Forssberg (Eds.), *Neurophysiology and neuropsychology of motor development* (pp. 232–256). London: MacKeith.

Newell, K. M., McDonald, P. V., & Baillargeon, R. (1993). Body scale and infant grip configurations. *Developmental Psychobiology, 26*, 195–205.

Newell, K. M., Scully, D. M., McDonald, P. V., & Baillargeon, R. (1989). Task constraints and infant grip configurations. *Developmental Psychobiology, 22*, 817–832.

Newell, K. M., Scully, D. M., Tenenbaum, F., & Hardiman, S. (1989). Body scale and the development of prehension. *Developmental Psychobiology, 22*, 1–13.

Newman, D. G., Pearn, J., Barnes, A., Young, C. M., Kehoe, M., & Newman, J. (1984). Norms for hand grip strength. *Archives of Disease in Childhood, 59*, 453–459.

Ng, C. L., Ho, D. D., & Chow, S. P. (1999). The Moberg pickup test: Results of testing with a standard protocol. *Journal of Hand Therapy, 12*, 309–312.

Nicoladoni, C. (1897). Daumenplastik [Plastic thumb]. *Weiner klinische Wochenschrift, 10*, 663–670.

Nicolelis, M. A. L., de Oliveira, L. M. O., Lin, R., & Chapin, J. (1996). Active tactile exploration influences the functional maturation of the somatosensory system. *Journal of Neurophysiology, 75*, 2192–2196.

Norkin, C. C., & White, D. J. (1985). *Measurement of joint motion: A guide to goniometry*. Philadelphia: Davis.

Norman, J. F., Norman, H. F., Clayton, A. M., Lianekhammy, J., & Zielke, G. (2004). The visual and haptic perception of natural object shape. *Perception & Psychophysics, 66*, 342–351.

Norsell, U., & Olausson, H. (1992). Human tactile directional sensibility and its peripheral origins. *Acta Physiologica Scandinavica, 144*, 155–161.

Nowak, D. A., & Hermsdörfer, J. (2003). Selective deficits of grip force control during object manipulation in patients with reduced sensibility of the grasping digits. *Neuroscience Research, 47*, 65–72.

Nowak, D. A., Hermsdörfer, J., Marquardt, C., & Fuchs, H.-H. (2002). Grip and load force coupling during discrete vertical arm movements with a grasped object in cerebellar atrophy. *Experimental Brain Research, 145*, 28–39.

Noyes, J. (1983). The QWERTY keyboard: A review. *International Journal of Man-Machine Studies, 18*, 265–281.

Obayashi, S., Suhara, T., Kawabe, K., Okauchi, T., Maeda, J., Akine, Y., Onoe, H., & Iriki, A. (2001). Functional brain mapping of monkey tool use. *NeuroImage, 14*, 853–861.

Ochiai, T., Mushiake, H., & Tanji, J. (2005). Involvement of the ventral premotor cortex in controlling image motion of the hand during performance of a target-capturing task. *Cerebral Cortex 15*, 929–937.

Ohtsuki, T. (1981a). Inhibition of individual fingers during grip strength exertion. *Ergonomics, 24*, 21–36.

Ohtsuki, T. (1981b). Decrease in grip strength induced by simultaneous bilateral exertion with reference to finger strength. *Ergonomics, 24*, 37–48.

Olausson, H., Lamarre, Y., Backlund, H., Morin, C., Wallin, B. G., Starck, G., Ekholm, S., Strigo, I., Worsley, K., Vallbo, A.B., & Bushnell, M.C. (2002). Unmyelinated tactile afferents signal touch and project to insular cortex. *Nature Neuroscience, 5*, 900–904.

Olausson, H., Wessberg, J., & Kakuda, N. (2000). Tactile directional sensibility: Peripheral neural mechanisms in man. *Brain Research, 866*, 178–187.

Oldfield, R. C. (1971). The assessment and analysis of handedness: The Edinburgh inventory. *Neuropsychologia, 9*, 97–114.

Oldfield, S. R., & Phillips, J. R. (1983). The spatial characteristics of tactile form perception. *Perception, 12*, 615–626.

Önne, L. (1962). Recovery of sensibility and sudomotor activity in the hand after nerve suture. *Acta Chirurgica Scandinavica 300*(Suppl.), 1–69.

Ouh-young, M., Beard, D. V., & Brooks, F. P. (1989). Force display performs better than visual display in a simple 6-D docking task. *Proceedings of the IEEE International Conference on Robotics and Automation, 3*, 1462–1466.

Pai, D. K., & Reissell, L.-M. (1996). Touching multiresolution curves. *Proceedings of ASME Dynamic Systems and Control Division,, 58*, 427–432.

Paillard, J., & Brouchon, M. (1968). Active and passive movements in the calibration of position sense. In S. J. Freedman (Ed.), *The neuropsychology of spa-*

tially oriented behavior (pp. 37–55). Homewood, IL: Dorsey.

Palmer, C. (1989). Mapping musical thought to musical performance. *Journal of Experimental Psychology: Human Perception and Performance, 15*, 331–346.

Palmer, C. (1997). Music performance. *Annual Reviews in Psychology, 48*, 115–138.

Palmer, C., & van de Sande, C. (1995). Range of planning in music performance. *Journal of Experimental Psychology: Human Perception and Performance, 21*, 947–962.

Palmer, C. F. (1989). The discriminating nature of infants' exploratory actions. *Developmental Psychology, 25*, 885–893.

Paré, M., Behets, C., & Cornu, O. (2003). Paucity of presumptive Ruffini corpuscles in the index finger pad of humans. *Journal of Comparative Neurology, 456*, 260–266.

Paré, M., Carnahan, H., & Smith, A. M. (2002). Magnitude estimation of tangential force applied to the fingerpad. *Experimental Brian Research, 142*, 342–348.

Paré, M., Smith, A. M., & Rice, F. L. (2002). Distribution and terminal arborizations of cutaneous mechanoreceptors in the glabrous finger pads of the monkey. *Journal of Comparative Neurology, 445*, 347–359.

Parrot, J. R., & Harrison, D. B. (1980). Surgically dividing pianists' hands [Letter to the editor]. *Journal of Hand Surgery, 5A*, 619.

Parsons, L. M., & Shimojo, S. (1987). Perceived spatial organization of cutaneous patterns on surfaces of the human body in various positions. *Journal of Experimental Psychology: Human Perception and Performance, 13*, 488–504.

Pascual-Leone, A., & Torres, F. (1993). Plasticity of the sensorimotor cortex representation of the reading finger in Braille readers. *Brain, 116*, 39–52.

Pascual-Leone, A., Wassermann, E. M., Sadato, N., & Hallett, M. (1995). The role of reading activity on the modulation of motor cortical outputs to the reading hand in Braille readers. *Annals of Neurology, 38*, 910–915.

Patel, M. R., & Bassini, L. (1999). A comparison of five tests for determining hand sensibility. *Journal of Reconstructive Microsurgery, 15*, 523–526.

Paul, R. L., Merzenich, M. M., & Goodman, H. (1972). Representation of slowly and rapidly adapting cutaneous mechanoreceptors of the hand in Brodmann's areas 3 and 1 of macaca mulatta. *Brain Research, 36*, 229–249.

Paulignan, Y., Jeannerod, M., MacKenzie, C., & Marteniuk, R. (1991). Selective perturbation of visual input during prehension movements: 2. The

effects of changing object size. *Experimental Brain Research, 87*, 407–420.

Paulignan, Y., MacKenzie, C., Marteniuk, R., & Jeannerod, M. (1991). Selective perturbation of visual input during prehension movements: 1. The effects of changing object position. *Experimental Brain Research, 83*, 502–512.

Pause, M., Kunesch, E., Binkofski, F., & Freund, H.-J. (1989). Sensorimotor disturbances in patients with lesions of the parietal cortex. *Brain, 112*, 1599–1625.

Pawluk, D. T. V., & Howe, R. D. (1999). Dynamic lumped element response of the human fingerpad. *Journal of Biomechanical Engineering, 121*, 178–183.

Pederson, D. R., Steele, D., & Klein, G. (1980). *Stimulus characteristics that determine infants' exploratory play.* Paper presented at the International Conference on Infant Studies, New Haven, CT.

Penfield, W., & Rasmussen, T. (1950). *The cerebral cortex of man.* New York: Macmillan.

Peters, M. (1980). Why the preferred hand taps more quickly than the non-preferred hand: Three experiments on handedness. *Canadian Journal of Psychology, 34*, 62–71.

Peters, M. (1981). Handedness: Effect of prolonged practice on between hand performance differences. *Neuropsychologia, 19*, 587–590.

Petersen, P., Petrick, M., Connor, H., & Conklin, D. (1989). Grip strength and hand dominance: Challenging the 10% rule. *American Journal of Occupational Therapy, 43*, 444–447.

Phillips, C. G. (1986). *Movements of the hand.* Liverpool, England: Liverpool University Press.

Phillips, J. R., Johansson, R. S., & Johnson, K. O. (1990). Representation of Braille characters in human nerve fibres. *Experimental Brain Research, 81*, 589–592.

Phillips, J. R., & Johnson, K. O. (1981). Tactile spatial resolution: II. Neural representation of bars, edges, and gratings in monkey primary afferents. *Journal of Neurophysiology, 46*, 1192–2003.

Piaget, J., & Inhelder, B. (1948). *The child's conception of space.* New York: Norton.

Picard, N., & Strick, P. L. (1996). Motor areas of the medial wall: A review of their location and functional activation. *Cerebral Cortex, 6*, 342–353.

Pick, H. L., Jr., & Pick, A. D. (1967). A developmental and analytic study of the size-weight illusion. *Journal of Experimental Child Psychology, 5*, 362–371.

Pick, H. L., Pick, A. D., & Klein, R. E. (1967). Perceptual integration in children. In L. P. Lipsitt & C. C. Spiker (Eds.), *Advances in child development and behavior* (Vol. 3, pp. 192–220). New York: Academic.

238   References

Piveteau, J. (1991). *La main et l'hominisation* [The hand and human evolution]. Paris: Masson.

Pohl, P. S., Winstein, C. J., & Fisher, B. E. (1996). The locus of age-related movement slowing: Sensory processing in continuous goal-directed aiming. *Journal of Gerontology: Psychological Sciences, 51B*, P94–P102.

Polanyi, M. (1958). *Personal knowledge towards a post-critical philosophy*. London: Routledge and Kegan Paul.

Pont, S. C., Kappers, A. M., & Koenderink, J. J. (1998). The influence of stimulus tilt on haptic curvature matching and discrimination by dynamic touch. *Perception, 27*, 869–880.

Poppen, N. K., McCarroll, H. R., Doyle, J. R., & Niebauer, J. J. (1979). Recovery of sensibility after suture of digital nerves. *Journal of Hand Surgery, 4*, 212–226.

Potts, R. O., Buras, E. M., & Chrisman, D. A. (1984). Changes with age in the moisture content of human skin. *Journal of Investigative Dermatology, 82*, 97–100.

Povel, D. J., & Collard, R. (1982). Structural factors in patterned finger tapping. *Acta Psychologica, 52*, 107–123

Poznanski, A. K. (1974). *The hand in radiologic diagnosis*. Philadelphia: Saunders.

Preuschoft, H., & Chivers, D. J. (Eds.). (1993). *Hands of primates*. New York: Springer-Verlag.

Prochazka, A. (1996). Proprioceptive feedback and movement regulation. In L. Rowell & J. T. Shepherd (Eds.), *Handbook of physiology: Sec. 12. Exercise: Regulation and integration of multiple systems* (pp. 89–127). New York: Oxford University Press.

Prochazka, A., & Hulliger, M. (1983). Muscle afferent function and its significance for motor control mechanisms during voluntary movements in cat, monkey, and man. In J. E. Desmedt (Ed.), *Motor control mechanisms in health and disease* (pp. 93–132). New York: Raven.

Prochazka, A., & Hulliger, M. (1998). The continuing debate about CNS control of proprioception. *Journal of Physiology, 513*, 315.

Proske, U., Schaible, H. G., & Schmidt, R. F. (1988). Joint receptors and kinesthesia. *Experimental Brain Research, 72*, 219–224.

Provins, K. A., & Glencross, D. J. (1968). Handwriting, typewriting and handedness. *Quarterly Journal of Experimental Psychology, 20*, 282–289.

Quilliam, T. A. (1978). The structure of finger print skin. In G. Gordon (Ed.), *Active touch: The mechanism of recognition of objects by manipulation: A multidisciplinary approach* (pp. 1–18). Oxford: Pergamon.

Quilliam, T. A., & Ridley, A. (1971). The receptor community in the fingertip. *Journal of Physiology, 216*, 15–17.

Rabin, E., & Gordon, A. M. (2004). Tactile feedback contributes to consistency of finger movements during typing. *Experimental Brain Research, 155*, 362–369.

Radwin, R. G., Oh, S., Jensen, T. R., & Webster, J. G. (1992). External finger forces in submaximal five-finger static pinch prehension. *Ergonomics, 35*, 275–288.

Ramachandran, V. S. (1993). Behavioral and magnetoencephalographic correlates of plasticity in the adult human brain. *Proceedings of the National Academy of Sciences (USA), 90*, 10413–10420.

Ramnani, N., & Miall, C. (2004). A system in the human brain for predicting the actions of others. *Nature Neuroscience, 7*, 85–90.

Ranganathan, V. K., Siemionow, V., Sahgal, V., & Yue, G. H. (2001). Effects of aging on hand function. *Journal of the American Geriatric Society, 49*, 1478–1484.

Rao, A. K., & Gordon, A. M. (2001). Contribution of tactile information to accuracy in pointing movements. *Experimental Brain Research, 138*, 438–445.

Rapp, B., Hendel, S. K., & Medina, J. (2002). Remodeling of somatosensory hand representations following cerebral lesions in humans. *NeuroReport, 13*, 207–211.

Reed, C. L., Caselli, R. J., & Farah, M. J. (1996). Tactile agnosia: Underlying impairment and implications for normal tactile object recognition. *Brain, 119*, 875–888.

Reed, C. L., Lederman, S. J., & Klatzky, R. L. (1990). Haptic integration of planar size with hardness, texture, and planar contour. *Canadian Journal of Psychology, 44*, 522–545.

Reed, C. M. (1995). Tadoma: An overview of research. In G. Plant & K.-E. Spens (Eds.), *Profound deafness and speech communication* (pp. 40–55). London: Whurr.

Reed, C. M., Delhorne, L. A., Durlach, N. I., & Fischer, S. D. (1995). A study of the tactual reception of sign language. *Journal of Speech and Hearing Research, 38*, 477–489.

Reed, C. M., & Durlach, N. I. (1998). Note on information transfer rates in human communication. *Presence, 7*, 509–518.

Reed, C. M., Durlach, N. I., Braida, L., & Schultz, M. C. (1989). Analytic study of the Tadoma method: Effects of hand position on segmental speech perception. *Journal of Speech and Hearing Research, 32*, 921–929.

Reed, C. M., Durlach, N. I., & Delhorne, L. A. (1992). Natural methods of tactual communication. In

I. R. Summers (Ed.), *Tactile aids for the hearing impaired* (pp. 218–230). London: Whurr.

Reed, C. M., Rabinowitz, W. M., Durlach, N. I., Braida, L. D., Conway-Fithian, S., & Schultz, M. C. (1985). Research on the Tadoma method of speech communication. *Journal of the Acoustical Society of America, 77,* 247–257.

Reed, C. M., Rabinowitz, W. M., Durlach, N. I., Delhorne, L. A., Braida, L. D., Pemberton, J.C., Mulcahey, B. D., & Washington, D. L. (1992). Analytic study of the Tadoma method: Improving performance through the use of supplementary tactual displays. *Journal of Speech and Hearing Research, 35,* 450–465.

Reid, D. A. C. (1960). Reconstruction of thumb. *Journal of Bone and Joint Surgery, 42B,* 444–465.

Reilly, K. T., & Hammond, G. R. (2000). Independence of force production by digits of the human hand. *Neuroscience Letters, 290,* 53–56.

Reilly, K. T., & Schieber, M. H. (2003). Incomplete functional subdivision of the human multitendoned finger muscle flexor digitorum profundus: An electromyographic study. *Journal of Neurophysiology, 90,* 2560–2570.

Reilmann, R., Gordon, A. M., & Henningsen, H. (2001). Initiation and development of fingertip forces during whole-hand grasping. *Experimental Brain Research, 140,* 443–452.

Rempel, D., Dennerlein, J., Mote, C. D., & Armstrong, T. (1994). A method of measuring fingertip loading during keyboard use. *Journal of Biomechanics, 27,* 1101–1104.

Rempel, D., Serina, E., Klinenberg, E., Martin, B. J., Armstrong, T. J., Foulke, J. A., & Natarajan, S. (1997). The effect of keyboard keyswitch make force on applied force and finger flexor muscle activity. *Ergonomics, 40,* 800–808.

Révész, G. (1950). *Psychology and art of the blind* (H. A. Wolff, Trans.). London: Longmans, Green.

Reynolds, H. M., Smith, N. P., & Hunter, P. J. (2004). Construction of an anatomically accurate geometric model of the forearm and hand musculoskeletal system. *Proceedings of the 26th Annual International Conference of the IEEE Engineering in Medicine and Biology Society, 3,* 1829–1832.

Rhodes, D., & Schwartz, G. (1981). Lateralized sensitivity to vibrotactile stimulation: Individual differences revealed by interaction of threshold and signal detection tasks. *Neuropsychologia, 19,* 831–835.

Richardson, B., & Wuillemin, D. (1981). Can passive tactile perception be better than active? *Nature, 292,* 90.

Robbins, F., & Reece, T. (1985). Hand rehabilitation after great toe transfer for thumb reconstruction. *Archives of Physical Medicine and Rehabilitation, 66,* 109–112.

Roberts, M. A., Andrews, G. R., & Caird, F. I. (1975). Skinfold thickness on the dorsum of the hand in the elderly. *Age and Ageing, 4,* 8–15.

Robertson, S. L., & Jones, L. A. (1994). Tactile sensory impairments and prehensile function in subjects with left-hemisphere cerebral lesions. *Archives of Physical Medicine and Rehabilitation, 75,* 1108–1117.

Robinson, H. B. (1964). An experimental examination of the size-weight illusion in young children. *Child Development, 35,* 91–107.

Robles-De-La-Torre, G., & Hayward, V. (2001). Force can overcome object geometry in the perception of shape through active touch. *Nature, 412,* 445–448.

Rochat, P. (1987). Mouthing and grasping in neonates: Evidence for the early detection of what hard and soft substances afford for action. *Infant Behavior and Development, 10,* 435–449.

Rochat, P. (1992). Self-sitting and reaching in 5- to 8-month-old infants: The impact of posture and its development on early eye-hand coordination. *Journal of Motor Behavior, 24,* 210–220.

Roland, P. E. (1987). Somatosensory detection of microgeometry, macrogeometry and kinesthesia after localized lesions of the cerebral hemispheres in man. *Brain Research Reviews, 12,* 43–94.

Roland, P. E., & Ladegaard-Pedersen, H. (1977). A quantitative analysis of sensations of tensions and of kinesthesia in man. *Brain, 100,* 671–692.

Roland, P. E., O'Sullivan, B., & Kawashima, R. (1998). Shape and roughness activate different somatosensory areas in the human brain. *Proceedings of the Natural Academy of Sciences (USA), 95,* 3295–3300.

Roland, P. E., & Zilles, K. (1996). Functions and structures of the motor cortices in humans. *Current Opinion in Neurobiology, 6,* 773–778.

Rollman, G. B., & Harris, G. (1987). The detectability, discriminability, and perceived magnitude of painful electrical shock. *Perception & Psychophysics, 42,* 257–268.

Romo, R., & Salinas, E. (2003). Flutter discrimination: Neural codes, perception, memory and decision making. *Nature Reviews Neuroscience, 4,* 203–218.

Rosch, E. (1978). Principles of categorization. In E. Rosch & B. Lloyd (Eds.), *Cognition and categorization* (pp. 27–48). Hillsdale, NJ: Erlbaum.

Rose, S. A., Gottfried, A. W., & Bridger, W. H. (1981). Cross-modal transfer in 6-month-old infants. *Developmental Psychology, 17,* 661–669.

Rosner, B. S. (1961). Neural factors limiting spatiotemporal discriminations. In W. A. Rosenblith (Ed.),

Sensory communication (pp. 725–738). New York: Wiley.

Ross, R. T. (1991). Dissociated loss of vibration, joint position and discriminatory tactile senses in disease of spinal cord and brain. *Canadian Journal of Neurological Sciences, 18,* 312–320.

Rouiller, E. M. (1996). Multiple hand representations in the motor cortical areas. In A. M. Wing, P. Haggard, & J. R. Flanagan (Eds.), *Hand and brain: The neurophysiology and psychology of hand movements* (pp. 99–124). San Diego, CA: Academic.

Rousseau, J.-J. (1961). *Emile* (B. Foxley, Trans.). New York: Dutton. (Original work published 1762)

Ruff, H. A. (1982). Role of manipulation in infants' responses to invariant properties of objects. *Developmental Psychology, 18,* 682–691.

Ruff, H. A. (1984). Infants' manipulative exploration of objects: Effects of age and object characteristics. *Developmental Psychology, 20,* 9–20.

Ruff, H. A., & Kohler, C. J. (1978). Tactual-visual transfer in six-month-old infants. *Infant Behavior and Development, 1,* 259–264.

Rumelhart, D. E., & Norman, D. A. (1982). Simulating a skilled typist: A study of skilled cognitive-motor performance. *Cognitive Science, 6,* 1–36.

Saels, P., Thonnard, J.-L., Detrembleur, C., & Smith, A. M. (1999). Impact of the surface slipperiness of grasped objects on their subsequent acceleration. *Neuropsychologia, 37,* 751–756.

Sainburg, R. L. (2002). Evidence for a dynamic-dominance hypothesis of handedness. *Experimental Brain Research, 142,* 241–258.

Sainburg, R. L., & Kalakanis, D. (2000). Differences in control of limb dynamics during dominant and nondominant arm reaching. *Journal of Neurophysiology, 83,* 2661–2675.

Salimi, I., Hollender, I., Frazier, W., & Gordon, A. M. (2000). Specificity of internal representations underlying grasping. *Journal of Neurophysiology, 84,* 2390–2397.

Salthouse, T. A. (1984). Effects of age and skill in typing. *Journal of Experimental Psychology: General, 113,* 345–371.

Salthouse, T. A. (1985). *A theory of cognitive aging.* Amsterdam: North-Holland.

Salthouse, T. A. (1986). Perceptual, cognitive, and motoric aspects of transcription typing. *Psychological Bulletin, 99,* 303–319.

Sanes, J. N., & Donoghue, J. P. (2000). Plasticity and primary motor cortex. *Annual Reviews of Neuroscience, 23,* 393–415.

Sanes, J. N., Donoghue, J. P., Thangaraj, V., Edelman, R. R., & Warach, S. (1995). Shared neural substrates controlling hand movements in human motor cortex. *Science, 268,* 1775–1777.

Santello, M., Flanders, M., & Soechting, J. F. (2002). Patterns of hand motion during grasping and the influence of sensory guidance. *Journal of Neuroscience, 22,* 1426–1435.

Santello, M., & Soechting, J. F. (1998). Gradual molding of the hand to object contours. *Journal of Neurophysiology, 79,* 1307–1320.

Santello, M., & Soechting, J. F. (2000). Force synergies for multifingered grasping. *Experimental Brain Research, 133,* 457–467.

Sasaki, S., Isa, T., Pettersson, L.-G., Alstermark, B., Naito, K., Yoshimura, K., Seki, K., & Ohki, Y. (2004). Dexterous finger movements in primate without monosynaptic corticomotoneuronal excitation. *Journal of Neurophysiology, 92,* 3142–3147.

Sathian, K., & Zangaladze, A. (1996). Tactile spatial acuity at the human fingertip and lip: Bilateral symmetry and interdigit variability. *Neurology, 46,* 1464–1466.

Saxe, J. G. (1936). The blind men and the elephant. In H. Felleman (Ed.), *The best loved poems of the American people* (pp. 521–522). New York: Doubleday.

Schellekens, J. M. H., Kalverboer, A. F., & Scholten, C. A. (1984). The micro-structure of tapping movements in children. *Journal of Motor Behavior, 16,* 20–39.

Schettino, L. F., Adamovich, S. V., & Poizner, H. (2003). Effects of object shape and visual feedback on hand configuration during grasping. *Experimental Brain Research, 151,* 158–166.

Schieber, M. H., & Hibbard, L. S. (1993). How somatotopic is the motor cortex hand area? *Science, 261,* 489–492.

Schieber, M., & Santello, M. (2004). Hand function: Peripheral and central constraints in performance. *Journal of Applied Physiology, 96,* 2293–2300.

Schiff, W., & Foulke, E. (1982). *Tactual perception: A sourcebook.* New York: Cambridge University Press.

Schiff, W., Kaufer, L., & Mosak, S. (1966). Informative tactile stimuli in the perception of direction. *Perceptual & Motor Skills, 23,* 1315–1335.

Schmidt, R. T., & Toews, J. V. (1970). Grip strength as measured by the Jamar dynamometer. *Archives of Physical Medicine and Rehabilitation, 66,* 69–74.

Schott, G. D. (1993). Penfield's homunculus: A note on cerebral cartography. *Journal of Neurology, Neurosurgery, and Psychiatry, 56,* 329–333.

Schultz, A. H. (1956). Postembryonic age changes. *Primatologia, 1,* 887–964.

Schultz, A. H. (1969). *The life of primates.* New York: Universe.

Schwartz, A. S., Marchok, P. L., Kreinick, C. J., & Flynn, R. E. (1979). The asymmetric lateralization of tactile extinction in patients with unilateral cerebral dysfunction. *Brain, 102,* 669–684.

Schwarzer, G., Kufer, I., & Wilkening, F. (1999). Learning categories by touch: On the development of holistic and analytic processing. *Memory and Cognition, 27,* 868–877.

Scott, R. N., Brittain, R. H., Caldwell, R. R., Cameron, A. B., & Dunfield, V. A. (1980). Sensory-feedback system compatible with myoelectric control. *Medical and Biological Engineering and Computing, 18,* 65–69.

Scott, R. N., & Parker, P. A. (1988). Myoelectric prostheses: State of the art. *Journal of Medical Engineering & Technology, 12,* 143–151.

Seashore, C. E. (1938). *Psychology of music.* New York: McGraw-Hill.

Sekiyama, K. (1991). Importance of head axes in perception of cutaneous patterns drawn on vertical body surfaces. *Perception & Psychophysics, 49,* 481–492.

Selzer, R. (1982). *Letters to a young doctor.* New York: Simon & Schuster.

Semjen, A., & Summers, J. J. (2002). Timing goals in bimanual coordination. *Quarterly Journal of Experimental Psychology, 55A,* 155–171.

Semmes, J. (1965). A non-tactual factor in astereognosis. *Neuropsychologia, 3,* 295–315.

SensAble Technologies. (2004, June 14). Available: http://www.sensable.com.

Sergio, L. E., & Scott, S. H. (1998). Hand and joint paths during reaching movements with and without vision. *Experimental Brain Research, 122,* 157–164.

Serina, E. R., Mockensturm, E., Mote, C. D., & Rempel, D. (1998). A structural model of the forced compression of the fingertip pulp. *Journal of Biomechanics, 31,* 639–646.

Serina, E. R., Mote, C. D., & Rempel, D. (1997). Force response of the fingertip pulp to repeated compression: Effects of loading rate, loading angle, and anthropometry. *Journal of Biomechanics, 30,* 1035–1040.

Shaffer, L. H. (1975). Control processes in typing. *Quarterly Journal of Experimental Psychology, 27,* 419–432.

Shaffer, L. H. (1976). Intention and performance. *Psychological Review, 83,* 375–393.

Shaffer, L. H. (1980). Analysing piano performance: A study of concert pianists. In G. E. Stelmach & J. Requin (Eds.), *Tutorials in motor behavior* (pp. 443–455). Amsterdam: North-Holland.

Shaffer, L. H. (1981). Performances of Chopin, Bach, and Bartok: Studies in motor programming. *Cognitive Psychology, 13,* 326–376.

Shannon, C. E., & Weaver, W. (1963). *The mathematical theory of communication.* Urbana: University of Illinois Press.

Sheridan, T. B. (1992). Defining our terms. *Presence, 1,* 272–274.

Sheridan, T. B., & Ferrell, W. R. (1974). *Man-machine systems: Information, control, and decision models of human performance.* Cambridge, MA: MIT Press.

Sherrick, C., & Cholewiak, R. (1986). Cutaneous sensitivity. In K. R. Boff, L. Kaufman, & J. P. Thomas (Eds.), *Handbook of perception and human performance* (Vol. 1, pp. 12-1–12-58). New York: Wiley.

Sherrick, C., & Rogers, R. (1966). Apparent haptic movement. *Perception & Psychophysics, 1,* 175–180.

Sherrington, C. S. (1900a). The muscular sense. In E. A. Shäfer (Ed.), *Textbook of physiology* (Vol. 2, pp. 1002–1025). Edinburgh: Pentland.

Sherrington, C. S. (1900b). The cutaneous sense. In E. A. Shäfer (Ed.), *Textbook of physiology* (Vol. 2, pp. 920–1001). Edinburgh: Pentland.

Sherrington, C. S. (1906). *The integrative action of the nervous system.* London: Constable.

Shima, K., Mushiake, H., Saito, N., & Tanji, J. (1996). Role for cells in the presupplementary motor area in updating motor plans. *Proceedings of the National Academy of Sciences, 93,* 8694–8698.

Shima, K., & Tanji, J. (1998). Both supplementary and presupplementary motor areas are crucial for the temporal organization of multiple movements. *Journal of Neurophysiology, 80,* 3247–3260.

Shinn, M. W. (1893). Notes on the development of a child. *University of California Public Education, 1,* 178–236.

Sholes, C. L., Glidden, C., & Soulé, S. W. (1868). Improvement in type writing machines. U.S. Patent No. 79,868.

Shrewsbury, M. M., & Johnson, R. K. (1983). Form, function, and evolution of the distal phalanx. *Journal of Hand Surgery, 8,* 475–479.

Silver, A. F., Montagna, W., & Karacan, I. (1965). The effect of age on human eccrine sweating. In W. Montagna (Ed.), *Advances in biology of skin* (Vol. 6, pp. 129–150). New York: Pergamon.

Sinclair, R. J., Kuo, J. J., & Burton, H. (2000). Effects on discrimination performance of selective attention to tactile features. *Somatosensory and Motor Research, 17,* 145–157.

Sloboda, J. A. (1974). The eye-hand span: An approach to the study of sight reading. *Psychology of Music, 2,* 4–10.

Sloboda, J. A. (1984). Experimental studies in music reading: A review. *Music Perception, 2,* 222–236.

Smeets, J. B. J., & Brenner, E. (1999). A new view on grasping. *Motor Control, 3,* 237–271.

Smith, A. M., Chapman, E., Deslandes, M., Langlais, J.-S., & Thibodeau, M.-P. (2002). Role of friction and tangential force variation in the subjective scaling of tactile roughness. *Experimental Brain Research, 144,* 211–223.

Smith, A. M., Gosselin, G., & Houde, B. (2002). Deployment of fingertip forces in tactile exploration. *Experimental Brain Research, 147,* 209–218.

Smith, A. M., & Scott, S. H. (1996). Subjective scaling of smooth surface friction. *Journal of Neurophysiology, 75,* 1957–1962.

Smith, C. D., Umberger, G. H., Manning, E. L., Slevin, J. T., Wekstein, D. R., Schmitt, F. A., Markesbery, W. R., Zhang, Z., Gerhardt, G. A., Kryscio, R. J., & Gash, D.M. (1999). Critical decline in fine motor hand movements in human aging. *Neurology, 53,* 1458–1461.

Smith, H. B. (1973). Smith hand function evaluation. *American Journal of Occupational Therapy, 27,* 244–251.

Smith, L. B., & Kemler, D. G. (1977). Developmental trends in free classification: Evidence for a new conceptualization of perceptual development. *Journal of Experimental Child Psychology, 24,* 279–298.

Soechting, J. F. (1984). Effect of target size on spatial and temporal characteristics of a pointing movement in man. *Experimental Brain Research, 54,* 121–132.

Soechting, J. F., & Flanders, M. (1992). Organization of sequential typing movements. *Journal of Neurophysiology, 67,* 1275–1290.

Soechting, J. F., & Flanders, M. (1997). Flexibility and repeatability of finger movements during typing: Analysis of multiple degrees of freedom. *Journal of Computational Neuroscience, 4,* 29–46.

Soechting, J. F., Gordon, A. M., & Engel, K. C. (1996). Sequential hand and finger movements: Typing and piano playing. In J. R. Bloedel, T. J. Ebner, & S. P. Wise (Eds.), *The acquisition of motor behavior in vertebrates* (pp. 343–360). Cambridge, MA: MIT Press.

Soechting, J. F., & Lacquaniti, F. (1981). Invariant characteristics of a pointing movement in man. *Journal of Neuroscience, 1,* 710–720.

Soechting, J. F., Tillery, S. I. H., & Flanders, M. (1990). Transformation from head- to shoulder-centered representation of target direction in arm movements. *Journal of Cognitive Neuroscience, 2,* 32–43.

Solomon, H. Y., Turvey, M. T., & Burton, G. (1989). Perceiving extents of rods by wielding: Haptic diagonalization and decomposition of the inertia tensor.

Journal of Experimental Psychology: Human Perception and Performance, 15, 58–68.

Spence, C., Pavani, F., & Driver, J. (2000). Crossmodal links between vision and touch in covert endogenous spatial attention. *Journal of Experimental Psychology: Human Perception and Performance, 26,* 1298–1319.

Sperry, R. W. (1950). Neural basis of the spontaneous optokinetic response produced by visual neural inversion. *Journal of Comparative and Physiological Psychology, 43,* 482–489.

Spray, D. C. (1986). Cutaneous temperature receptors. *Annual Reviews of Physiology, 48,* 625–638.

Spreen, O., & Gaddes, W. H. (1969). Developmental norms for 15 neuropsychological tests age 6 to 15. *Cortex, 5,* 170–191.

Srinivas, K., Greene, A. J., & Easton, R. D. (1997a). Implicit and explicit memory for haptically experienced two-dimensional patterns. *Psychological Science, 8,* 243–246.

Srinivas, K., Greene, A. J., & Easton, R. D. (1997b). Visual and tactile memory for 2-D patterns: Effects of changes in size and left-right orientation. *Psychonomic Bulletin & Review, 4,* 535–540.

Srinivasan, M. (1989). Surface deflection of primate fingertip under line load. *Journal of Biomechanics, 22,* 343–349.

Srinivasan, M. A., & LaMotte, R. H. (1987). Tactile discrimination of shape: Responses of slowly and rapidly adapting mechanoreceptive afferents to a step indented into the monkey fingerpad. *Journal of Neuroscience, 7,* 1682–1697.

Srinivasan, M. A., & LaMotte, R. H. (1995). Tactual discrimination of softness. *Journal of Neurophysiology, 73,* 88–101.

Srinivasan, M. A., & LaMotte, R. H. (1996). Tactual discrimination of softness: Abilities and mechanisms. In O. Franzen, R. Johansson, & L. Terenius (Eds.), *Somesthesis and the neurobiology of the somatosensory cortex* (pp. 123–136). Basel: Birkhäuser Verlag.

Srinivasan, M. A., Whitehouse, J. M., & LaMotte, R. H. (1990). Tactile detection of slip: Surface microgeometry and peripheral neural codes. *Journal of Neurophysiology, 63,* 1323–1332

Stark, B., Carlstedt, T., Hallin, R. G., & Risling, M. (1998). Distribution of human Pacinian corpuscles in the hand: A cadaver study. *Journal of Hand Surgery, 23B,* 370–372.

Starkes, J. L., Payk, I., & Hodges, N. J. (1998). Developing a standardized test for the assessment of suturing skill in novice microsurgeons. *Microsurgery, 18,* 19–22.

Starkes, J. L., Payk, I., Jennen, P., & Leclair, D. (1993). A stitch in time: Cognitive issues in microsurgery. In

J. L. Starkes & F. Allard (Eds.), *Cognitive issues in motor expertise* (pp. 225–240). Amsterdam: North-Holland Elsevier.

Steele, D., & Pederson, D. R. (1977). Stimulus variables which affect the concordance of visual and manipulative exploration in six-month-old infants. *Child Development, 48,* 104–111.

Steenhuis, R. E. (1996). Hand preference and performance in skilled and unskilled activities. In D. Elliott & E. A. Roy (Eds.), *Manual asymmetries in motor performance* (pp. 123–142). Boca Raton, FL: CRC.

Stetson, R. H., & McDill, J. A. (1923). Mechanism of the different types of movement. *Psychological Monographs, 32,* 18–40.

Stevens, J. C. (1979). Thermal intensification of touch sensation: Further extensions of the Weber phenomenon. *Sensory Processes, 3,* 240–248.

Stevens, J. C. (1991). Thermal sensibility. In M. Heller and W. Schiff. (Eds.), *The psychology of touch* (pp. 61–90). Hillsdale, NJ: Erlbaum.

Stevens, J. C. (1992). Aging and spatial acuity of touch. *Journal of Gerontology: Psychological Science, 47,* 35–40.

Stevens, J. C., & Choo, K. K. (1996). Spatial acuity of the body surface over the life span. *Somatosensory and Motor Research, 13,* 153–166.

Stevens, J. C., & Choo, K. K. (1998). Temperature sensitivity of the body surface over the life span. *Somatosensory and Motor Research, 15,* 13–28.

Stevens, J. C., & Green, B. G. (1978). Temperature-touch interaction: Weber's phenomenon revisited. *Sensory Processes, 2,* 206–209.

Stevens, J. C. & Marks, L. E. (1971). Spatial summation and the dynamics of warmth sensation. *Perception & Psychophysics, 9,* 391–398.

Stevens, J. C., Okulicz, W. C., & Marks, L. E. (1973). Temporal summation at the warmth threshold. *Perception & Psychophysics, 14,* 307–312.

Stevens, J. C., & Patterson, M. Q. (1995). Dimensions of spatial acuity in the touch sense: Changes over the life span. *Somatosensory and Motor Research, 12,* 29–47.

Stevens, S. S. (1959). Tactile vibration: Dynamics of sensory intensity. *Journal of Experimental Psychology, 57,* 210–218.

Stevens, S. S. (1975). *Psychophysics: Introduction to its perceptual, neural and social prospects.* New York: Wiley.

Stevens, S. S., Carton, A. S., & Shickman, G. M. (1958). A scale of apparent intensity of electric shock. *Journal of Experimental Psychology, 56,* 328–338.

Stevens, S. S., & Harris, J. (1962). Scaling of roughness and smoothness. *Journal of Experimental Psychology, 64,* 489–494.

Stoeckel, M. C., Weder, B., Binkofski, F., Buccino, G., Shah, N. J., & Seitz, R. J. (2003). A fronto-parietal circuit for tactile object discrimination: An event-related fMRI study. *NeuroImage, 19,* 1103–1114.

Streri, A. (1987). Tactile discrimination of shape and intermodal transfer in 2- to 3-month-old infants. *British Journal of Developmental Psychology, 5,* 213–220.

Streri, A., & Gentaz, E. (2003). Cross-modal recognition of shape from hand to eyes in human newborns. *Somatosensory and Motor Research, 20,* 13–18.

Streri, A., Lhote, M., & Dutilleul, S. (2000). Haptic perception in newborns. *Developmental Science, 3,* 319–327.

Streri, A., & Molina, M. (1993). Visual and tactual transfer between objects and pictures in 2-month-old infants. *Perception, 22,* 1299–1318.

Streri, A., & Spelke, E. (1988). Haptic perception of objects in infancy. *Cognitive Psychology, 20,* 1–23.

Streri, A., Spelke, E., & Rameix, E. (1993). Modality-specific and amodal aspects of object perception in infancy: The case of active touch. *Cognition, 47,* 251–279.

Summers, C., & Lederman, S. J. (1990). Perceptual asymmetries in the somatosensory system. *Cortex, 26,* 201–226.

Swanson, A. B. (1964). Evaluation of impairment of function in the hand. *Surgical Clinics of North America, 44,* 925–940.

Swanson, A. B., De Groot Swanson, G., & Göran-Hagert, G. (1990). Evaluation of impairment of hand function. In J. M. Hunter, L. H. Scheider, E. J. Mackin, & A. D. Callahan (Eds.), *Rehabilitation of the hand: Surgery and therapy* (3d ed., pp. 109–138). St. Louis, MO: Mosby.

Swash, M., & Fox, K. P. (1972). Muscle spindle innervation in man. *Journal of Anatomy, 112,* 61–80.

Szabo, R. M., Gelberman, R. H., Williamson, R. V., Dellon, A. L., Yaru, N. C., & Dimick, M. P. (1984). Vibratory sensory testing in acute peripheral nerve compression. *Journal of Hand Surgery, 9A,* 104–109.

Talbot, W. H., Darian-Smith, I., Kornhuber, H. H., & Mountcastle, V. B. (1968). The sense of flutter-vibration: Comparison of the human capacity with response patterns of mechanoreceptive afferents from the monkey hand. *Journal of Neurophysiology, 31,* 301–334.

Tallis, R. (2003). *The hand: A philosophical inquiry into human being.* Edinburgh: Edinburgh University Press.

Tan, H. Z., Durlach, N. I., Beauregard, M. G., & Srinivasan, M. A. (1995). Manual discrimination of compliance using active pinch grasp: The roles of force and work cues. *Perception & Psychophysics, 57,* 495–510.

Tan, H. Z., Durlach, N. I., Rabinowitz, W. M., Reed, C. M., & Santos, J. R. (1997). Reception of Morse code through motional, vibrotactile, and auditory stimulation. *Perception & Psychophysics, 59,* 1004–1017.

Tan, H., Lim, A., & Traylor, R. (2000). A psychophysical study of sensory saltation with an open response paradigm. *Proceedings of the ASME Dynamic Systems and Control Division, 69,* 1109–1115.

Tanji, J. (1996). New concepts of the supplementary motor area. *Current Opinion in Neurobiology, 6,* 782–787.

Taub, E., & Berman, A. J. (1968). Movement and learning in the absence of sensory feedback. In S. J. Freedman (Ed.), *The neuropsychology of spatially oriented behavior* (pp. 173–192). Homewood, IL: Dorsey.

Taus, R. H., Stevens, J. C., & Marks, L. E. (1975). Spatial localization of warmth. *Perception & Psychophysics, 17,* 194–196.

Tawney, G. (1895). The perception of two points is not the space-threshold. *Psychological Review, 2,* 585–593.

Taylor, C. L., & Schwarz, R. J. (1970). The anatomy and mechanics of the human hand. In *Selected Articles From Artificial Limbs, January 1954–Spring 1966* (pp. 49–62). Huntington, NY: Kreiger.

Taylor, M. M., & Lederman, S. J. (1975). Tactile perception of grooved surfaces: A model and the effect of friction. *Perception & Psychophysics, 17,* 23–36.

Taylor, N., Sand, P. L., & Jebsen, R. H. (1973). Evaluation of hand function in children. *Archives of Physical Medicine and Rehabilitation, 54,* 129–135.

Taylor, N. L., Raj, A. D., Dick, H. M., & Solomon, S. (2004). The correction of ulnar claw fingers: A follow-up study comparing the extensor-to-flexor with the palmaris longus 4-tailed tendon transfer in patients with leprosy. *Journal of Hand Surgery, 29,* 595–604.

Teghtsoonian, R., & Teghtsoonian, M. (1965). Seen and felt length. *Psychonomic Science, 3,* 465–466.

Teghtsoonian, R., & Teghtsoonian, M. (1970). Two varieties of perceived length. *Perception & Psychophysics, 8,* 389–392.

Tenner, E. (2004). *Our own devices.* New York: Vintage.

Teraoka, T. (1979). Studies on the peculiarity of grip strength in relation to body positions and aging. *Kobe Journal of Medical Science, 25,* 1–17.

Terzuolo, C. A., & Viviani, P. (1980). Determinants and characteristics of motor patterns used for typing. *Neuroscience, 5,* 1085–1103.

Thelen, E., & Smith, L. B. (1994). *A dynamical system approach to the development of cognition and action.* London: Bradford.

Thomine, J.-M. (1981). The skin of the hand. In R. Tubiana (Ed.), *The hand* (Vol. 1, pp. 107–115). Philadelphia: Saunders.

Thornbury, J. M., & Mistretta, C. M. (1981). Tactile sensitivity as a function of age. *Journal of Gerontology, 36,* 34–39.

Thorngren, K. G., & Werner, C. O. (1979). Normal grip strength. *Acta Orthopaedica Scandinavica, 50,* 255–259.

Tiffin, J. (1968). *Purdue pegboard examiner manual.* Chicago: Science Research Associates.

Todor, J. I., & Kyprie, P. M. (1980). Hand differences in the rate and variability of rapid tapping. *Journal of Motor Behavior, 12,* 57–62.

Todor, J. I., & Smiley, A. L. (1985). Performance differences between the hands: Implications for studying disruption to limb praxis. In E. A. Roy (Ed.), *Neuropsychological studies of apraxia and related disorders* (pp. 309–344). Amsterdam: Elsevier.

Todor, J. I., & Smiley-Oyen, A. L. (1987). Force modulation as a source of hand differences in rapid finger tapping. *Acta Psychologica, 65,* 65–73.

Tower, S. (1940). Pyramid lesion in the monkey. *Brain, 63,* 36–90

Tremblay, F., Wong, K., Sanderson, R., & Cote, L. (2003). Tactile spatial acuity in elderly persons: Assessment with grating domes and relationship with manual dexterity. *Somatosensory and Motor Research, 20,* 127–132.

Turner-Stokes, L., & Reid, K. (1999). Three-dimensional motion analysis of upper limb movement in the bowing arm of string-playing musicians. *Clinical Biomechanics, 14,* 426–433.

Turvey, M. T. (1996). Dynamic touch. *American Psychologist, 51,* 1134–1152.

Turvey, M. T., Burton, G., Amazeen, E. L., Butwill, M., & Carello, C. (1998). Perceiving the width and height of a hand-held object by dynamic touch. *Journal of Experimental Psychology: Human Perception and Performance, 24,* 35–48.

Tuttle, R. H. (1969). Quantitative and functional studies on the hands of the Anthropoidea: 1. The Hominoidea. *Journal of Morphology, 128,* 309–364.

Twitchell, T. E. (1970). Reflex mechanisms and development of prehension. In K. J. Connolly (Ed.), *Mechanisms of motor skill development* (pp. 25–38). New York: Academic.

Umiker-Sebeok, J., & Sebeok, T. A. (Eds.). (1987). *Monastic sign languages.* New York: Mouton de Gruyter.

Uno, Y., Kawato, M., & Suzuki, R. (1989). Formation and control of optimal trajectory in human multijoint arm movement: Minimum torque change model. *Biological Cybernetics, 61,* 89–101.

U.S. National Research Council. (1995). *Currency features for visually impaired people.* Washington, DC: National Academy of Sciences.

Vallbo, A. (1974). Afferent discharge from human muscle spindles in non-contracting muscles: Steady state impulse frequency as a function of joint angle. *Acta Physiologica Scandinavica, 90,* 303–318.

Vallbo, A. (1995). Somatic sensation. In M. Gazzaniga (Ed.), *The cognitive neurosciences* (pp. 237–252). Cambridge, MA: MIT Press.

Vallbo, A. B., & Johansson, R. S. (1978). The tactile sensory innervation of the glabrous skin of the human hand. In G. Gordon (Ed.), *Active touch: The mechanism of recognition of objects by manipulation: A multidisciplinary approach* (pp. 29–54). Oxford: Pergamon.

Vallbo, A. B., & Johansson, R. S. (1984). Properties of cutaneous mechanoreceptors in the human hand related to touch sensation. *Human Neurobiology, 3,* 3–14.

Vallbo, A. B., Olausson, H., & Wessberg, J. (1999). Unmyelinated afferents constitute a second system coding tactile stimuli of the human hairy skin. *Journal of Neurophysiology, 81,* 2753–2763.

Vallbo, A. B., Olausson, H., Wessberg, J., & Kakuda, N. (1995). Receptive field characteristics of tactile units with myelinated afferents in hairy skin of human subjects. *Journal of Physiology, 483,* 783–795.

Van Boven, R. W., & Johnson, K. O. (1994). A psychophysical study of the mechanisms of sensory recovery following nerve injury in humans. *Brain, 117,* 149–167.

van der Meer, A. L., van der Weel, F. R., & Lee, D. N. (1995). The functional significance of arm movements in neonates. *Science, 267,* 693–695.

Van Doren, C. L., Gescheider, G. A., & Verrillo, R. T. (1990). Vibrotactile temporal gap detection as a function of age. *Journal of the Acoustical Society of America, 87,* 2201–2206.

van Galen, G., & Wing, A. M. (1984). The sequencing of movements. In M. M. Smyth & A. M. Wing (Eds.), *The psychology of human movement* (pp. 153–181). London: Academic.

van Santvoord, A. A. M., & Beek, P. J. (1996). Spatiotemporal variability in cascade juggling. *Acta Psychologica, 91,* 131–151.

Varney, N. R. (1986). Somesthesis. In H. J. Hannay (Ed.), *Experimental techniques in human neuropsychology* (pp. 212–237). New York: Oxford University Press.

Vega-Bermudez, F., & Johnson, K. O. (1999). SA1 and RA receptive fields, response variability, and population responses mapped with a probe array. *Journal of Neurophysiology, 81,* 2701–2710.

Vega-Bermudez, F., & Johnson, K. O. (2001). Differences in spatial acuity between digits. *Neurology, 56,* 1389–1391.

Vega-Bermudez, F., Johnson, K. O., & Hsiao, S. S. (1991). Human tactile pattern recognition: Active vs. passive touch, velocity effects, and patterns of confusion. *Journal of Neurophysiology, 65,* 531–546.

Venkataraman, S. T., & Iberall, T. (Eds.). (1990). *Dextrous robot hands.* New York: Springer.

Verrillo, R. T. (1963). Effect of contactor area on vibrotactile threshold. *Journal of the Acoustical Society of America, 35,* 1962–1966.

Verrillo, R. T. (1965). Temporal summation in vibrotactile sensitivity. *Journal of the Acoustical Society of America, 37,* 843–846.

Verrillo, R. T. (1968). A duplex mechanism of mechanoreception. In D. R. Kenshalo (Ed.), *The skin senses* (pp. 139–159). Springfield, IL: Thomas.

Verrillo, R. T. (1971). Vibrotactile thresholds measured at the finger. *Perception & Psychophysics, 9,* 329–330.

Verrillo, R. T. (1977). Comparison of child and adult vibrotactile thresholds. *Bulletin of the Psychonomic Society, 9,* 197–200.

Verrillo, R. T. (1985). Psychophysics of vibrotactile stimulation. *Journal of the Acoustical Society of America, 77,* 225–232.

Verrillo, R. T., & Bolanowski, S. J., Jr. (1986). The effects of skin temperature on the psychophysical responses to vibration on glabrous and hairy skin. *Journal of the Acoustical Society of America, 80,* 528–532.

Verrillo, R. T., Bolanowski, S. J., & Gescheider, G. A. (2002). Effect of aging on the subjective magnitude of vibration. *Somatosensory and Motor Research, 19,* 238–244.

Verrillo, R. T., Bolanowski, S. J., & McGlone, F. P. (1999). Subjective magnitude estimate of tactile roughness. *Somatosensory and Motor Research, 16,* 352–360.

Verrillo, R. T., & Capraro, A. J. (1975). Effect of extrinsic noise on vibrotactile information processing channels. *Perception & Psychophysics, 18,* 88–94.

Verrillo, R. T., & Gescheider, G. A. (1975). Enhancement and summation in the perception of two successive vibrotactile stimuli. *Perception & Psychophysics, 18,* 128–136.

Verrillo, R. T., & Gescheider, G. A. (1977). Effect of prior stimulation on vibrotactile thresholds. *Sensory Processes, 1,* 292–300.

Victor Raj, D., Ingty, K., & Devanandan, M. S. (1985). Weight appreciation in the hand in normal subjects and in patients with leprous neuropathy. *Brain, 108,* 95–102.

Viemeister, N. F., & Wakefield, G. H. (1991). Temporal integration and multiple looks. *Journal of the Acoustical Society of America, 90,* 858–865.

Viitasalo, J. T., Era, P., Leskinen, A. L., & Heikkinen, E. (1985). Muscular strength profiles and anthropometry in random samples of men aged 31–35, 51–55 and 71–75 years. *Ergonomics, 28,* 1563–1574.

Voisin, J., Benoit, G., & Chapman, C. E. (2002). Haptic discrimination of object shape in humans: Two-dimensional angle discrimination. *Experimental Brain Research, 145,* 239–250.

Voisin, J., Lamarre, Y., & Chapman, C. E. (2002). Haptic discrimination of object shape in humans: Contribution of cutaneous and proprioceptive inputs. *Experimental Brain Research, 145,* 251–260.

Vorberg, D., & Hambuch, R. (1978). On the temporal control of rhythmic performance. In J. Requin (Ed.), *Attention and performance* (Vol. 12, pp. 535–555). Hillsdale, NJ: Erlbaum.

Voss, V. H. (1971). Tabelle der absoluten und relativen Muskelspindelzahlen der menschlichen Skelettmuskulatur [Table of the absolute and relative number of muscle spindles in skeletal muscles]. *Anatomischer Anzeiger, 129,* 562–572.

Vrbova, G. (1995). *Nerve-muscle interaction* (2d ed.). New York: Chapman & Hall.

Wall, J. T., Kaas, J. H., Sur, M., Nelsen, R. J., Felleman, D. J., & Merzenich, M. M. (1986). Functional reorganization in somatosensory cortical areas 3b and 1 of adult monkeys after median nerve repair: Possible relationships to sensory recovery in humans. *Journal of Neuroscience, 6,* 218–233.

Wallace, S. A., Weeks, D. L., & Kelso, J. A. S. (1990). Temporal constraints in reaching and grasping behavior. *Human Movement Science, 9,* 69–93.

Wang, X., Merzenich, M. M., Sameshima, K., & Jenkins, W. M. (1995). Remodeling of hand representation in adult cortex determined by timing of tactile stimulation. *Nature, 378,* 71–75.

Ward, T. B. (1989). Analytic and holistic modes of processing in category learning. In B. E. Shepp & S. Ballesteros (Eds.), *Object perception: Structure and process* (pp. 387–419). Hillsdale, NJ: Erlbaum.

Ward, T. B., & Scott, J. (1987). Analytic and holistic modes of learning family-resemblance concepts. *Memory and Cognition, 15,* 42–54.

Weber, E. H. (1978). *The sense of touch* (H. E. Ross & D. J. Murray, Eds. and Trans.). London: Academic. (Original work published 1834)

Wei, F. C., Carver, N., Lee, Y. H., Chuang, D. C., & Cheng, S. L. (2000). Sensory recovery and Meissner corpuscle number after toe-to-hand transplantation. *Plastic and Reconstructive Surgery, 105,* 2405–2411.

Weinstein, S. (1962). Tactile sensitivity of the phalanges. *Perceptual and Motor Skills, 14,* 351–354.

Weinstein, S. (1968). Intensive and extensive aspects of tactile sensitivity as a function of body part, sex, and laterality. In D. R. Kenshalo (Ed.), *The skin senses* (pp. 195–222). Springfield, IL: Thomas.

Weinstein, S., & Sersen, E. A. (1961). Tactual sensitivity as a function of handedness and laterality. *Journal of Comparative and Physiological Psychology, 54,* 665–669.

Welch, R. B., & Warren, D. H. (1986). Intersensory interactions. In K. R. Boff, L. Kaufman, & J. P. Thomas (Eds.), *Handbook of perception and human performance* (Vol. 1, pp. 25-1–25-36). New York: Wiley Interscience.

Welford, A. T. (1980). Choice reaction time: Basic concepts. In A. T. Welford (Ed.), *Reaction times* (pp. 73–128). New York: Academic.

Welford, A. T. (1988). Reaction time, speed of performance, and age. *Annals of the New York Academy of Sciences, 515,* 1–17.

Werner, G., & Mountcastle, V. B. (1965). Neural activity in mechanoreceptive cutaneous afferents: Stimulus-response relations, Weber functions, and information transmission. *Journal of Neurophysiology, 28,* 359–397.

West, L. J., & Sabban, Y. (1982). Hierarchy of stroking habits at the typewriter. *Journal of Applied Psychology, 67,* 370–376.

Westling, G., & Johansson, R. S. (1984). Factors influencing the force control during precision grip. *Experimental Brain Research, 53,* 277–284.

Westling, G., & Johansson, R. S. (1987). Responses in glabrous skin mechanoreceptors during precision grip in humans. *Experimental Brain Research, 66,* 128–140.

Whitsel, B. L., Favorov, O. V., Kelly, D. G., & Tommerdahl, M. (1991). Mechanisms of dynamic peri- and intra-columnar interactions in somatosensory cortex: Stimulus-specific contrast enhancement by NMDA receptor activation. In O. Franzen & J. Westman (Eds.), *Information processing in the somatosensory system* (pp. 353–370). New York: Stockton.

Whitsel, B. L., Favorov, O. V., Tommerdahl, M., Diamond, M., Juliano, S., & Kelly, D. G. (1989). Dynamic processes govern the somatosensory cortical response to natural stimulation. In J. S. Lund (Ed.), *Sensory processing in the mammalian brain* (pp. 84–116). New York: Oxford University Press.

Wiesendanger, M. (1999). Manual dexterity and the making of tools: An introduction from an evolutionary perspective. *Experimental Brain Research, 128,* 1–5.

Wiesendanger, M., Kazennikov, O., Perrig, S., & Kaluzny, P. (1996). Two hands, one action: The problem of bimanual coordination. In A. M. Wing, P. Haggard, & J. R. Flanagan (Eds.), *Hand and brain: The neurophysiology and psychology of hand movements* (pp. 283–300). New York: Academic.

Wilkes, G. L., Brown, I. A., & Wildnauer, R. H. (1973). The biomechanical properties of skin. *CRC Critical Reviews in Bioengineering, 1,* 453–495.

Wilska, A. (1954). On the vibrational sensitivity in different regions of the body surface. *Acta Physiologica Scandinavica, 31,* 285–289.

Wilson, B. C., Iacoviello, J. M., Wilson, J. J., & Risucci, D. (1982). Purdue pegboard performance of normal preschool children. *Journal of Clinical Neuropsychology, 4,* 19–26.

Wilson, F. R. (1998). *The hand.* New York: Pantheon.

Wing, A. M., & Fraser, C. (1983). The contribution of the thumb to reaching movements. *Quarterly Journal of Experimental Psychology, 35A,* 297–309.

Wing, A. M., Haggard, P., & Flanagan, J. R. (Eds.). (1996). *Hand and brain: The neurophysiology and psychology of hand movements.* San Diego, CA: Academic.

Wing, A. M., & Kristofferson, A. B. (1973). Response delays and the timing of discrete motor responses. *Perception & Psychophysics, 14,* 5–12.

Wing, A. M., & Lederman, S. L. (1998). Anticipating load torques produced by voluntary movements. *Journal of Experimental Psychology: Human Perception and Performance, 24,* 1571–1581.

Winges, S. A., Weber, D. J., & Santello, M. (2003). The role of vision on hand preshaping during reach to grasp. *Experimental Brain Research, 152,* 489–498.

Wise, S. P., di Pellegrino, G., & Boussaoud, D. (1992). Primate premotor cortex: Dissociation of visuomotor from sensory signals. *Journal of Neurophysiology, 68,* 969–972.

Witney, A. G. (2004). Internal models for bi-manual tasks. *Human Movement Science, 23,* 747–770.

Wolff, P. H., & Hurwitz, I. (1976). Sex differences in finger tapping: A developmental study. *Neuropsychologia, 14,* 35–41.

Wong, T. S. (1975). The respective role of limb and eye movements in the haptic and visual Müller-Lyer illusion. *Quarterly Journal of Experimental Psychology, 27,* 659–666.

Wong, T.S. (1977). Dynamic properties of radial and tangential movements as determinants of the haptic horizontal-vertical illusion with an L-figure. *Journal of Experimental Psychology: Human Perception and Performance, 3,* 151–164.

Wong, T. S., Ho, R., & Ho, J. (1974). Influence of shape of receptor organ on the horizontal-vertical illu-

sion in passive touch. *Journal of Experimental Psychology, 103,* 414–419.

Woo, T. L., & Pearson, K. (1927). Dextrality and sinistrality of hand and eye. *Biometrika, 19,* 165–199.

Wood-Jones, F. (1944). *The principles of anatomy as seen in the hand* (2d ed.). London: Baillière, Tindall.

Woodworth, R. S. (1899). The accuracy of voluntary movement. *Psychological Review, 13*(Suppl.), 1–119.

Wu, G., Ekedahl, R., & Hallin, R. G. (1998). Clustering of slowly adapting type II mechanoreceptors in human peripheral nerve and skin. *Brain, 121,* 265–279.

Wynn Parry, C. B., & Salter, M. (1976). Sensory re-education after median nerve lesions. *Hand, 8,* 250–257.

Yan, J. H., Thomas, J. R., & Stelmach, G. E. (1998). Aging and rapid aiming arm movement control. *Experimental Aging Research, 24,* 155–168.

Yoblick, D. A., & Salvendy, G. (1970). Influence of frequency on the estimation of time for auditory, visual, and tactile modalities: The Kappa effect. *Journal of Experimental Psychology, 86,* 157–164.

Yoshioka, T., Gibb, B., Dorsch, A. K., Hsiao, S. S., & Johnson, K. O. (2001). Neural coding mechanisms underlying perceived roughness of finely textured surfaces. *Journal of Neuroscience, 21,* 6905–6916.

Young, R. W. (2003). Evolution of the human hand: The role of throwing and clubbing. *Journal of Anatomy, 202,* 165–174.

Yu, W., & Brewster, S. (2002). Comparing two haptic interfaces for multimodal graph rendering. In *IEEE Proceedings of the 10th International Symposium on Haptic Interfaces for Virtual Environment and Teleoperator Systems,* pp. 3–9.

Zatsiorsky, V. M., Gao, F., & Latash, M. L. (2003). Prehension synergies: Effects of object geometry and prescribed torques. *Experimental Brain Research, 148,* 77–87.

Zelaznik, H. N., Hawkins, B., & Kisselburgh, L. (1983). Rapid visual feedback processing in single-aiming movements. *Journal of Motor Behavior, 15,* 217–236.

Zimny, M. L. (1988). Mechanoreceptors in articular tissues. *American Journal of Anatomy, 182,* 16–32.

Zimny, M. L., DePaolo, C., & Dabezies, E. (1989). Mechano-receptors in the flexor tendons of the hand. *Journal of Hand Surgery, 14B,* 229–231.

Zuidhoek, S., Kappers, A. M., van der Lubbe, R. H., & Postma, A. (2003). Delay improves performance on a haptic spatial matching task. *Experimental Brain Research, 149,* 320–330.

AUTHOR INDEX

SUBJECT INDEX